Spacecraft Maximum Allowable Concentrations for Selected Airborne Contaminants

Volume 2

Subcommittee on Spacecraft Maximum
Allowable Concentrations

Committee on Toxicology

Board on Environmental Studies
and Toxicology

Commission on Life Sciences

National Research Council

NATIONAL ACADEMY PRESS
Washington, D.C., 1996

NATIONAL ACADEMY PRESS 2101 Constitution Ave., N.W., Washington, D.C. 20418

NOTICE: The project that is the subject of this report was approved by the Governing Board of the National Research Council, whose members are drawn from the councils of the National Academy of Sciences, the National Academy of Engineering, and the Institute of Medicine. The members of the committee responsible for the report were chosen for their special competences and with regard for appropriate balance.

This report has been reviewed by a group other than the authors according to procedures approved by a Report Review Committee consisting of members of the National Academy of Sciences, the National Academy of Engineering, and the Institute of Medicine.

The National Academy of Sciences is a private, nonprofit, self-perpetuating society of distinguished scholars engaged in scientific and engineering research, dedicated to the furtherance of science and technology and to their use for the general welfare. Upon the authority of the charter granted to it by the Congress in 1863, the Academy has a mandate that requires it to advise the federal government on scientific and technical matters. Dr. Bruce Alberts is president of the National Academy of Sciences.

The National Academy of Engineering was established in 1964, under the charter of the National Academy of Sciences, as a parallel organization of outstanding engineers. It is autonomous in its administration and in the selection of its members, sharing with the National Academy of Sciences the responsibility for advising the federal government. The National Academy of Engineering also sponsors engineering programs aimed at meeting national needs, encourages education and research, and recognizes the superior achievements of engineers. Dr. Harold Liebowitz is president of the National Academy of Engineering.

The Institute of Medicine was established in 1970 by the National Academy of Sciences to secure the services of eminent members of appropriate professions in the examination of policy matters pertaining to the health of the public. The Institute acts under the responsibility given to the National Academy of Sciences by its congressional charter to be an adviser to the federal government and, upon its own initiative, to identify issues of medical care, research, and education. Dr. Kenneth I. Shine is president of the Institute of Medicine.

The National Research Council was organized by the National Academy of Sciences in 1916 to associate the broad community of science and technology with the Academy's purposes of furthering knowledge and advising the federal government. Functioning in accordance with general policies determined by the Academy, the Council has become the principal operating agency of both the National Academy of Sciences and the National Academy of Engineering in providing services to the government, the public, and the scientific and engineering communities. The Council is administered jointly by both Academies and the Institute of Medicine. Dr. Bruce Alberts and Dr. Harold Liebowitz are chairman and vice chairman, respectively, of the National Research Council.

The project was supported by the National Aeronautics and Space Administration Grant No. NAGW-2239.

Library of Congress Catalog Card Number 95-73151
International Standard Book Number 0-309-05478-8

Additional copies of this report are available from
National Academy Press
2101 Constitution Ave., N.W.
Box 285
Washington, D.C. 20055
800-624-6242 or 202-334-3313 (in the Washington Metropolitan Area)

B-718

Copyright 1996 by the National Academy of Sciences. All rights reserved.

Printed in the United States of America

Subcommittee on Spacecraft Maximum Allowable Concentrations

DONALD E. GARDNER *(Chair)*, Consultant, Raleigh, N.C.
JOSEPH V. BRADY, The Johns Hopkins University School of Medicine, Baltimore, Md.
RICHARD J. BULL, Washington State University, Pullman, Wash.
GARY P. CARLSON, Purdue University, West Lafayette, Ind.
CHARLES E. FEIGLEY, University of South Carolina, Columbia, S.C.
MARY E. GAULDEN, University of Texas, Southwestern Medical Center, Dallas, Tex.
WILLIAM E. HALPERIN, National Institute for Occupational Safety and Health, Cincinnati, Ohio
ROGENE F. HENDERSON, Lovelace Biomedical and Environmental Research Institute, Albuquerque, N.Mex.
E. MARSHALL JOHNSON, Thomas Jefferson Medical College, Philadelphia, Pa.
RALPH L. KODELL, National Center for Toxicological Research, Jefferson, Ark.
ROBERT SNYDER, Environmental and Occupational Health Sciences Institute, Piscataway, N.J.
BERNARD M. WAGNER, Bernard M. Wagner Associates, Millburn, N.J.
G. DONALD WHEDON, Consultant, Clearwater Beach, Fla.
GAROLD S. YOST, University of Utah, Salt Lake City, Utah

Staff

KULBIR S. BAKSHI, Project Director
MARGARET E. MCVEY, Project Officer
RUTH E. CROSSGROVE, Editor
SHARON L. HOLZMANN, Administrative Associate
CATHERINE M. KUBIK, Senior Program Assistant
LUCY V. FUSCO, Project Assistant

Sponsor: National Aeronautics and Space Administration

Committee on Toxicology

ROGENE F. HENDERSON *(Chair)*, Lovelace Biomedical and Environmental Research Institute, Albuquerque, N.Mex.
DONALD E. GARDNER *(Vice-Chair)*, Raleigh, N.C.
DEBORAH A. CORY-SLECHTA, University of Rochester, Rochester, N.Y.
ELAINE M. FAUSTMAN, University of Washington, Seattle, Wash.
CHARLES E. FEIGLEY, University of South Carolina, Columbia, S.C.
DAVID W. GAYLOR, U.S. Food and Drug Administration, Jefferson, Ark.
WALDERICO M. GENEROSO, Oak Ridge National Laboratory, Oak Ridge, Tenn.
IAN A. GREAVES, University of Minnesota, Minneapolis, Minn.
SIDNEY GREEN, U.S. Food and Drug Administration, Laurel, Md.
LOREN D. KOLLER, Oregon State University, Corvallis, Oreg.
MICHELE A. MEDINSKY, Chemical Industry Institute of Toxicology, Research Triangle Park, N.C.
JOHN L. O'DONOGHUE, Eastman Kodak Company, Rochester, N.Y.
ROBERT SNYDER, Environmental and Occupational Health Sciences Institute, Piscataway, N.J.
BAILUS WALKER, JR., Howard University, Washington, D.C.
ANNETTA P. WATSON, Oak Ridge National Laboratory, Oak Ridge, Tenn.
HANSPETER R. WITSCHI, University of California, Davis, Calif.
GERALD N. WOGAN, Massachusetts Institute of Technology, Cambridge, Mass.
GAROLD S. YOST, University of Utah, Salt Lake City, Utah

Staff

KULBIR S. BAKSHI, Program Director
MARVIN A. SCHNEIDERMAN, Senior Staff Scientist
MARGARET E. MCVEY, Staff Officer
RUTH E. CROSSGROVE, Editor
CATHERINE M. KUBIK, Senior Program Assistant
LUCY V. FUSCO, Project Assistant

Board on Environmental Studies and Toxicology

PAUL G. RISSER *(Chair)*, Miami University, Oxford, Ohio
MICHAEL J. BEAN, Environmental Defense Fund, Washington, D.C.
EULA BINGHAM, University of Cincinnati, Cincinnati, Ohio
PAUL BUSCH, Malcolm Pirnie, Inc., White Plains, N.Y.
EDWIN H. CLARK, II, Clean Sites, Inc., Alexandria, Va.
ALLAN H. CONNEY, Rutgers University, Piscataway, N.J.
ELLIS COWLING, North Carolina State University, Raleigh, N.C.
GEORGE P. DASTON, Procter & Gamble Co., Cincinnati, Ohio
DIANA FRECKMAN, Colorado State University, Fort Collins, Colo.
ROBERT A. FROSCH, Harvard University, Cambridge, Mass.
RAYMOND C. LOEHR, University of Texas, Austin, Tex.
GORDON ORIANS, University of Washington, Seattle, Wash.
GEOFFREY PLACE, Hilton Head, S.C.
DAVID P. RALL, Washington, D.C.
LESLIE A. REAL, Indiana University, Bloomington, Ind.
KRISTIN SHRADER-FRECHETTE, University of South Florida, Tampa, Fla.
BURTON H. SINGER, Princeton University, Princeton, N.J.
MARGARET STRAND, Bayh, Connaughton and Malone, Washington, D.C.
GERALD VAN BELLE, University of Washington, Seattle, Wash.
BAILUS WALKER, JR., Howard University, Washington, D.C.
TERRY F. YOSIE, E. Bruce Harrison Co., Washington, D.C.

Staff Program Directors of the Board on Environmental Studies and Toxicology

JAMES J. REISA, Director
DAVID J. POLICANSKY, Associate Director and Program Director for Natural Resources and Applied Ecology
CAROL A. MACZKA, Program Director for Toxicology and Risk Assessment
LEE R. PAULSON, Program Director for Information Systems and Statistics
RAYMOND A. WASSEL, Program Director for Environmental Sciences and Engineering

Commission on Life Sciences

THOMAS D. POLLARD *(Chair),* The Johns Hopkins University, Baltimore, Md.
FREDERICK R. ANDERSON, Cadwalader, Wickersham & Taft, Washington, D.C.
JOHN C. BAILAR III, University of Chicago, Chicago, Ill.
JOHN E. BURRIS, Marine Biological Laboratory, Woods Hole, Mass.
MICHAEL T. CLEGG, University of California, Riverside, Calif.
GLENN A. CROSBY, Washington State University, Pullman, Wash.
URSULA W. GOODENOUGH, Washington University, St. Louis, Mo.
SUSAN E. LEEMAN, Boston University, Boston, Mass.
RICHARD E. LENSKI, Michigan State University, East Lansing, Mich.
THOMAS E. LOVEJOY, Smithsonian Institution, Washington, D.C.
DONALD R. MATTISON, University of Pittsburgh, Pittsburgh, Pa.
JOSEPH E. MURRAY, Wellesley Hills, Mass.
EDWARD E. PENHOET, Chiron Corp., Emergyville, Calif.
EMIL A. PFITZER, Research Institute for Fragrance Materials, Hackensack, N.J.
MALCOLM C. PIKE, University of Southern California, Los Angeles, Calif.
HENRY C. PITOT III, University of Wisconsin, Madison, Wisc.
JONATHAN M. SAMET, The Johns Hopkins University, Baltimore, Md.
HAROLD M. SCHMECK, JR., North Chatham, Mass.
CARLA J. SHATZ, University of California, Berkeley, Calif.
JOHN L. VANDEBERG, Southwestern Foundation for Biomedical Research, San Antonio, Tex.

PAUL GILMAN, Executive Director

Other Recent Reports

Board on Environmental Studies and Toxicology

Upstream: Salmon and Society in the Pacific Northwest (1996)
Science and the Endangered Species Act (1995)
Wetlands: Characteristics and Boundaries (1995)
Biologic Markers (Urinary Toxicology (1995), Immunotoxicology (1992), Environmental Neurotoxicology (1992), Pulmonary Toxicology (1989), Reproductive Toxicology (1989))
Review of EPA's Environmental Monitoring and Assessment Program (three reports, 1994-1995)
Science and Judgment in Risk Assessment (1994)
Ranking Hazardous Sites for Remedial Action (1994)
Measuring Lead Exposure in Infants, Children, and Other Sensitive Populations (1993)
Pesticides in the Diets of Infants and Children (1993)
Issues in Risk Assessment (1993)
Setting Priorities for Land Conservation (1993)
Protecting Visibility in National Parks and Wilderness Areas (1993)
Dolphins and the Tuna Industry (1992)
Hazardous Materials on the Public Lands (1992)
Science and the National Parks (1992)
Animals as Sentinels of Environmental Health Hazards (1991)
Assessment of the U.S. Outer Continental Shelf Environmental Studies Program, Volumes I-IV (1991-1993)
Human Exposure Assessment for Airborne Pollutants (1991)
Monitoring Human Tissues for Toxic Substances (1991)
Rethinking the Ozone Problem in Urban and Regional Air Pollution (1991)
Decline of the Sea Turtles (1990)
Tracking Toxic Substances at Industrial Facilities (1990)

Committee on Toxicology

Permissible Exposure Levels for Selected Military Fuel Vapors (1996)
Nitrate and Nitrite in Drinking Water (1995)
Guidelines for Chemical Warfare Agents in Military Field Drinking Water (1995)
Review of the U.S. Naval Medical Research Institute's Toxicology Program (1994)
Health Effects of Permethrin-Impregnated Army Battle-Dress Uniforms (1994)
Spacecraft Maximum Allowable Concentrations for Selected Airborne Contaminants, Volume 1 (1994)
Health Effects of Ingested Fluoride (1993)
Guidelines for Developing Community Emergency Exposure Levels for Hazardous Substances (1993)
Guidelines for Developing Spacecraft Maximum Allowable Concentrations for Space Station Contaminants (1992)
Review of the U.S. Army Environmental Hygiene Agency Toxicology Division (1991)
Permissible Exposure Levels and Emergency Exposure Guidance Levels for Selected Airborne Contaminants (1991)

These reports may be ordered from the National Academy Press:

(800) 624-6242 or (202) 334-3313

Preface

The National Aeronautics and Space Administration (NASA) is aware of the potential toxicological hazards to humans that might be associated with prolonged spacecraft missions. Despite major engineering advances in controlling the atmosphere within spacecraft, some contamination of the air appears inevitable. NASA has measured numerous airborne contaminants during space missions. As the missions increase in duration and complexity, ensuring the health and well-being of astronauts traveling and working in this unique environment becomes increasingly difficult.

As part of its efforts to promote safe conditions aboard spacecraft, NASA requested the National Research Council (NRC) to develop guidelines for establishing spacecraft maximum allowable concentrations (SMACs) for contaminants, and to review SMACs for various spacecraft contaminants to determine whether NASA's recommended exposure limits are consistent with the guidelines recommended by the subcommittee.

In response to NASA's request, the NRC organized the Subcommittee on Guidelines for Developing Spacecraft Maximum Allowable Concentrations for Space Station Contaminants within the Committee on Toxicology (COT). In the first phase of its work, the subcommittee developed the criteria and methods for preparing SMACs for spacecraft contaminants. The subcommittee's report, entitled *Guidelines for Developing Spacecraft Maximum Allowable Concentrations for Space Station Contaminants,* was published in 1992. The executive summary of that report is reprinted as Appendix A of this volume.

In the second phase of the study, the Subcommittee on Spacecraft Maximum Allowable Concentrations reviewed reports prepared by NASA scientists and contractors recommending SMACs for approximately 35 spacecraft contaminants. The subcommittee sought to determine whether the SMAC reports were consistent with the 1992 guidelines. Appendix B of this volume contains the SMAC reports for 12 chemical contaminants that have been reviewed for their application of the guidelines developed in the first phase of this activity and approved by the subcommittee. This report is the second volume in the series *Spacecraft Maximum Allowable Concentrations for Space Station Contaminants*. The first volume was published in 1994.

The subcommittee gratefully acknowledges the valuable assistance provided by the following personnel from NASA and its contractors: Dr. John James, Dr. Martin Coleman, Dr. Lawrence Dietlein, Mr. Jay Perry, Mr. Kenneth Mitchell (all from NASA), Mr. James Hyde (Jet Propulsion Laboratory), Dr. King Lit Wong (U.S Department of Commerce, Patent and Trademark Office), Dr. Hector Garcia, Dr. Chiu Wing Lam (both from Krug International), and Mr. Donald Cameron (Boeing Company). The subcommittee is grateful to astronauts Drs. Shannon Lucid, Drew Gaffney, Mary Cleave, and Martin Fettman for sharing their experiences. The subcommittee also acknowledges the valuable assistance provided by the Johnson Space Center, Houston, Texas, the Marshall Space Flight Center, Huntsville, Alabama, the Kennedy Space Center, Cape Canaveral, Florida, and the Space Station Freedom Program Office, Reston, Virginia, for providing tours of their facilities. No effort of this kind can be accomplished without the hard work and dedication of Sharon Holzmann, Catherine Kubik, and Margaret McVey. Lucy Fusco was the project assistant. Ruth Crossgrove edited the report. The subcommittee particularly acknowledges Dr. Kulbir Bakshi, project director for the subcommittee, whose hard work and expertise were most effective in bringing the report to completion.

Donald E. Gardner, *Chair*
Subcommittee on Spacecraft Maximum
Allowable Concentrations

Rogene F. Henderson, *Chair*
Committee on Toxicology

Contents

SPACECRAFT MAXIMUM ALLOWABLE CONCENTRATIONS FOR SELECTED AIRBORNE CONTAMINANTS   **1**
   Introduction, **1**
   Summary of Report on Guidelines for Developing SMACS, **3**
   Review of SMAC Reports, **4**
   References, **5**

APPENDIX A
   Guidelines for Developing Spacecraft Maximum Allowable Concentrations for Space Station Contaminants: Executive Summary   **7**

APPENDIX B
   Reports on Spacecraft Maximum Allowable Concentrations for Selected Airborne Contaminants   **17**
      B1 Acrolein, **19**
      B2 Benzene, **39**
      B3 Carbon Dioxide, **105**
      B4 2-Ethoxyethanol, **189**
      B5 Hydrazine, **213**
      B6 Indole, **235**
      B7 Mercury, **251**
      B8 Methylene Chloride, **277**
      B9 Methyl Ethyl Ketone, **307**
      B10 Nitromethane, **331**
      B11 2-Proponol, **351**
      B12 Toluene, **373**

Spacecraft Maximum Allowable Concentrations for Selected Airborne Contaminants

INTRODUCTION

The space station—a multinational effort—is expected to be launched in 1997 and, in its present configuration, is expected to carry a crew of four to eight astronauts for up to 180 days. Because the space station will be a closed and complex environment, some contamination of its internal atmosphere is unavoidable. Several hundred chemical contaminants are likely to be found in the closed-loop atmosphere of the space station, most at very low concentrations. Important sources of atmospheric contaminants include off-gassing of cabin materials, operation of equipment, and metabolic waste products of crew members. Other potential sources of contamination are releases of toxic chemicals from experiments and manufacturing activities performed on board the space station and accidental spills and fires. The water recycling system has also been shown to produce chemical contaminants that can enter the cabin air. Therefore, the astronauts potentially can be chronically exposed to low levels of airborne contaminants and occasionally to high levels of contaminants in the event of a leak, spill, or fire.

The National Aeronautics and Space Administration (NASA) seeks to ensure the health, safety, and functional abilities of astronauts and seeks to prevent the exposure of astronauts to toxic levels of spacecraft contaminants. Consequently, exposure limits need to be established for continuous exposure of astronauts to spacecraft contaminants for up to 180 days (for normal space-station operations) and for short-term (1-24 hr) emergency exposures to high levels of chemical contaminants.

Federal regulatory agencies, such as the U.S. Occupational Safety and Health Administration (OSHA) and the U.S. Environmental Protection Agency (EPA), have not promulgated exposure limits for the unique environment of spacecraft, nor are their existing standards appropriate for this environment. In 1972, the National Research Council's Committee on Toxicology (COT) first recommended maximum levels for continuous and emergency exposures to spacecraft contaminants (NRC, 1972). However, that early report did not provide documentation of toxicity data or the rationale for the recommended exposure levels. Toxicity data for most of the compounds were not well developed at that time, and the risk-assessment methods were rudimentary. Over the past several years, COT has recommended emergency exposure guidance levels (EEGLs) and continuous exposure guidance levels (CEGLs) for several hundred chemical substances for the U.S. Department of Defense (NRC, 1984a,b,c,d; 1985a,b; 1986; 1987; 1988). However, EEGLs and CEGLs are not available for most spacecraft contaminants. Because of the experience of COT in recommending EEGLs and CEGLs, NASA requested that COT establish guidelines for developing spacecraft maximum allowable concentrations (SMACs) that could be used uniformly by scientists involved in preparing SMACs for airborne contaminants and review the SMACs for individual contaminants to ascertain whether they are consistent with the guidelines.

SMACs are intended to provide guidance on chemical exposures during normal operations of spacecraft as well as emergency situations. Short-term SMACs refer to concentrations of airborne substances (such as a gas, vapor, or aerosol) that will not compromise the performance of specific tasks by astronauts during emergency conditions or cause serious or permanent toxic effects. Such exposures might cause reversible effects, such as mild skin or eye irritation, but they are not expected to impair judgment or interfere with proper responses to emergencies. Long-term SMACs are intended to avoid adverse health effects (either immediate or delayed) and to prevent decremental change in crew performance under continuous exposure to chemicals in the closed environment of the space station for as long as 180 days.

In response to NASA's request to establish guidelines for developing SMACs and to review SMAC documents for selected spacecraft contaminants, COT organized the Subcommittee on Guidelines for Developing Spacecraft Maximum Allowable Concentrations for Space Station Contaminants. The subcommittee comprised experts in toxicology, epidemiology, medicine, physiology, biochemistry, pathology, pharmacology, neurotoxicology, industrial hygiene, statistics, and risk assessment. In the first phase of the study, the subcommittee prepared *Guidelines for Developing Spacecraft Maximum Allowable Concentrations for Space Station Contaminants* (NRC, 1992). That report provided guidance for deriving SMACs from available toxicological and epidemiological data. It also provided guidance on what data to use, how to evaluate the data for appropriateness, how to perform risk assessment for carcinogenic and noncarcinogenic effects, and how to consider the effects of physiological changes induced by microgravity that might enhance the susceptibility of astronauts to certain spacecraft contaminants. The executive summary of that report is contained in Appendix A of this volume.

SUMMARY OF REPORT ON GUIDELINES FOR DEVELOPING SMACS

As described in Appendix A, the first step in establishing SMACs for a chemical is to collect and review all relevant information available on a compound. Various types of evidence are assessed in establishing SMAC values for a chemical contaminant. These include information from (1) chemical–physical characterizations, (2) structure-activity relationships, (3) in vitro toxicity studies, (4) animal toxicity studies, (5) human clinical studies, and (6) epidemiological studies. For chemical contaminants, dose-response data from human exposure are most applicable and are used when available in preference to data from animal studies and in vitro studies. Toxicity data from inhalation exposures are most useful because inhalation is the most likely route of exposure.

For most chemicals, actual human toxicity data are not available. Therefore, toxicity data from studies conducted in animals are extrapolated to estimate the potential toxicity in humans. Extrapolation requires experienced scientific judgment. The toxicity data from animal species most representative of humans in terms of pharmacodynamic and pharmacokinetic properties are used for determining SMACs. If data are not available on which species best represents humans, the data from the most sensitive animal species are used to set SMACs. Safety or uncertainty factors are commonly used when animal data are extrapolated to a safe level for humans. The magnitude of uncertainty factors depends on the quality of the animal data used to determine the no-observed-adverse-effect level (NOAEL). Conversion from animals to humans is done on a body-weight or surface-area basis. When available, pharmacokinetic data on tissue doses are considered for use in species interconversion.

Based on the review of the toxicity data and the use of appropriate safety factors, SMACs for different periods are developed, and a rationale is provided for each recommendation. One- or 24-hr emergency SMACs are usually derived from acute exposure toxicity studies. Development of 1- or 24-hr SMACs usually begins with providing a SMAC for the shortest exposure of 1 hr. Values for 24-hr SMACs might necessitate using Haber's rule ($C \times T = k$) when applicable. Detoxification or recovery and data available on 24-hr exposures are taken into account in modifying Haber's rule.

When data from chronic exposure studies are available, they are used to derive 7-, 30-, or 180-day SMACs, and safety factors are applied as needed. For substances that affect several organ systems or have multiple effects, all end points—including reproductive (in both sexes), developmental, carcinogenic, neurotoxic, respiratory, and other organ-related effects—are evaluated, the most important or most sensitive effects receiving the major attention. With carcinogenic chemicals, quantitative carcinogenic assessment is done, and the SMAC is set so that the estimated lifetime risk of a neoplasm is no more than 1 in 10,000 exposed persons. When a substance is known to cause an effect that will be aggravated by microgravity, additional safety factors are used.

REVIEW OF SMAC REPORTS

In the second phase of the study, the Subcommittee on Spacecraft Maximum Allowable Concentrations reviewed reports for approximately 35 spacecraft contaminants to determine whether the recommended exposure limits were consistent with the 1992 guidelines (see Appendix A). The SMAC reports were prepared solely by NASA scientists or contractors.

These SMAC reports are intended for use by engineers in developing design criteria for the space station. The SMAC reports will also be applicable to the space shuttle, because the recommended SMACs will cover the exposure times that are of interest to the space-shuttle program—1-hr and 24-hr SMACs for emergencies and 7-day and 30-day SMACs for continuous exposures.

The subcommittee's review of the SMAC reports prepared by NASA, NASA's contractors, and members of the subcommittee involved both oral and written presentations to the subcommittee by the authors of the reports. The subcommittee provided advice and recommendations for revisions. The authors of the SMAC reports presented their revised reports at subsequent meetings until the subcommittee agreed that the reports followed the guidelines developed in the first phase of the study (NRC, 1992).

The subcommittee recognizes that many factors, such as the alterations in normal human physiological and biochemical processes associated with spaceflight, are not fully understood and could warrant revisions of proposed SMAC values as additional scientific data become available. Because of the enormous amount of data presented in the SMAC reports, the subcommittee could not verify all the data. The subcommittee relied on NASA scientists for the accuracy and completeness of the toxicity data cited in the SMAC reports. Although individual data points were not verified by the subcommittee, the subcommittee agrees with the proposed SMAC values.

This report is the second volume in the series *Spacecraft Maximum Allowable Concentrations for Selected Airborne Contaminants*. It contains SMAC reports on 12 spacecraft contaminants. The subcommittee concludes that these reports, presented in Appendix B, are consistent with the 1992 NRC guidelines. The first report in the series, *Spacecraft Maximum Allowable Concentrations for Selected Airborne Contaminants, Volume 1*, published in 1994, contains SMAC reports on 11 substances. SMAC reports for additional spacecraft contaminants will be presented in subsequent volumes.

REFERENCES

NRC (National Research Council). 1968. Atmospheric Contaminants in Spacecraft. Washington, D.C.: National Academy of Sciences.

NRC (National Research Council). 1972. Atmospheric Contaminants in Manned Spacecraft. Washington, D.C.: National Academy of Sciences.

NRC (National Research Council). 1984a. Emergency and Continuous Exposure Limits for Selected Airborne Contaminants, Vol. 1. Washington, D.C.: National Academy Press.

NRC (National Research Council). 1984b. Emergency and Continuous Exposure Limits for Selected Airborne Contaminants, Vol. 2. Washington, D.C.: National Academy Press.

NRC (National Research Council). 1984c. Emergency and Continuous Exposure Limits for Selected Airborne Contaminants, Vol. 3. Washington, D.C.: National Academy Press.

NRC (National Research Council). 1984d. Toxicity Testing: Strategies to Determine Needs and Priorities. Washington, D.C.: National Academy Press.

NRC (National Research Council). 1985a. Emergency and Continuous Exposure Guidance Levels for Selected Airborne Contaminants, Vol. 4. Washington, D.C.: National Academy Press.

NRC (National Research Council). 1985b. Emergency and Continuous Exposure Guidance Levels for Selected Airborne Contaminants, Vol. 5. Washington, D.C.: National Academy Press.

NRC (National Research Council). 1986. Emergency and Continuous Exposure Guidance Levels for Selected Airborne Contaminants, Vol. 6. Washington, D.C.: National Academy Press.

NRC (National Research Council). 1987. Emergency and Continuous Exposure Guidance Levels for Selected Airborne Contaminants, Vol. 7. Washington, D.C.: National Academy Press.

NRC (National Research Council). 1988. Emergency and Continuous Exposure Guidance Levels for Selected Airborne Contaminants, Vol. 8. Washington, D.C.: National Academy Press.

NRC (National Research Council). 1992. Guidelines for Developing Spacecraft Maximum Allowable Concentrations for Space Station Contaminants, Washington, D.C.: National Academy Press.

NRC (National Research Council). 1994. Spacecraft Maximum Allowable Concentrations for Selected Airborne Contaminants, Volume 1, Washington, D.C.: National Academy Press.

Appendix A

Guidelines for Developing Spacecraft Maximum Allowable Concentrations for Space Station Contaminants: Executive Summary[1]

---

[1] NRC (National Research Council). 1992. Guidelines for Developing Spacecraft Maximum Allowable Concentrations for Space Station Contaminants. Washington, D.C.: National Academy Press.

Guidelines for Developing Spacecraft Maximum Allowable Concentrations for Space Station Contaminants: Executive Summary

The National Aeronautics and Space Administration (NASA) is preparing to launch a manned space station by the mid-1990s. Because the space station will be a closed complex environment, some contamination of its atmosphere is inevitable. Several hundred chemicals are likely to be found in the closed atmosphere of the space station, most in very low concentrations. Important sources of atmospheric contaminants include metabolic waste products of crew members and off-gassing of cabin materials and equipment. Release of chemicals from experiments performed on board the space station is also a possible source of contamination, and the water reclamation system has the potential to introduce novel compounds into the air. NASA is concerned about the health, safety, and functional abilities of crews exposed to these contaminants.

This report, prepared by the Committee on Toxicology of the National Research Council's Board on Environmental Studies and Toxicology, is in response to a request from NASA for guidelines to develop spacecraft maximum allowable concentrations (SMACs) for space-station contaminants. SMACs are used to provide guidance on allowable chemical exposures during normal operations and emergency situations. Short-term SMACs refer to concentrations of airborne substances (such as gas, vapor, or aerosol) that will not compromise the performance of specific tasks during emergency conditions lasting up to 24 hr. Exposure to 1- or 24-hr SMACs will not cause serious or permanent effects but may cause reversible effects that do not impair judgment or interfere with proper responses to emergencies such as fires or accidental releases.

Long-term SMACs are intended to avoid adverse health effects (either immediate or delayed) and to avoid degradation in crew performance with continuous exposure in a closed space-station environment for as long as 180 days. Chemical accumulation, detoxification, excretion, and repair of toxic insults are thus important in determining 180-day SMACs.

ENVIRONMENTAL CONTROL AND LIFE-SUPPORT SYSTEM

The environmental control and life-support system (ECLSS) of the space station is designed to control temperature, humidity, and composition of space-station air, including CO_2 removal; recover water; dispose of waste; and detect and suppress fires. Fires are a great potential hazard and much attention has been given to suppressing them. A fire suppression system is available, but if all else fails, an escape vehicle can be used. A subsystem of the ECLSS, the atmosphere revitalization system, which includes a mass spectrometer called the major constituent analyzer, will analyze cabin air for O_2, N_2, H_2, CO, H_2O, and CH_4 in all areas of the habitation and laboratory modules. A design criterion for the atmosphere revitalization subsystem is the maintenance of space-station exposure levels below the 180-day SMACs under normal conditions.

MODIFICATION OF CONTAMINANT TOXICITY BY ENVIRONMENTAL FACTORS

The special conditions of the space environment must be taken into account in defining spacecraft contaminant exposure limits. Deposition of particles is clearly different and lung function and the toxic potential of inhaled particles may be different under microgravity conditions than under full gravity conditions, as on earth.

Astronauts will be physically, physiologically, and psychologically compromised for the following reasons: loss of muscle and bone mass, altered immune system, cardiovascular changes, decreased red-blood-cell mass, altered nutritional requirements, behavioral changes from stress, fluid shift in the body, altered hormonal status, and altered drug metabolism. These changes could be important factors in disease susceptibility.

The physiological changes noted in spaceflight to date demonstrate that the astronaut is in an altered homeostatic state and thus may be more susceptible to toxic chemicals. How this altered state modifies reactions to chemicals in the space-station environment is not fully known. The physiological changes induced in the space crew are important and their impact must be taken into account in developing SMAC values for various contaminants.

SOURCES AND TYPES OF DATA FOR ESTABLISHMENT OF SMACS

The subcommittee recommends the use of data derived from a number of sources in establishing SMAC values. These sources provide information on a variety of health effects including mortality, morbidity, clinical signs and symptoms, pulmonary effects, neurobehavioral effects, immunotoxicity, reproductive and developmental toxicity, pathology, mutagenicity, carcinogenicity, and biochemical and enzyme changes.

Chemical–Physical Characteristics of Toxicants

The chemical and physical characteristics of a substance provide valuable information on potential tissue dosimetry of the compound within the body and on its likely toxic effects. Preliminary estimates of the toxic potential of new chemicals also may be derived from known toxicities of structurally similar, well-investigated compounds. However, additional uncertainty (safety) factors must be applied to arrive at safe levels for those congeners that have no dose-response data from intact animals.

In Vitro Toxicity Studies

Useful information can be obtained from studies conducted to investigate adverse effects of chemicals on cellular or subcellular systems in vitro. Systems in which toxicity data have been collected include isolated organ systems, single-cell systems, and tissue cultures from multicellular organisms maintained under defined conditions or from functional units derived from whole cells. In vitro studies can be used to elucidate the toxic effects of chemicals and to study their mechanism of action.

Animal Toxicity Studies

The data necessary to evaluate the relationship between exposure to a toxic chemical and its effects on people are frequently not available from human experience; therefore, animal toxicity studies must be relied on to provide information on responses likely to occur in humans.

The usefulness of animal data depends in part on the route of exposure and species used. Although inhalation studies are most relevant in assessing the toxicity of atmospheric contaminants, data from skin absorption, ingestion, and parenteral studies are also useful. The relevance of animal data to humans may be limited by the absence of information on affected target organs and knowledge of metabolic pathways and pharmacokinetics in animals and humans.

Clinical and Epidemiological Observations

In establishing SMACs for chemicals, dose-response data from human exposure should be used whenever possible. Data from clinical inhalation exposures are most useful because inhalation is the most likely route of exposure. Human toxicity data also are available from epidemiological studies of long-term industrial exposures, from short-term high-level exposures following accidents, or from therapeutic uses of some pharmaceutical agents. Some of these data provide a basis for estimating a dose-response relationship.

Epidemiological studies have contributed to our knowledge of the health effects of many airborne chemical hazards. The limitations of epidemiology stem from its use of available data. The accuracy of data on health outcomes varies with the source of the information, and records documenting historical exposure levels are often sparse. For example, mortality information derived from death certificates is sometimes inaccurate, and exposure information collected from administrative purposes is limited. Despite these limitations, if the populations studied are large enough and have been exposed to high enough doses over a sufficient period to allow for the expression of disease, epidemiological studies usually provide valuable information on the effects of exposure in humans without resorting to cross-species extrapolation or to exposing humans in an experimental situation to possible injuries from chemical hazards.

Pharmacokinetics and Metabolism

Evaluation of the health effects of any chemical in a given environment is greatly facilitated by an understanding of its physiological disposition in the body. Many chemicals require some form of metabolic activation to exert their toxic effects. The formation of reactive metabolites may depend on the level of exposure and the pharmacokinetics of the chemical. Modern pharmacokinetic studies can provide physiologically based models describing disposition of chemicals within organs and tissues in the body. The space station is a closed system with limited capacity to clear the air of chemical vapors; the crew contributes to the removal of the chemicals from the air through sequestration and metabolism.

Toxic metabolites may be highly reactive chemically. These metabolites are biologically reactive intermediates that may covalently bind to nucleic acids or proteins that in turn, may alter DNA replication or transcription. In addition to formation of reactive metabolites, metabolic activity also may lead to formation of species of active oxygen that may damage nucleic acids or proteins or cause lipid peroxidation. The resulting health effects may range from direct, short-term target-organ toxicity to carcinogenesis.

Biological Markers

Biological markers are indicators of change within an organism that link exposure to a chemical to subsequent development of adverse health effects. Biological markers within an exposed individual can indicate the degree of exposure to a pollutant and may provide evidence of the initial structural, functional, or biochemical changes induced by the exposure and, ultimately, the biochemical or physiological changes associated with adverse health effects.

Biological markers can be divided into three classes:

1. Biological markers of exposure to pollutants may be thought of as "footprints" that the chemical leaves behind upon interaction with the body. Such markers contain the chemical itself or a metabolic fragment of the chemical and thus are usually chemical-specific.

2. Biological markers of the effects of exposure include the totality of subclinical and clinical signs of chemically induced disease states. The markers of greatest interest are those that are early predictors of serious effects or late-occurring effects. Such markers would be useful in determining what levels of pollutants in the space station can be tolerated without causing irreversible deleterious health effects.

3. Biological markers of increased susceptibility to the effects of exposure to pollutants could be used to predict which persons are most likely to be at excess risk as space-station crew members.

RISK ASSESSMENT
(DEVELOPMENT OF EXPOSURE CRITERIA)

The assessment of toxicants that do not induce carcinogenic or mutagenic effects traditionally has been based on the concept that an adverse health effect will not occur below a certain level of exposure, even if exposure continues over a lifetime. Given this assumption, a reference dose intended to avoid toxic effects may be established by dividing the no-observed-adverse-effect level by an appropriate uncertainty factor or set of factors.

For carcinogenic effects, especially those known to be due to direct mutagenic events, no threshold dose may exist. However, when carcinogenesis is due to epigenetic or nongenotoxic mechanisms, a threshold dose may be considered. Attempts to estimate carcinogenic risks associated with levels of exposure have involved fitting mathematical models to experimental data and extrapolating from these models to predict risks at doses that are usually well below the experimental range. The multistage model of Armitage and Doll is used most frequently for low-dose extrapolation. According to multistage theory, a malignant cancer cell develops from a single stem cell as a result of a number of biological events (e.g., mutations) that must occur in a specific order. Recently, a two-stage model that explicitly provides for tissue growth and cell kinetics also has been used in carcinogenic risk assessment.

The multistage model, characterized by low-dose linearity, forms the basis for setting SMACs for carcinogens. Low-dose linearity is generally assumed for chemical carcinogens that act through direct interaction with genetic material.

ISSUES IN MAKING RECOMMENDATIONS FOR THE ESTABLISHMENT OF SMACS

A number of issues need to be considered in developing recommendations for establishing SMACs. These issues include (1) translating animal toxicity data to predict toxicities in humans; (2) determining 30- or 180-day SMACs for carcinogens based on toxicological or epidemiological studies that often involve long-term or lifetime exposure; (3) considering limits set by the Occupational Safety and Health Administration, the American Conference of Governmental Industrial Hygienists, and the National Research Council in developing SMACs; (4) evaluating the toxicities of mixtures; and (5) modifying risk assessments based on the altered environment in the microgravity of space.

Appendix B

Reports on Spacecraft Maximum Allowable Concentrations for Selected Airborne Contaminants

B1 Acrolein

King Lit Wong, Ph.D.
Johnson Space Center Toxicology Group
Biomedical Operations and Research Branch
Houston, Texas

PHYSICAL AND CHEMICAL PROPERTIES

Acrolein is a colorless or yellowish, volatile, flammable liquid with an extremely sharp, irritating odor (Sax, 1984).

Synonyms:	2-Propenal, acrylic aldehyde
Formula:	CH_2CHCHO
CAS number:	107-02-08
Molecular weight:	56
Boiling point:	52.7°C
Melting point:	-87°C
Vapor pressure:	214 mm Hg at 20°C
Conversion factors at 25°C, 1 atm:	1 ppm = 2.29 mg/m^3 1 mg/m^3 = 0.44 ppm

OCCURRENCE AND USE

Acrolein is not used in the spacecraft, but acrolein is a potential atmospheric contaminant in spacecraft because it was found to be off-gassed from the hardware of two Spacelab missions at a rate of 0.007 mg/d (Geiger, 1984).

PHARMACOKINETICS AND METABOLISM

In dogs, inhaled acrolein is primarily retained by the upper respiratory tract (Bowes and Cater, 1968). In the rat, the evidence suggests that acrolein is oxidized by two pathways. The liver and lung microsomes oxidize acrolein in vitro to its epoxide metabolite, glycidaldehyde (Patel et al., 1980), which is converted to glyceraldehyde by epoxide hydrase (Hayakawa et al., 1975). Acrolein is also oxidized to acrylic acid probably involving aldehyde dehydrogenase (Patel et al., 1980). In rats given acrolein orally, S-carboxyethylmercapturic acid is the metabolite found in urine (Draminski et al., 1983).

TOXICITY SUMMARY

The major toxicity of acrolein is mucosal irritation.

Acute and Short-Term Toxicity

Mucosal Irritation

The NRC's Committee on Toxicology cited a report by the Shell Chemical Corporation that acrolein produced moderate mucosal irritation at a concentration as low as 0.25 ppm in humans, but no information on the exposure duration was given (Shell Chemical Corp., 1958). In comparison, Weber-Tschopp et al. (1977) reported that an exposure to acrolein at 0.3 ppm resulted in only a little eye discomfort in 1.5 min and moderate eye irritation in 1 h. Darley et al. (1960) demonstrated that a 5-min exposure to acrolein at 1.3-1.6 ppm was found to produce only moderate eye irritation in college students. In contrast, Sim and Pattle (1957) found that 1.2 ppm was extremely irritating to the mucosal surfaces in 5 min. Because of these variations in human test results, comparisons of the qualitative descriptions of acrolein's irritancy in different studies should be made with care.

Some studies show that acrolein's eye irritancy tends to increase progressively within the first 40 min of the exposure. Stephens et al.

(1961) showed that an exposure to acrolein at 0.5 ppm resulted in eye irritation in 10-35% of subjects within 5 min and in 91% within 12 min (Stephens et al., 1961). Exposure to acrolein at 0.3 ppm produced mild eye irritation in human subjects in 10 min, mild-to-moderate eye irritation in 20 min, and moderate eye irritation in 40 min (Weber-Tschopp et al., 1977). From 40 to 60 min, the degree of eye, nose, and throat irritation stayed constant (Weber-Tschopp et al., 1977).

The relative sensitivity of human eyes and nose toward acrolein's irritation depends on the exposure concentration and duration. For exposures lasting about an hour, acrolein is more irritating to the eyes than the nose; the reverse is true for short exposures at below 0.3 ppm (Weber-Tschopp et al., 1977). A 40-60 min exposure to acrolein at 0.3 ppm caused moderate eye irritation and slight nose irritation (Weber-Tschopp et al., 1977). In a 1.5-min exposure of human volunteers to acrolein, 0.6 ppm produced the same degree of irritation in the eyes and the nose (somewhat slight irritation) (Weber-Tschopp et al., 1977). However, in a similar exposure, 0.15 ppm caused a little bit of nose discomfort, although it was not irritating to the eyes (Weber-Tschopp et al., 1977).

There are some data on acrolein's dose-response relationship in the literature. Darley et al. (1960) showed that in a 5-min exposure of humans to acrolein, the eye irritation was moderate at a concentration of 1.3-1.6 ppm and moderate to severe at 2.0-2.3 ppm. Weber-Tschopp et al. (1977) reported that in human volunteers exposed to acrolein for 1.5 min, mild irritation to the eyes and nose was detected at 0.60 ppm, a little bit of discomfort to the eyes and nose was detected at 0.30 ppm, and a little bit of discomfort to the nose and no effect on the eye was detected at 0.15 ppm (Weber-Tschopp et al., 1977). Probably because of the little bit of discomfort to the nose, some of the test subjects expressed a desire to leave the room when questioned in a survey, and they characterized the air quality as acceptable, versus good by the control group (Weber-Tschopp et al., 1977).

The exposure concentration seems to be more important than the exposure duration in causing acute irritating effects in humans. Sim and Pattle (1957) found that an exposure to acrolein at 0.8 ppm produced lacrimation in humans in 20 s. However, at a slightly higher concentration of 1.2 ppm, acrolein produced lacrimation in only 5 s. Had acrolein's mucosal irritancy followed Haber's rule, 1.2 ppm would not be expected to produce lacrimation until 13 s into an exposure. That means $C \times T$ was not constant in that study.

Respiratory Effects

Like most sensory irritants, acrolein also affects respiration. In about half of the human volunteers exposed to acrolein at 0.3 ppm, respiratory rates were decreased by 10% after 10 min of exposure (Weber-Tschopp et al., 1977). A 10-min exposure of Swiss-Webster mice to acrolein at 0.22 ppm also resulted in a decrease of approximately 25% in the respiratory rate (Steinhagen and Barrow, 1984). Even an exposure as low as 0.04 ppm resulted in a decrease of about 10% in the respiratory rate in the mouse (Steinhagen and Barrow, 1984). In guinea pigs exposed to acrolein at 0.4-1.0 ppm for 2 h, there were decreases in the respiratory rate and increases in the total pulmonary resistance (Murphy et al., 1963). The respiratory effects of acrolein in both the mouse and guinea pig are readily reversible when the exposure ends (Steinhagen and Barrow, 1984; Murphy et al., 1963). In a recent study, a 2-h exposure of guinea pigs to acrolein at 0.3 ppm increased the pulmonary resistance and bronchial responsiveness (Leikauf et al., 1989). In the lung lavage fluid, the exposure also increased the amount of thromboxane B_2 and prostaglandin F_2 immediately after exposure and the number of neutrophils 24 h after exposure (Leikauf et al., 1989).

Miscellaneous Effects

Acrolein readily reacts with sulfhydryl groups. As a result, a 3-h inhalation exposure of rats to acrolein at 2.5 ppm decreased mucosal glutathione in the respiratory region of the nose (Lam et al., 1985). A 6-h exposure of rats to acrolein at 2 ppm did not cause DNA-protein crosslinks in the respiratory mucosa of the nose (Lam et al., 1985).

Finally, acrolein exposures at a sufficiently high concentration could be lethal. The 6-h LC_{50} of acrolein in mice was 66 ppm (Philippin et al., 1970) and one man died 10 min after exposure at 150 ppm (Prentiss, 1937).

Subchronic and Chronic Toxicity

Mucosal Irritation and Effects on the Respiratory System

Subchronic exposures to acrolein generally produced signs of mucosal irritation and pathology of the respiratory system. An exposure of mice to acrolein at 6 ppm for 2 w, 6 h/d, 5 d/w caused atelectasis, inflammation, and edema in the lung and body-weight reduction (Philippin et al., 1970). When the acrolein exposure was extended to 6 w, 8 h/d, 5 d/w, Lyon et al. (1970) found that 3.7 ppm produced eye irritation, squamous metaplasia, and basal-cell hyperplasia in the trachea of monkeys. Dogs, rats, and guinea pigs were similarly exposed in the same study, but the rats and guinea pigs appeared to be less susceptible to acrolein than the monkeys; the rats and guinea pigs did not show any gross sign of irritation, tracheal metaplasia, or tracheal hyperplasia (Lyon et al., 1970). Only two dogs were used in each exposure group, and in general the clinical signs of eye and nose irritation as well as lung injury in the dogs were similar to those in the monkeys. An exposure of acrolein at 0.7 ppm for 6 w, 8 h/d, 5 d/w caused chronic peribronchial inflammation in all exposed animals, indicating chronic bronchial irritation (Lyon et al., 1970).

Unfortunately, the nose was not examined microscopically in any of the experimental animals in either of the subchronic studies reviewed above. Feron et al. (1978) were the first to look for any nasal histopathology produced by acrolein. In an exposure of hamsters, rats, and rabbits to acrolein for 13 w, 6 h/d, 5 d/w, Feron et al. found that the rat was the most susceptible species, and the rabbit appeared to be the least susceptible. The 13-w exposure at 1.4 ppm resulted in metaplasia and inflammation of the nasal mucosa in the rats, only minimal nasal inflammation in the hamsters, and no adverse effects in the rabbits (Feron et al., 1978). The no-observed-adverse-effect level (NOAEL) was 0.4 ppm for this exposure of hamsters and rats to acrolein for 13 w, 6 h/d, 5 d/w (Feron et al., 1978).

In a 6-w exposure of mice, acrolein at 2 ppm for 6 h/d had no effect on the body-weight gain and the wet-to-dry lung-weight ratio in the mice (Northrop, 1985). However, it is of interest that a 6-w exposure of mice to acrolein at 4 ppm for 3 h/d or at 8 ppm for 1.5 h/d also did not affect the wet-to-dry lung-weight ratio, but it did reduce the body-weight gain (Northrop, 1985). These results show that, in subchronic exposures to acrolein, the exposure concentrations seem to be more important than the exposure times. As discussed above, a similar conclusion was drawn for the acute mucosal irritancy of acrolein in humans.

Lyon et al. (1970) exposed monkeys, dogs, rats, and guinea pigs to acrolein continuously for 90 d, but no nasal microscopic examinations were conducted. Lyon et al. did observe that exposure to acrolein at 1.0 ppm produced ocular and nasal discharge, squamous metaplasia, and basal-cell hyperplasia in the tracheas of the monkeys. The effects from the exposure on the internal organs of the monkeys might not be reliable, because parasitic infestation was found in the lungs, livers, hearts, and brains of these monkeys (Lyon et al., 1970). The 1.0-ppm exposure did not cause any exposure-related histopathological changes in the tracheas or lungs of the guinea pigs and rats (Lyon et al., 1970). Lyon et al. used only two dogs in the 1.0-ppm exposure group, and the dogs reacted to acrolein exposure similarly to the monkeys. A 90-d continuous exposure of four dogs to acrolein at 0.22 ppm led to emphysema and lung congestion in two dogs.

Miscellaneous Effects

In the 90-d continuous exposure of four dogs to acrolein conducted by Lyon et al. (1970), an exposure at 0.22 ppm resulted in splenic hemorrhage in two dogs and thyroid hyperplasia in the other two dogs. However, no exposure-related histopathological changes were detected in the internal organs of the monkeys, guinea pigs, and rats. (Lyon et al. did not examine the trachea in the 0.22-ppm exposure groups.) A similar exposure at 1.0 ppm was found to cause focal liver necrosis in guinea pigs and rats. The meaning of the finding of liver necrosis is uncertain, because a 90-d continuous exposure at a higher concentration of 1.8 ppm failed to induce liver necrosis in the guinea pigs and rats (Lyon et al., 1970).

In addition to the Lyon et al. investigation, Sinkuvene (1970) investigated the toxic effects of a continuous subchronic exposure to acrolein. The latter study showed that a 16-d continuous exposure of rats to acrolein at 0.056 ppm produced some changes, but they were not specified

(Sinkuvene, 1970). A similar exposure at 0.011 ppm failed to change the blood cholinesterase activity, the "chronaxy" of antagonistic muscles, and body-weight gain (Sinkuvene, 1970).

Carcinogenicity

The U.S. Environmental Protection Agency (EPA) classified acrolein as a possible human carcinogen on the basis of limited animal carcinogenicity data, mutagenicity in bacteria, and structural similarity with two probable human carcinogens, formaldehyde and acetaldehyde (EPA, 1990). A 1-y exposure of hamsters to acrolein at 4 ppm failed to cause any increase in tumor incidence (Feron and Kruysse, 1977). However, the hamster study did not prove that acrolein is noncarcinogenic in laboratory animals for two reasons. First, it is likely that 1 y is not of sufficient duration for exposure. Second, the hamster has not been shown to be the most sensitive test species, at least on the basis of noncarcinogenic end points, to the toxicity of acrolein in subchronic exposures (Feron et al., 1978). Lijinski and Reuber (1987) found an increased incidence of adrenal cortical adenomas in female rats given acrolein at a concentration of 625 ppm in drinking water for 5 d/w for 100 w, but the increase was not statistically significant. (Evidence of the carcinogenicity of acrolein in this study is inconclusive, however, because of the p difference and the small number of rats, 20 in each group used.) It should be noted that there is no evidence of the carcinogenicity of acrolein in humans (EPA, 1990). The International Agency for Research on Cancer (IARC) characterized acrolein as a compound with inadequate evidence of carcinogenicity in both humans and animals (IARC, 1987).

Genotoxicity

There are some indications that acrolein might be genotoxic. It induced mutations in *Salmonella typhimurium* strain TA104 but not in strains TA98, TA1535, TA1537, and TA1538 (Lutz et al., 1982; Hales, 1982; Marnett et al., 1985). Acrolein induced recessive lethal mutations in *Drosophila melanogaster* (Zimmering et al., 1985) and sister chromatid exchange in Chinese hamster ovary cells (Au et al., 1980). However, acrolein failed to affect the dominant lethal assay in mice after an intraperitoneal injection (Epstein et al., 1972).

TABLE 1-1 Toxicity Summary[a]

Concentration, ppm	Exposure Duration	Species	Effects	Reference
0.15	1.5 min	Human	Some desire to leave the room. The air quality was acceptable (versus good for the controls). No effect on the eye. A little bit of nose discomfort.	Weber-Tschopp et al., 1977
0.25	N.S.	Human	Moderate mucosal irritation.	Schell Cemical Corp., 1958
0.3	1.5 min	Human	Some desire to leave the room. The air quality was acceptable (versus good for the controls). A little bit of eye and nose discomfort.	Weber-Tschopp et al., 1977
0.3	10 min	Human	Moderate eye irritation in 18% and severe eye irritation in 3% of the subjects (mild eye irritation on the average). Respiratory rate decreased by 10% in 47% of the subjects. 50% of the subjects desired to leave the room.	Weber-Tschopp et al., 1977
0.3	20 min	Human	Moderate eye irritation in 35% of the subjects and severe eye irritation in 18% of the subjects (mild-to-moderate eye irritation on the average). Respiratory rate decreased in 60% of the subjects. 72% of the subjects desired to leave the room.	Weber-Tschopp et al., 1977
0.3	40-60 min	Human	On the average, moderate eye irritation, mild nose irritation, and very little throat irritation. Increased eye-blinking frequency; decreased respiratory rate.	Weber-Tschopp et al., 1977
0.5	5-12 min	Human	Eye irritation in 10-35% of the subjects within 5 min and in 91% of the subjects within 12 min.	Stephens et al., 1961
0.6	1.5 min	Human	Some desire to leave the room. The air quality was semi-acceptable, semi-poor. Mild eye and nose irritation.	Weber-Tschopp et al., 1977

TABLE 1-1 (Continued)

Concentration, ppm	Exposure Duration	Species	Effects	Reference
0.8	10 min	Human	Lacrimation developed within 20 s. The mucosal irritation was just tolerable in 10 min.	Sim and Pattle, 1957
1.2	5 min	Human	Lacrimation developed in 5 s. An exposure of more than 5 min would have been extremely distressing because of mucosal irritation.	Sim and Pattle, 1957
1.3-1.6	5 min	Human	Moderate eye irritation.	Darley et al., 1960
2.0-2.3	5 min	Human	Moderate-to-severe eye irritation	Darley et al., 1960
140	15 s	Human	Threshold of eye irritation.	Douglas and Coe, 1987
150	10 min	Human	Death.	Prentiss, 1937
0.011	24 h/d, 16 d	Rat	No adverse effect.	Sinkuvene, 1970
0.04	10 min	Mouse	About 10% decrease in respiratory rate	Steinhagen and Barrow, 1984
0.056	24 h/d, 16 d	Rat	Changes that the investigator did not specify.	Sinkuvene, 1970
0.22	10 min	Mouse	About 25% decrease in respiratory rate.	Steinhagen and Barrow, 1984
0.22	24 h/d, 90 d	Monkey, dog, guinea pig, rat	2/4 dogs developed lung emphysema and congestion and spleen hemorrhage; 2/4 developed thyroid hyperplasia. Monkeys, guinea pigs, and rats developed no exposure-related histopathology in internal organs. One of 17 monkeys developed infection of one eye and died. All animals appeared normal and gained weight normally.	Lyon et al., 1970
0.3	24 h/d, 16 d	Rat	Reduced body-weight gain and changes in cholinesterase activity in blood.	Sinkuvene, 1970

0.31	2 h	Guinea pig	Increases in pulmonary resistance and bronchial responsiveness. Increases in thromboxane B_2 and PGF_2 immediately after exposure. Increases in neutrophils 24 h after exposure.	Murphy et al., 1963
0.4	6 h/d, 5 d/w, 62 d	Rat	No adverse effect.	Kutzman et al., 1985
0.4	6 h/d, 5 d/w, 13 w	Rabbit, hamster, rat	No adverse effect in rabbits and hamsters. 1/12 rats developed metaplasia and inflammation of the nasal mucosa.	Feron et al., 1978
0.7	8 h/d, 5 d/w, 6 w	Monkey, rat, guinea pig	Chronic peribronchial inflammation and mild, focal emphysema in the lung.	Lyon et al., 1970
1.0	24 h/d 90 d	Monkey, guinea pig, rat	Monkeys developed ocular and nasal discharge, keeping the eyes closed; parasitic infestation in the lung, liver, brain, and heart. 1 of the 8 monkeys was biten and died from the infection. Guinea pigs appeared normal, with pulmonary inflammation and focal liver necrosis. Rats appeared normal, with occasional lung hemorrhage and focal liver necrosis and reduced body-weight gain.	Lyon et al., 1970
1.4	6 h/d, 5 d/w, 62 d	Rat	3/31 rats showed some pulmonary histopathology. Increase in collagen concentration in the lung.	Kutzman et al., 1985
1.4	6 h/d, 5 d/w, 13 w	Rabbit, hamster, rat	No adverse effect in rabbits. Minimal nasal inflammation in hamsters. Metaplasia and inflammation of the nasal mucosa in rats.	Feron et al., 1978
1.7	6 h/d, 5 d	Mouse	Squamous metaplasia, inflammation, exfoliation, and ulceration of the nasal mucosa.	Buckley et al., 1985
1.8	24 h/d, 90 d	Monkey, guinea pig, rat	Monkeys had excessive salivation and ocular discharge, squamous metaplasia, and basal-cell hyperplasia of the trachea. Rats had reduced body-weight gain.	Lyon et al., 1970

TABLE 1-1 (Continued)

Concentration, ppm	Exposure Duration	Species	Effects	Reference
2	6 h/d 6 w	Mouse	No effects on body-weight gain and wet-to-dry lung-weight ratio. No adverse clinical signs.	Northrop, 1985
2.5	3 h	Rat	Depletion of glutathione in the respiratory mucosa of the nose.	Lam et al., 1985
3	N.S.[b]	Mouse	Less effective in inactivation of inhalation challenges of *Staphylococcus aureus*.	Astry and Jakab, 1983
3	6 h/d, 5 d/w, 3 w	Rat	Exfoliation, necrosis, erosion, and metaplasia of nasal mucosa, and reduction in spleen and body weight. No effect on the number of antibody plaque-forming cells in the lung-associated lymph nodes, lymphocyte blastogenesis, resistance to *Listeria* challenge, pulmonary clearance of inhaled *Kiebsiella pneumoniae*, and the number of cells lavaged from the lung.	Sherwood et al., 1986; Leach et al., 1987
3.7	8 h/d, 5 d/w, 6 w	Monkey, rat, guinea pig	Monkeys had excessive salivation, blinked their eyes frequently, and kept them closed most of the time; had squamous metaplasia and basal-cell hyperplasia of the trachea, necrotizing bronchitis, and bronchiolitis obliterans. 2/9 monkeys died, one of which had parasitic worms in the large intestine. Monkeys and rats had focal calcification of renal tubular epithelium. Reduced body-weight gain in all species.	Lyon et al., 1970
4	3 h/d, 6 w	Mouse	Reduced body-weight gain. No effect on wet-to-dry lung weight ratio. No adverse clinical signs.	Northrop, 1985

4	6 h/d, 5 d/w, 62 d	Rat	32/57 rats died. Bronchiolar necrosis, bronchiolar and lung edema. Increases in lung weight, connective tissue and collagen concentrations.	Kutzman et al., 1985
4	7 h/d, 5 d/w, 52 w	Hamster	No increase in tumor incidence.	Feron and Kruysse, 1977
6	6 h/d, 5 d/w, 2 w	Mouse	Atelectasis, inflammation, and edema in the lung. Body-weight reduction.	Philippin et al., 1970
8	1.5 h/d, 6 w	Mouse	Reduced body-weight gain. No effect on wet-to-dry lung-weight ratio. No adverse clinical signs.	Northrop, 1985
8.3	4 h	Rat	Half of the rats died.	Ballantyne et al., 1986
12	5 min	Baboon	Mucosal irritation. No effect on the time the baboon took to press the correct lever to open the door of the exposure chamber to escape.	Kaplan et al. 1986
26	1 h	Rat	Half of the rats died.	Ballantyne et al., 1986
66	6 h	Mouse	Half of the mice died.	Lam et al., 1985
200-300	10 min	Dog	Airway damage and lung edema	Hales et al., 1988

[a]Only the most important results are included.
[b]N.S. = not specified.

TABLE 1-2 Exposure Limits Set by Other Organizations

Organization	Concentration, ppm
ACGIH's TLV	0.1 (TWA)
ACGIH's STEL	0.3
OSHA's PEL	0.1 (TWA)
NIOSH's IDLH	5
NRC's 10-min EEGL	0.1
NRC's 60-min EEGL	0.05
NRC's 24-h EEGL	0.01
NRC's 90-d CEGL	0.01

TLV = threshold limit value. TWA = time-weighted average. STEL = short-term exposure limit. PEL = permissible exposure limit. IDLH = immediately dangerous to life and health. EEGL = emergency exposure guidance level. CEGL = continuous exposure guidance level.

TABLE 1-3 Spacecraft Maximum Allowable Concentrations

Duration	ppm	mg/m^3	Target Toxicity
1 h	75	170	Mucosal irritation
24 h	35	80	Mucosal irritation
7 d[a]	15	30	Mucosal irritation
30 d	15	30	Mucosal irritation
180 d	15	30	Mucosal irritation

[a]Former 7-d SMAC = 50 ppm.

RATIONALE FOR ACCEPTABLE CONCENTRATIONS

Mucosal irritation is the most important toxic end point to be used in setting SMACs for acrolein because mucosal irritation was detected in humans and mice after exposure to concentrations lower than those that produced histopathological changes in the respiratory systems of various laboratory animal species (Weber-Tschopp et al., 1977; Steinhagen and Barrow, 1984; Lyon et al., 1970; Feron et al., 1978). The eyes and nose differ in their sensitivities to acrolein's irritation. Weber-Tschopp et al. found that a 1-h exposure of human volunteers to acrolein at 0.3 ppm caused moderate eye irritation and only slight nose irritation

(Weber-Tschopp et al., 1977). Because the eye appears to be more sensitive than the nose, SMACs for acrolein are set on the basis of eye irritation.

The 1-h and 24-h SMACs are designed for contingency scenarios, so they are aimed at preventing irreversible injuries and significant performance decrements. It is acceptable that these short-term SMACs might not protect against slight mucosal irritation.

1-h SMAC

The goal is to find an exposure concentration that would produce only slight eye irritation in 1 h. A 1-h exposure to acrolein at 0.3 ppm produced, on the average, moderate eye irritation in humans (Weber-Tschopp et al., 1977). Darley et al. (1960) showed that, after a 5-min exposure to acrolein, eye irritancy in humans decreased by half a grade, from moderate-to-severe to moderate, when the concentration was reduced 35-40% (from 2.0-2.3 ppm to 1.3-1.6 ppm). According to Figure 4 in the report of Weber-Tschopp et al. (1977), as the concentration of acrolein was reduced from 0.6 ppm to 0.3 ppm in a 1.5-min exposure of human volunteers to acrolein, the effect on the eye was reduced from being mild irritation to a little bit of discomfort. There was no effect on the eye when the concentration was further reduced to 0.15 ppm. From the data of Darley et al. and Weber-Tschopp et al., lowering the 1-h exposure concentration of 0.3 ppm, which was moderately irritating to the eye, fourfold should result in a concentration that is only mildly irritating to the eye.

1-h SMAC based on eye irritation
  = 1-h moderately irritating concentration $\times$ 1/extrapolation factor
  = 0.3 ppm $\times$ 1/4
  = 75 ppb.

24-h SMAC

Because Weber-Tschopp et al. (1977) showed that the mucosal irritancy caused by exposure to acrolein at 0.3 ppm stayed constant from

40 min to 60 min, the irritancy concentration should remain unchanged when extending the exposure from 1 h to 24 h. Therefore, theoretically, the 24-h SMAC could be set equal to the 1-h SMAC. However, to reduce the degree of mucosal irritation that the astronauts have to endure in a 24-h emergency, the 24-h SMAC is derived by dividing the 1-h SMAC by two. The factor of two is selected because Weber-Tschopp et al. (1977) demonstrated that, when the concentration of acrolein in a 1.5-min exposure was lowered twofold (from 0.6 ppm to 0.3 ppm), the severity of the eye irritation was reduced from slight irritation to only a little bit of discomfort. As a result, dividing the mildly irritating 1-h SMAC by two should yield a concentration that will cause only a little eye discomfort.

24-h SMAC based on eye irritation
= 1-h slightly irritating concentration × 1/extrapolation factor
= 75 ppb × 1/2
= 35 ppb.

7-d, 30-d, and 180-d SMACs

If the irritancy of acrolein at 24 h is likely to be the same as that at 1 h, it stands to reason that the irritancy would not worsen when exposure is extended to 180 d. That is because mucosal irritation is a surface phenomenon and is generally not considered to be exposure-duration-dependent after the initial exposure period. Acute eye-irritation effects of exposure to acrolein in humans (Sim and Pattle, 1957) and subchronic depressive effects on body-weight gain in mice (Northrop, 1985) are more dependent on the exposure concentration than on the exposure duration (Northrop, 1985). Moreover, in dogs exposed repetitively or continuously to acrolein, the signs of mucosal irritation were found to diminish after the first week, indicating the development of reduced susceptibility as acrolein exposure is lengthened (Lyon et al., 1970). Therefore, long-term SMACs could be established by basing an estimate for a nonirritating concentration of acrolein on data of 1-h exposures.

As discussed above, a 1-h exposure at 75 ppb is expected to be only slightly irritating to the eye. To estimate a 1-h exposure concentration that is nonirritating to the eye, an extrapolation factor of 4 is applied on 75 ppb. This factor of 4 was derived from the data of Weber-Tschopp et al. (1977), who showed that, in a 1.5-min exposure of human subjects to acrolein, 0.60 ppm produced a slight eye irritation, 0.30 ppm caused a little eye discomfort, and 0.15 ppm resulted in no adverse effects on the eye (going from 0.60 ppm to 0.15 ppm is a factor of 2 × 2, or 4). With a lack of data, the concentration-response relationship of acrolein exposure and mucosal irritation in a 1-h exposure is assumed to be the same as that in a 1.5-min exposure. An additional safety factor of 10/(square root of n) is applied for the potential differences among individuals in a human population.

7-d, 30-d, and 180-d SMACs based on eye irritation
 = 1-h slightly irritating concentration × 1/extrapolation factor
   × 1/small n factor
 = 75 ppb × 1/4 × (square root of n)/10
 = 75 ppb × 1/4 × (square root of 53)/10
 = 15 ppb.

The Establishment of SMAC Values

Consequently, the 1-h, 24-h, 7-d, 30-d, and 180-d SMACs are set at 75, 35, 15, 15, and 15 ppb, respectively, to prevent mucosal irritation. Because mucosal irritation is not expected to be significantly influenced by physiological changes caused by microgravity, the SMAC values are not adjusted for any microgravity-induced physiological changes.

REFERENCES

Astry, C.L., and G.J. Jakab. 1983. The effects of acrolein exposure on pulmonary antibacterial defenses. Toxicol. Appl. Pharmacol. 67: 49-54.

Au, W., O.I. Sokova, B. Kopnin, and F.E. Arrighi. 1980. Cytogenetic toxicity of cyclophosphamide and its metabolites in vitro. Cytogenet. Cell Genet. 26:108-116.

Ballantyne, B., D.E. Dodd, I.M. Pritts, D.J. Nachreiner, and E.H.

Fowler. 1989. Acute vapor inhalation toxicity of acrolein and its influence as a trace contaminant in 2-methoxy-3,4-dihydro-^{2}H-pyran. Hum. Toxicol. 8:229-235.

Bowes, J.H., and C.W. Cater. 1968. The interaction of aldehydes with collagen. Biochim. Biophys. Acta 168:341-352.

Buckley, L.A., X.Z. Jiang, R.A. James, K.T. Morgan, and C.S. Barrow. 1984. Respiratory tract lesions induced by sensory irritants at the RD_{50} concentration. Toxicol. Appl. Pharmacol. 74:417-429.

Darley, E.F., J.T. Middleton, and M.J. Garber. 1960. Plant damage and eye irritation from ozone-hydrocarbon reactions. J. Agr. Food Chem. 8:484-485.

Douglas, R.B., and J.E. Coe. 1987. The relative sensitivity of the human eye and lung to irritant gases. Ann. Occup. Hyg. 31:265-267.

Draminski, W., E. Eder, and D. Henschler. 1983. A new pathway of acrolein metabolism in rats. Arch. Toxicol. 52:243-247.

EPA. 1990. Acrolein. In Integrated Risk Information System. Office of Health and Environmental Assessment, U.S. Environmental Protection Agency, Washington, D.C.

Epstein, S.S., E. Arnold, J. Andrea, W. Bass, and Y. Bishop. 1972. Detection of chemical mutagens by the dominant lethal assay in the mouse. Toxicol. Appl. Pharmacol. 23:288-325.

Feron, V.J., and A. Kruysse. 1977. Effects of exposure to acrolein vapor in hamsters simultaneously treated with benzo(*a*)pyrene or diethylnitrosamine. J. Toxicol. Environ. Health 3:379-394.

Feron, V.J., A. Kruysse, H.P. Til, and H.R. Immel. 1978. Repeated exposure to acrolein vapour: Subacute studies in hamsters, rats, and rabbits. Toxicology 9:47-57.

Geiger, T. 1984. P. 11 in Spacelab Mission 3 Aggregate Trace Contaminant Assessment. Publ. No. EP45(84-148). NASA, Marshall Space Flight Center, Huntsville, Ala.

Hales, B. 1982. Comparison of the mutagenicity and teratogenicity of cyclophosphamide and its active metabolites, 4-hydroxycyclophosphamide, phosphoramide mustard, and acrolein. Cancer Res. 42: 3016-3021.

Hales, C.A., P.W. Barkin, W. Jung, E. Trautman, D. Lamborghini, N. Herrig, and J. Burke. 1988. Synthetic smoke with acrolein but not HCl produces pulmonary edema. J. Appl. Physiol. 64:1121-1133.

Hayakawa, T., S. Udenfriend, H. Yagi, and D.M. Jerina. 1975. Substrates and inhibitors of hepatic glutathione-S-epoxide transferase. Arch. Biochem. Biophys. 170:438-451.

IARC. 1987. Acrolein. P. 78 in IARC Monographs on the Evaluation of Carcinogenic Risks to Humans. Overall Evaluations of Carcinogenicity: An Updating of IARC Monographs, Vols. 1-42, Suppl. 7. International Agency for Research on Cancer, Lyon, France.

Kaplan, H.L., A.F. Grand, W.G. Switzer, D.S. Mitchell, W.R. Rogers, and G.E. Hartzell. 1986. Effects of combustion gases on escape performance of the baboon and the rat. Dang. Prop. Ind. Mat. Rep. 6:2-12.

Kutzman, R.S., E.A. Popenoe, M. Schmaeler, and R.T. Drew. 1985. Changes in rat lung structure and composition as a result of subchronic exposure to acrolein. Toxicology 34:139-151.

Lam, C.-W., M. Casanova, and H. d'A. Heck. 1985. Depletion of nasal mucosal glutathione by acrolein and enhancement of formaldehyde-induced DNA-protein cross-linking by simultaneous exposure of acrolein. Arch. Toxicol. 58:67-71.

Leach, C.L., N.S. Hatoun, H.V. Ratajczak, and J.M. Gerhart. 1987. The pathologic and immunologic effects of inhaled acrolein in rats. Toxicol. Lett. 39:189-198.

Leikauf, G.D., L.M. Leming, J.R. O'Donnell, and C.A. Doupnik. 1989. Bronchial responsiveness and inflammation in guinea pigs exposed to acrolein. J. Appl. Physiol. 66:171-178.

Lijinsky, W., and M.D. Reuber. 1987. Chronic carcinogenesis studies of acrolein and related compounds. Toxicol. Ind. Health 3:337-345.

Lutz, D., E. Eder, T. Neudecker, and O. Henschler. 1982. Structure-mutagenicity relationships in 2,8-unsaturated carbonylic compounds and their corresponding allylic alcohols. Mutat. Res. 93:303-315.

Lyon, J.P., L.J. Jenkins, Jr., R.A. Jones, R.A. Coon, and J. Siegel. 1970. Repeated and continuous exposure of laboratory animals to acrolein. Toxicol. Appl. Pharmacol. 17:726-732.

Marnett, L.J., H.K. Hurd, M.C. Hollstein, D.E. Levin, H. Esterbaure, and B.N. Ames. 1985. Naturally occurring carbonyl compounds are mutagenic in *Salmonella* tester strain TA104. Mutat. Res. 148:25-34.

Murphy, S.D., D.A. Klinghirn, and E.E. Ulrich. 1963. Respiratory response of guinea pig during acrolein inhalation and its modification by drugs. J. Pharmacol. Exp. Ther. 141:79-83.

Northrop. 1985. Pulmonary Toxicity of Nitrogen Dioxide and Acrolein. Final Report. Publ. No. SP-4220-85-49. Northrop Environmental Sciences, Research Triangle Park, N.C.

Patel, J.M., J.C. Wood, and K.C. Leibman. 1980. The biotransformation of allyl alcohol and acrolein in rat liver and lung preparations. Drug Metab. Dispos. 8:305-308.

Philippin, C., A. Gilgen, and E. Grandjean. 1970. Toxicological and physiological investigation on acrolein inhalation in the mouse. Int. Arch. Arbeitsmed. 26:281-305.

Prentiss, A.M. 1937. Pp. 139-140 in Chemicals in War. A Treatise on Chemical Warfare. New York: McGraw-Hill.

Sax, I. 1984. P. 127 in Dangerous Properties of Industrial Materials. New York: Van Nostrand Reinhold.

Shell Chemical Corp. 1958. Toxicity Data Sheet: Acrolein. SC: 57-76. Ind. Hyg. Bull. 4 pp.

Sherwood, R.L., C.L. Leach, N.S. Hatoum, and C. Aranyi. 1986. Effects of acrolein on macrophage functions in rats. Toxicol. Lett. 32:41-49.

Sinkuvene, D.S. 1970. [Hygienic evaluation of acrolein as an air pollutant.] Hyg. Sanit. (USSR) 35:325-329.

Sim, V.M., and R.E. Pattle. 1957. Effect of possible smog irritants on human subjects. J. Am. Med. Assoc. 165:1908-1913.

Steinhagen, W.H., and C.S. Barrow. 1984. Sensory irritation structure-activity study of inhaled aldehydes in B6C3F1 and Swiss-Webster mice. Toxicol. Appl. Pharmacol. 72:495-503.

Stephens, E.R., E.F. Darley, O.C. Taylor, and W.E. Scott. 1961. Photochemical reaction products in air pollution. J. Air Pollut. 4:79-100.

Weber-Tschopp, A., T. Fischer, R. Geier, and E. Grandjean. 1977. Experimentally induced irritating effects of acrolein on men. Int. Arch. Occup. Environ. Health 40:117-130.

Zimmering, S., J.M. Mason, R. Valencia, and R.C. Woodruff. 1985. Chemical mutagenesis testing in *Drosophila*. II. Results of 20 coded compounds tested for the National Toxicology Program. Environ. Mutagen. 7:87-100.

B2   *Benzene*

John T. James, Ph.D., and Harold L. Kaplan, Ph.D.
Johnson Space Center Toxicology Group
Biomedical Operations and Research Branch
Houston, Texas

PHYSICAL AND CHEMICAL PROPERTIES

Benzene is a clear, colorless, highly flammable liquid with an odor characteristic of aromatic hydrocarbons (Sandmeyer, 1981; ATSDR, 1989). The odor threshold is 4-5 ppm (ATSDR, 1989).

Synonym:	Benzol
Chemical structure:	(benzene ring)
Formula:	C_6H_6
CAS number:	71-43-2
Molecular weight:	78.11
Boiling point:	80.1 °C
Melting point:	5.5 °C
Specific gravity:	0.88 (20 °C)
Vapor pressure:	95 torr (25 °C)
Solubility:	Slightly soluble in water, very soluble in organic solvents
Conversion factors at 25 °C, 1 atm:	1 ppm = 3.26 mg/m^3 1 mg/m^3 = 0.31 ppm

OCCURRENCE AND USE

Benzene is a natural constituent of crude oil, coal tar, and other fossil fuels (Sandmeyer, 1981). Most of the millions of gallons of benzene that are used in the United States each year are produced by petroleum refining. A major use of benzene is as a component in gasoline, particularly in unleaded fuels, because of its antiknock properties (ATSDR, 1989). Its content in gasoline is estimated to range from 1% to 2% in the United States and up to 5% in European countries. Large quantities are also used to synthesize chemicals for the manufacture of various plastics, resins, elastomers, dyes, and pesticides (Marcus, 1987). Minimal amounts are now used as a solvent in paints, cements, adhesives, and paint removers. Sources of atmospheric contamination include fugitive emissions from gasoline handling, thermal degradation of plastics, solid waste gasification, and tobacco smoke (Sandmeyer, 1981; Marcus, 1987).

Benzene has been detected in approximately 10% of recent air samples in the space-shuttle cabin and in Spacelab at concentrations of 0.01-0.1 mg/m3 (James et al., 1994). Benzene has not been used as a payload or system chemical aboard the space shuttle; hence, the low concentrations observed are due to materials out-gassing. Benzene has been found as a pyrolysis product of electronic components identical to ones that failed in data-display units aboard STS-35 (J. Boyd, NASA, unpublished data).

PHARMACOKINETICS AND METABOLISM

Absorption

In humans, benzene vapor is rapidly absorbed by the lungs in amounts equivalent to about 50% of the doses inhaled over several hours of exposure to concentrations of 50-100 ppm (Nomiyama and Nomiyama, 1974a,b; Sato and Nakajima, 1979; R. Snyder et al., 1981; IARC, 1982). In men and women exposed to 52-62 ppm for 4 h, respiratory uptake was 47%, with little difference between the sexes (Nomiyama and Nomiyama, 1974a,b; IARC, 1982). Absorption was greatest during the first 5 min of exposure and reached a constant level between 15 min (Srbova et al., 1950) and 3 h (Nomiyama and Nomiyama, 1974a,b; IARC, 1982). Respiratory retention (the difference between respiratory uptake and excretion) was estimated as 30% of the inhaled dose (Nomiyama and Nomiyama, 1974a,b; IARC, 1982).

Benzene can also be absorbed through the skin, but the rate of absorption is lower than that for inhalation exposure (ATSDR, 1989). It has been calculated that an adult working in ambient air containing benzene at 10 ppm would absorb 7.5 μL/h from inhalation versus 1.5 μL/h from whole-body dermal absorption (Blank and McAuliffe, 1985). Absorption of benzene vapor by animals also is rapid, but retention of absorbed benzene might be affected by exposure concentration. In rats and mice, the percentage of inhaled vapor that was retained decreased from 33% to 15% during a 6-h exposure and from 50% to 10%, respectively, as the concentration was increased from approximately 10 ppm to 1000 ppm (Sabourin et al., 1987).

Distribution

Once benzene is absorbed into the blood, it is rapidly distributed to tissues; the relative uptake is dependent on the perfusion rate of tissues (ATSDR, 1989). Because of its high lipophilicity, benzene tends to accumulate in fatty tissues. In experimental human exposures, lower blood concentrations and slower elimination in females than in males were attributed primarily to relatively higher fat content of females (Sato et al., 1975). Tissue levels of benzene in accidental- or intentional-exposure victims are variable but generally indicate higher concentrations in brain, fat, and liver (Winek et al., 1967; Winek and Collom, 1971). About 60% of the absorbed benzene was found in bone marrow, adipose tissue, and liver of humans exposed to unspecified concentrations (Duvoir et al., 1946).

Distribution of benzene in animals also is rapid; the relative uptake and accumulation in tissues appear to be dependent on perfusion rate and lipid content (Schrenk et al., 1941; Ghantous and Danielsson, 1986). Following a 10-min inhalation exposure of mice, benzene was present in well-perfused tissues, such as liver and kidney, and in lipid-rich tissues, such as brain and fat (Ghantous and Danielsson, 1986). In rats exposed to 500 ppm, steady-state concentrations were highest in fat, bone marrow, and blood (15:5.5:1 ratio) and lower in kidney, lung, liver, brain, and spleen (Rickett et al., 1979). Female rats and male rats with large body fat content stored benzene longer and eliminated it more slowly than lean animals (Sato et al., 1974).

Excretion

Following inhalation, benzene is eliminated from the body by humans and animals in unchanged form in the exhaled air and in metabolized form in the urine (ATSDR, 1989). Estimates of the fraction of absorbed benzene excreted in the expired air of humans range from 12% to 50% (Srbova et al., 1950; Teisinger et al., 1952; Nomiyama and Nomiyama, 1974a,b; IARC, 1982). The respiratory elimination is described as triphasic—an initial fast component having a half-life of 0.9 h and two slower components having half-lives of 3 and 15 h, respectively (Nomiyama and Nomiyama, 1974a,b; IARC, 1982). No differences in respiratory elimination were observed between men and women (Nomiyama and Nomiyama, 1974a,b; IARC, 1982).

In rats exposed to 500 ppm for 6 h, a biphasic pattern of respiratory elimination of benzene was observed, with half-lives of 0.7 and 13.1 h (Rickert et al., 1979). Respiratory elimination might be increased as a result of saturation of metabolic pathways by high doses of benzene. At lower concentrations (10-130 ppm), less than 6% of inhaled ^{14}C-benzene was exhaled by rats and mice, whereas at concentrations of 260 and 870 ppm (rats) and 990 ppm (mice), exhaled radioactivity increased from 11% to 48% (Sabourin et al., 1987).

Metabolism

The metabolism of benzene is complex and not completely elucidated. It is well established that most of the absorbed benzene is metabolized through a variety of major and minor pathways in humans and animals and excreted as metabolites in the urine (Snyder, 1987). The major site of metabolism is the liver, although mixed-function oxidases that catalyze the oxidation of benzene also occur in bone marrow, the target organ of benzene toxicity (Snyder, 1987). Benzene is metabolized mostly by oxidation to the major metabolites—phenol, catechol, and hydroquinone, which are excreted in the urine as sulfate and glucuronide conjugates (Snyder, 1987). Many minor metabolites also are formed, of which phenylmercapturic acid and *trans,trans*-muconic acid are the most important.

In the initial metabolic step, benzene is thought to be oxidized by hepatic microsomal mixed-function oxidase to a reactive intermediate, benzene oxide (Erexson et al., 1985; Yardley-Jones et al., 1991). Most of the benzene oxide rearranges spontaneously to phenol. Phenol is mostly conjugated and excreted as sulfate ester and glucuronide, but some can be further oxidized to catechol and hydroquinone (Erexson et al., 1985; ATSDR, 1989). The latter compound spontaneously oxidizes to form 1,4-benzoquinone. The remaining benzene oxide can be conjugated with glutathione to produce phenylmercapturic acid, which is excreted in the urine, or it can be converted to benzene glycol (ATSDR, 1989; Erexson et al., 1985). The latter compound can undergo dehydrogenation to form catechol or further oxidation and ring breakage to produce *trans,trans*-muconic acid. Most of the catechol is conjugated and excreted, but a small amount is oxidized to the trihydroxybenzene 1,2,4-benzenetriol.

In humans, the major urinary metabolite of benzene is phenol (Teisinger et al., 1952). Most of the phenol is excreted as sulfate ester (Teisinger et al., 1952), but significant amounts of glucuronide can be formed, especially after exposure to high concentrations of benzene (Sherwood, 1972). In an inhalation study with human subjects, 28.8% of the absorbed benzene was excreted in the urine as phenol, 2.9% as catechol, and 1.1% as hydroquinone (Teisinger et al., 1952). Urinary excretion was highest within the first 24 h following exposure and was essentially complete within 48 h.

In workers exposed for 7 h to benzene at 1-76 ppm, the correlation between exposure concentration and urinary phenol excretion was 0.891 (Inoue et al., 1986). A urinary phenol concentration of 75 mg/L indicates an 8-h (time-weighted average) exposure at 10 ppm (NIOSH, 1974) and a concentration of 100 mg/L indicates an 8-h exposure at 25 ppm (Sandmeyer, 1981). The ACGIH biological exposure index (BEI) for benzene exposure is a urinary concentration of phenol at 50 mg/g of creatinine at the end of a workshift, but ACGIH notes that phenol is usually present in unexposed individuals and also might result from exposure to other chemicals (ACGIH, 1991). Recently, *S*-phenylmercapturic acid in urine (Stommel et al., 1989) and benzene hemoglobin adducts in blood (Sun et al, 1990) were evaluated as possible biomarkers of benzene exposure.

In a study in which radiolabeled benzene was administered to rabbits by oral intubation, 43% of the radioactivity was recovered as exhaled, unmetabolized benzene and 1.5% was recovered as carbon dioxide (Parke and Williams, 1953). Urinary metabolites (representing 35% of the dose) were mainly in the form of phenolic sulfates and glucuronides and included phenol (23%), hydroquinone (4.8%), catechol (2.2%), *trans,trans*-muconic acid (1.3%), phenylmercapturic acid (0.5%), and 1,2,4-trihydroxybenzene (0.3%). This same general profile was also found in rats (Cornish and Ryan, 1965), mice (Longacre et al., 1981), and cats and dogs (Oehme, 1969).

Benzene metabolism appears to be qualitatively similar but quantitatively different among species. Mice metabolized the largest fraction of benzene (67%) to hydroquinone conjugates and muconic acid metabolites, followed by monkeys (31%), rats (17%), and chimpanzees (14%) (Sabourin et al., 1992). Urinary-metabolite data from workers exposed for about 7 h to benzene at 50 ppm suggest that the metabolism of benzene to hydroquinone compounds in humans is quantitatively comparable to that in mice, whereas the metabolism to muconic acid is comparable to that in rats and one-third of that in mice (Henderson et al., 1992).

Metabolic pathways leading to putative toxic metabolites, such as hydroquinone and muconic acid metabolites, have been designated "toxification pathways" in contrast to "detoxification pathways," which lead to the less toxic metabolites, such as phenyl conjugates and phenylmercapturic acid products (Henderson et al., 1992). In mice, rats, and monkeys, the toxification pathways appear to be low-capacity, high-affinity pathways that become saturated at relatively low concentrations, resulting in proportionately less formation of hydroquinone and muconic acid at higher concentrations (Henderson et al., 1992).

Stimulation or inhibition of hepatic mixed-function oxidase activity by benzene, other chemicals, or dietary factors might alter the rate of metabolism of benzene. Exposure of mice to benzene enhanced the in vitro metabolism of benzene by hepatic microsomes from these animals, but exposure to phenolic metabolites did not (Dean, 1978; Gonasun et al., 1973). In contrast, repeated inhalation exposure of mice and rats to benzene at 600 ppm for 6 h/d, 5 d/w, for 3 w had minimal effects on urinary-metabolite profiles (Sabourin et al., 1990). Ethanol ingestion as well as food deprivation and carbohydrate restriction enhanced the metabolism of benzene in rats (Sato and Nakajima, 1985).

Three physiologically based pharmacokinetic (PBPK) models were recently proposed to describe the pharmacokinetics and metabolism of benzene in animals (Medinsky et al., 1989; Woodruff et al., 1989; Travis et al., 1990). The models are herein arbitrarily referred to by the name of the first author as the Medinsky model (Medinsky et al., 1989), the Travis model (Travis et al., 1990), and the Woodruff model (Woodruff et al., 1989). The Medinsky model was adjusted to data obtained experimentally with mice and rats (Medinsky et al., 1989), the Travis model with data on mice, rats, and humans (Travis et al., 1990), and the Woodruff model with data on rats (Woodruff et al., 1989). The models have similar structures but differ in the parameter values used for the same species. In a comparison and evaluation of the models, the investigators concluded that PBPK models are useful for investigating the mechanism of toxicity of benzene but not for risk assessment of cancer (Bois et al., 1991).

TOXICITY SUMMARY

Acute and Short-Term Toxicity

Neurotoxicity

In humans, acute inhalation of benzene produces CNS effects, including euphoria, giddiness, nausea and drowsiness at lower concentrations, and ataxia, narcosis, delirium, convulsions, unconsciousness, and even death at high concentrations (Sandmeyer, 1981). Recovery is usually rapid, but, in some cases, symptoms have persisted for weeks.

The symptoms, and their severity, vary with concentration and duration of exposure. It is estimated (without supporting data) that exposure at 25 ppm for 8 h has no effects; 50-150 ppm for 5 h produces headache, lassitude, and weakness, symptoms that are exaggerated at 500 ppm; 3000 ppm for 0.5-1.0 h can be tolerated; 7500 ppm for 30 min is dangerous to life; and 19,000-20,000 ppm can be fatal in 5-10 min (Gerarde, 1962; Von Oettingen, 1940). The rapid development of CNS effects, including death in some cases, suggests that benzene, not a metabolite, is responsible for the acute toxicity and that the cause of sudden death is asphyxiation, respiratory arrest, CNS depression, or cardiac arrhythmia (Sandmeyer, 1981).

Acute inhalation exposure to benzene also causes CNS effects in animals. In mice, exposure at 2200 ppm produced narcosis, and at 4600 ppm and 11,800 ppm produced narcosis in 51 min and 8 min, respectively (Von Oettingen, 1940). At 11,800 ppm, deaths occurred in 38-295 min, and at 24,000 ppm, deaths occurred in 50 min. About 4000 ppm was the narcotic concentration threshold for laboratory animals, and more than 10,000 ppm was fatal after several hours of exposure (Leong, 1977). At 35,000-45,000 ppm, anesthesia occurred in about 4 min, with excitation and tremors after 5 min, loss of pupillary reflexes after 6.5 min, involuntary blinking after 15 min, and death after 22-71 min (Carpenter, et al., 1944).

Lethality data of rats confirm the low potential for benzene to cause death via inhalation. A 4-h exposure at 16,000 ppm resulted in the deaths of four of six rats (Smyth et al., 1962). An LC_{50} value of 13,700 ppm was determined for a 4-h exposure of rats (Drew and Fouts, 1974). Respiratory paralysis followed by ventricular fibrillation was observed in male rats exposed to lethal concentrations (Sandmeyer, 1981).

A limited number of animal studies measured electroencephalographic and behavioral changes to investigate the CNS effects of benzene. In cats, exposure to benzene at a concentration of 12,000 ppm for 10 min caused restlessness, rapid respiration, and head nodding, accompanied by hypersynchronous amygdaloid EEG activity (Contreras et al., 1979). Ataxia and postural collapse occurred when concentrations were increased to 52,000 ppm. With repeated daily 10-min exposures, a 3-Hz spike-wave activity in the gyrus cinguli developed and generalized tonic-clonic seizures developed after a sensitization period. Behavioral disturbances, characterized by increased milk-licking, were evident in C57BL mice after the first week of exposure to 100 or 300 ppm (Dempster et al., 1984). Less sensitive parameters, home-cage food intake and hind-limb grip strength, were reduced at 1000 and 3000 ppm, but not at 100 or 300 ppm, even when exposure durations were adjusted to yield a minimum Ct (concentration × time) product of 3000 ppm.

Cardiac Sensitization

Acute inhalation of high concentrations of benzene by cats and monkeys induced ventricular dysrhythmias, which were abolished by removal of the adrenals and the stellate ganglia, and were restored by injections of epinephrine (Nahum and Hoff, 1934). The effects were attributed to the sensitization of the myocardium to epinephrine by benzene.

In Wistar rats, previous inhalation of benzene at 3000 and 7000 ppm, but not at 1500 ppm, for 15 min increased the number of ectopic ventricular beats induced by coronary ligation or intravenous administration of aconitine (Magos et al., 1990). With an increased dose of aconitine, ventricular fibrillation developed rapidly at 7000 ppm, and progressed to asystole and death after 16 min.

Hematotoxicity and Immunotoxicity

Although benzene-induced hematotoxicity and immunotoxicity are generally associated with prolonged exposure, abnormal hematological parameters have been observed in some workers exposed to low concentrations for short periods (ATSDR, 1989). These observations are consistent with the results of animal studies showing hematological changes after short-term, and even acute, exposures. After an 8-h inhalation exposure of mice to 4680 ppm, a significant depletion of bone-marrow colony-forming cells was evident in an in vitro cell culture (Uyeki et al., 1977). In mice, continuous exposure at 100 ppm for 2 d produced leukocytopenia (Gill et al., 1980) and a 1-w exposure (6 h/d, 5 d/w) at 300 ppm decreased peripheral blood erythrocyte and lymphocyte counts (Snyder et al., 1978). Continuous exposure of NMRI mice at a concentration of 21 ppm for 4-10 d significantly decreased cellularity (number of nucleated cells) and colony-forming granulopoietic stem cells (CFU-C) in tibia bone marrow (Toft et al., 1982). Intermittent exposure (8 h/d, 5 d/w) for 2 w at 21 ppm reduced the number of CFU-C cells.

In female Wistar rats exposed 8 h/d for 7 d, peripheral leukocyte counts were depressed significantly after exposures at 50-300 ppm but not at 20 ppm (Li et al., 1986). Leukocyte alkaline phosphatase (LAP) concentrations were significantly increased at 300 ppm, marginally increased at 100 ppm, and not affected at 20 or 50 ppm.

Short-term exposures of animals at low concentrations might produce hematological changes that can affect immune-associated processes. In male C57BL mice exposed at 10 ppm for 6 h/d for 6 d, femoral lipopolysaccharide (LPS)-induced B-lymphocyte-colony-forming ability was significantly depressed, but total numbers of B lymphocytes were not (Rozen et al., 1984). At 30 ppm, splenic phytohemagglutinin (PHA)-induced blastogenesis was significantly depressed, but there was no concomitant significant depression in numbers of T lymphocytes. After six exposures at 300 ppm, mitogen-induced proliferation of bone marrow and splenic B and T lymphocytes was depressed, and numbers of T lymphocytes in thymus and spleen were reduced (Rozen and Snyder, 1985).

Genotoxicity

Benzene is not mutagenic in most in vitro test systems, including both *Salmonella typhimurium* (five strains) and *Saccharomyces cerevisiae* with and without metabolic activation, *Drosophila melanogaster,* mouse lymphoma cells, various human, mouse, and Chinese hamster cells, and others (ATSDR, 1989; Marcus, 1987). In vitro studies of chromosomal aberrations and other genotoxic effects of benzene yielded positive, negative, or mixed results, depending on the end point and test system. Positive results were obtained in studies of DNA binding in rabbit bone marrow and rat liver mitoblasts; negative results were obtained in studies of DNA breaks in rat hepatocytes, Chinese hamster V79 cells, and mouse L5178Y cells; and mixed results were obtained in studies of chromosomal aberrations and sister chromatid exchange (SCE) in human lymphocytes (ATSDR, 1989). Benzene did not increase SCE frequency in human lymphocytes stimulated by phytohemagglutinin (Morimoto and Wolff, 1980) or in human lymphocytes incubated without rat liver S-9 (Morimoto, 1983). Delaying addition of benzene, however, to 24 h after mitogen stimulation produced significant concentration-related increases in SCE frequency, decreases in mitotic indices, and inhibition of cell-cycle kinetics without S-9 (Erexson et al., 1985).

In contrast to in vitro results, benzene-induced cytogenetic effects, including chromosomal and chromatid aberrations, SCE, and micronuclei, were consistently found in in vivo animal studies (ATSDR, 1989). Acute inhalation studies have shown cytogenetic effects in animals, even at low exposure concentrations. Exposure of mice at 10 ppm for 6 h induced SCE in peripheral blood lymphocytes and bone marrow as well as micronuclei in bone-marrow polychromatic erythrocytes (Erexson et al., 1985). Exposure of DBA/2 mice at 3100 ppm for 4 h significantly increased SCE frequency in bone-marrow cells in both sexes and inhibited marrow cellular proliferation in males only, but did not affect the frequency of chromosomal aberrations (Tice et al., 1980).

Subchronic and Chronic Toxicity

Prolonged inhalation of benzene by humans can result in CNS, hematotoxic, myelotoxic, immunotoxic, genotoxic, or carcinogenic effects. These effects are well established for chronic exposure, but less is known about some of these effects as a result of subchronic exposure. Limited information exists on the potential of benzene to cause adverse effects on reproductive function and pre- and postnatal development in humans.

Neurotoxicity

Involvement of the CNS might be an important effect of chronic inhalation exposure of humans and animals to benzene, but it can be masked by other more-visible effects (Sandmeyer, 1981). Workers exposed even to low concentrations (e.g., 50 ppm) reported symptoms of headaches, dizziness, fatigue, anorexia, dyspnea, and visual disturbances (Sandmeyer, 1981). Some workers also exhibited signs of CNS lesions, such as abnormal caloric labyrinth irritability and impairment of hearing. Although there are reports of polyneuritis associated with exposure to benzene, other chemicals were also involved (Sandmeyer, 1981). Exposure of rats for 5.5 mo to 20 ppm resulted in a delay in conditioned reflex response time; however, the effect was not seen at 4 ppm (Novikov, 1956).

Hematotoxicity and Myelotoxicity

The effects of benzene on the hematopoietic system have been known for many years. Prolonged exposure causes hypoplasia and depressed function of bone marrow, resulting in leukopenia, anemia, or thrombocytopenia (Sandmeyer, 1981; ATSDR, 1989). With continued exposure, bone-marrow aplasia results in pancytopenia and aplastic anemia; bone-marrow aplasia might progress and develop into myelogenous leukemia or other types of leukemia. These are not distinct diseases but rather are a continuum of changes reflecting the severity of damage to the bone marrow.

The cytopenias, which can occur as a group or in various combinations, can manifest themselves as specific adverse health effects (ATSDR, 1989). For example, thrombocytopenia induces capillary fragility, petechiae, and hemorrhage, which might result in death. Decreased circulating granulocytes decrease defenses against infection, which might have been responsible for deaths of some benzene-exposed individuals. Also, lymphocytopenia and eosinophilia, which might be related to impaired immune function, were observed in some workers.

A high prevalence and wide range of hematological responses to benzene are evident in the numerous epidemiological studies and case reports of occupationally exposed workers. Of 332 rotogravure workers exposed to benzene at 11-1060 ppm for 6-60 mo, 23 had severe cytopenia (23 of 23, leukopenia; 15 of 23, erythropenia; 18 of 23, thrombocytopenia) (Goldwater, 1941). In a rubber factory, 25 of 1104 workers exposed at up to 500 ppm (100 ppm average) developed severe pancytopenia (9 of the 25 were hospitalized), and 83 others had mild hematological disorders (Wilson, 1942). The relationship between pancytopenia, preleukemia and acute leukemia was reviewed by Goldstein (1977). Pancytopenia also was diagnosed in 6 of 217 apparently healthy shoe-factory workers exposed at 30-210 ppm for 3 mo to 17 y; 45 others had some hematological abnormalities (Aksoy et al., 1971). There also are numerous reports of aplastic anemia in occupationally exposed workers (Aksoy et al., 1972; Vigliani and Forni, 1976). Of 32 cases of aplastic anemia among workers exposed at 150-650 ppm for 4 mo to 15 y, there were eight deaths due to thrombocytopenic hemorrhage and infection (Aksoy et al., 1972). In a 10-y followup of 216 workers in a study of 282 workers, four had persistent cytopenias and one died of aplastic anemia 9 y after cessation of exposure (Guberan and Kocher, 1971). The exposure level associated with development of noncarcinogenic hematological effects of benzene has not been established (ATSDR, 1989). A threshold of about 10 ppm for cytopenia was suggested on the basis of observations of minimal hematotoxicity in workers exposed at 20 ppm (Chang, 1972).

There is evidence that benzene-induced pancytopenia or aplastic anemia is associated with the later development of leukemia (ATSDR, 1989). In 44 patients with pancytopenia (exposure at 150-650 ppm for 4 mo to 15 y), six developed leukemia within 6 y of followup (Aksoy and Erdem, 1978). Leukemia also occurred in workers with aplastic anemia either during exposure to high concentrations or shortly after cessation of exposure; however, in a few cases the latency period was long (Aksoy et al., 1976; Aksoy, 1978). Benzene-induced leukemia is discussed in more detail in the section on carcinogenicity.

The hematotoxic effects observed in humans have been reproduced experimentally in animals; however, the response depends on species, strain, sex, and intermittent vs. continuous exposure, in addition to exposure concentration and duration (ATSDR, 1989).

Exposure of rats to benzene at concentrations of 831, 65, or 61 ppm produced a significant leukopenia within 2 to 4 w, and a less severe leukopenia after 5-8 w at concentrations of 47 or 44 ppm (Deichmann et al., 1963). Leukocyte counts were not affected by 31 ppm for 4 mo, 29 ppm for 3 mo, or 15 ppm for 7 mo. In CD-1 mice, 300 ppm for 6 h/d, 5 d/w for 13 w increased mean cell volume and mean cell hemoglobin and decreased hematocrit, hemoglobin, erythrocyte, leukocyte, and platelet counts and percentage of lymphocytes (Ward et al., 1985). Histopathological changes, including bone-marrow hypoplasia, lymphoid depletion in lymph nodes and tissue, and increased splenic extramedullary hematopoiesis, were more prevalent and severe in males than females. No effects were evident at 1, 10, or 30 ppm.

Sprague-Dawley rats were less severely affected by the same exposure regimen than mice. The rats exhibited no effects at 1, 10, or 30 ppm and significant decreases only in leukocyte counts and percentage of lymphocytes at a concentration of 300 ppm (Ward et al., 1985). The only histological lesion was slightly decreased femoral-marrow cellularity. Repeated inhalation exposures at 80-85 ppm for 136 exposures of rats, 175 exposures of rabbits, and 193 exposures of guinea pigs induced leukopenia, increased spleen weights, and histopathological changes in bone marrow in the rats, guinea pigs, or rabbits (Wolf et al., 1956).

Several animal studies indicate that benzene induces its hematotoxicity by acting on early progenitor cells in the bone marrow and spleen. In CD-1 mice, 103 ppm and higher exposures for 6 h/d for 5 d reduced marrow and spleen cellularity and decreased granulocyte macrophage colony-forming units (GM-CFU-C) in spleen but not in marrow (Green et al., 1981a). At concentrations of 302 ppm for 26 w, marrow and spleen cellularity, colony-forming units in spleen (CFU-S), and marrow GM-CFU-S were decreased. Depression of CFU-S also was reported in C57BL mice exposed at 400 ppm for 6 h/d intermittently for 9 d or consecutively for 11 d (Harigaya et al., 1981). Bone-marrow cellularity and pluripotential stem cells were significantly reduced in C57BL mice exposed for 2 w at 100 ppm, but not at 10 or 25 ppm (Cronkite et al., 1985). At 300 ppm, 2 w were required for recovery of stem-cell numbers after 2- or 4-w exposures, and 25 w were required for recovery to 92% of control values after a 16-w exposure. Peripheral blood lymphocyte counts were not affected at 10 ppm (2 w), but exhibited a dose-related decrease at 25-400 ppm.

Other investigators also observed depletions in pluripotential stem-cell numbers (Gill et al., 1980) and reductions in granulocyte and macrophage progenitor cells (C.A. Snyder et al., 1981) in mice. In studies of the erythroid cell line, repeated exposures to benzene at 10 ppm reduced the number of progenitor red blood cells, i.e., erythroid colony-forming units (CFU-E) in mice (Baarson et al., 1984; Valle-Paul and Snyder, 1986). The effects of benzene at concentrations of 100, 300, and 900 ppm (6 h/d, 5 d/w, for up to 16 w) on hematopoietic stem-cell compartments were investigated in a series of studies with female BDF1 mice (Seidel et al., 1989a,b, 1990). The CFU-E was the most sensitive compartment, showing significant concentration-dependent decreases and a marked decrease at 100 ppm as early as 1 w after the start of exposure (Seidel et al., 1990). The BFU-E (burst-forming units), CFU-S, and CFU-C compartments of progenitor cells showed dose-related decreases at 300 and 900 ppm. Recovery of stem-cell compartments was slow, requiring 73-185 d after exposure at 300 ppm.

Potential mechanisms for the development of pancytopenia and its variants in humans and animals exposed to benzene include destruction of bone-marrow stem cells, impairment of the differentiation of these cells, and destruction of more mature hematopoietic cell precursors and circulating cells (Goldstein, 1977). Numerous studies have shown that benzene-induced bone-marrow depression is the result of inhibitory effects on proliferation, maturation, or replication of pluripotential stem cells or early proliferating committed cells in either the erythroid or myeloid lines (ATSDR, 1989). Several molecular mechanisms have been proposed to explain the hematotoxicity and myelotoxicity, as well as other toxicities, of benzene. These include suppression of RNA and DNA synthesis, alkylation of cellular sulfhydryl groups, disruption of the cell cycle, oxygen activation (or free radical formation), and covalent building of benzene metabolites to cellular macromolecules (ATSDR, 1989). These mechanisms will be reviewed in greater detail in the section on carcinogenicity.

Although the exact mechanism is unknown, it is generally accepted that benzene must be metabolized before its toxic effects (other than neurotoxicity and cardiac sensitization) are manifest (ATSDR, 1989). Evidence for a primary role for benzene metabolites in inducing myelosuppression and hematotoxicity is provided by studies showing that agents that alter benzene metabolism modify its toxicity. Partial hepatectomy alleviated benzene-induced depression of erythropoiesis, and decreased urinary levels of benzene metabolites and covalent binding of reactive metabolites to bone-marrow protein (Sammett et al., 1979). The apparent decrease in toxicity provided by phenobarbital and Arochlor-1254 was attributed to increased detoxifying metabolism of benzene in the liver (Ikeda and Ohtsuji, 1971; ATSDR, 1989); the decrease provided by toluene and aminotriazole was attributed to inhibition of metabolism resulting in a decreased rate of toxic metabolite formation (Hirokawa and Nomiyama, 1962). On the other hand, ethanol ingestion generally increases benzene-induced hematotoxicity, possibly by increasing the rate of formation of toxic metabolites (Driscoll and Snyder, 1984).

Also supporting a causative role of benzene metabolites in bone-marrow suppression and hematotoxicity are studies showing the accumulation of metabolites in bone marrow. Although benzene can be metabolized in bone marrow, mixed-function oxidase activity appears to be insufficient to account for the high concentrations of phenol, hydroquinone, and catechol in bone marrow (ATSDR, 1989). It is postulated, therefore, that one or more metabolites formed in the liver are transported to the bone marrow where they accumulate and produce a metabolic impairment expressed as bone-marrow depression (Marcus, 1987). Finally, metabolites of benzene, including benzene oxide, hydroquinone, catechol, and *trans,trans*-muconaldehyde (a precursor of muconic acid), have been shown to be hematotoxic to animals (Marcus, 1987; ATSDR, 1989).

Immunotoxicity

It has long been suspected that benzene might adversely affect human immune functions. Studies in the early 1900s demonstrated an increased susceptibility of benzene-treated rabbits to tuberculosis and pneumonia (Marcus, 1987). Later, depression of lymphocytes and increased susceptibility to infection became increasingly associated with exposure of workers to benzene. With advances in immunology, alterations in serum immunoglobulin and complement levels were detected in occupationally exposed workers (Marcus, 1987). In 35 painters exposed to benzene, along with toluene and xylene, at concentrations of 3.4-48 ppm, serum IgG and IgA levels were significantly decreased compared with controls, and IgM levels were increased (Lange et al., 1973). Leukocyte agglutinins also were increased in some workers, leading to the suggestion that benzene might cause an allergic blood dyscrasia in some individuals. Other findings in workers, including eosinophilia and leukocyte agglutination associated with granulocytopenia, also suggest that autoimmunity or allergy is responsible for benzene-induced effects on immune function (Goldstein, 1977). Furthermore, auto-immune phenomena and reticulosis were implicated in the pathogenesis of bone-marrow disease.

In C57BL mice, 300-ppm exposures for 6 h/d for 6, 30, or 115 d reduced mitogen-induced proliferation of bone-marrow and splenic B and T lymphocytes and markedly reduced the number of B lymphocytes in bone marrow and spleen and the number of T lymphocytes in thymus and spleen (Rozen and Snyder, 1985). Increased bone-marrow cellularity and numbers of thymic T cells between the 6th and 30th exposure suggested a compensating proliferative response. Thymic lymphoma was observed after 115 exposures. Significant suppression of the primary antibody response to fluid (FTT) and adsorbed (APTT) tetanus toxoid was observed in Swiss albino mice exposed at 200 ppm for 6 h/d for 10-20 d but was not observed at 50 ppm (Stoner et al., 1981). At concentrations of 400 ppm for 5, 12, or 22 exposures, the primary antibody response to FTT was reduced by 74-89% and that to APTT by 8%, 36%, and 85%, respectively. The secondary antibody response was unaffected at 50, 200, or 400 ppm.

In a study of cell-mediated immunity, host resistance to *Listeria monocytogenes* was measured in mice exposed to benzene for either 5 d prior to infection (pre-exposure regimen) or for 5 d prior to and 7 d after infection (continuous regimen) (Rosenthal and Snyder, 1985). The pre-exposure regimen at 300 ppm increased splenic bacterial counts (730% of controls) on day 4 but had no effect at 10, 30, or 100 ppm. With the continuous-exposure regimen, bacterial counts were increased at 30, 100, and 300 ppm to 490%, 750%, and 720% of controls, respectively, and were unaffected at 10 ppm. On day 7, bacterial counts were not increased by either regimen, indicating recovery of the immune response. At concentrations of 30 ppm and higher, a concentration-dependent decrease in T and B lymphocytes was observed, with B lymphocytes showing a greater decrease. Tumor resistance, another parameter of cell-related immunity, also is adversely affected by benzene. In male C57BL mice exposed at 100 ppm for 6 h/d, 5 d/w, for 20 exposures and then injected with cells from a virus-induced tumor, 90% developed lethal tumors, compared with 30% of controls (Rosenthal and Snyder, 1986).

Several metabolites of benzene are suspect in benzene's immunotoxicity, but the identity of the causative agent or agents and the mechanism of action have not been established. Catechol, hydroquinone, benzoquinone, and 1,2,4-benzenetriol are cytotoxic to spleen cells, reduce the number of progenitor cells from the spleen and bone marrow, or suppress T- and B-lymphocyte mitogen responses (Irons and Neptun, 1980; Pfeifer and Irons, 1981). Suppression of cell growth and function in the lymphoid system, as in the bone marrow, correlates with the concentration of hydroquinone and catechol, which accumulate in lymphoid tissue following exposure to benzene (Greenlee et al., 1981; Wierda and Irons, 1982; Irons et al., 1982). Also, hydroquinone, benzoquinone, phenol, and catechol suppress microtubule assembly and progenitor cells (Kalf et al., 1987). Inhibition of microtubule function might result in suppression of phytohemagglutinin-stimulated lymphocyte activation; the inactivation correlates with the ability of the metabolites to undergo sulfhydryl-dependent autoxidation (Irons and Neptun, 1980; Pfeifer and Irons, 1981). It has been suggested that hydroquinone or its terminal oxidation product, *p*-benzoquinone, might be responsible for these effects (Irons and Neptun, 1980).

Mutagenicity and Genotoxicity

Evidence that benzene is genotoxic to humans comes from epidemiological studies of occupationally exposed workers. These studies show that workers with benzene-induced blood disorders consistently exhibited an increased prevalence of chromosomal aberrations; in workers who were without overt signs of toxicity or were exposed to benzene at low concentrations, the results were less consistent (ATSDR, 1989).

There are extensive reviews of epidemiological studies and case reports of benzene-induced chromosomal aberrations in workers (Snyder et al., 1977; White et al., 1980; Van Raalte and Grasso, 1982; Dean, 1985). A significantly higher number of lymphocytes with unstable chromosomal aberrations were found in 20 men (many with neutropenia) exposed to benzene for 1-20 y than were found in unexposed controls (1.4% exposed vs. 0.6% controls) (Tough and Brown, 1965). In rotogravure workers exposed at 125-532 ppm for 1-22 y, unstable and stable chromosomal aberrations in lymphocytes were significantly increased, compared with controls (Forni et al., 1971a). Similar results were reported in 25 persons (13 men and 12 women), even after recovery from hemopathy (Forni et al., 1971b).

Even at low concentrations, chromosomal aberrations were increased; for example, increases were found in 52 workers exposed at <10 ppm (estimated time-weighted-average exposure 2.1 ppm) for 5 y (Picciano, 1979) and in 22 healthy workers exposed at 13 ppm for 11 y (Sarto et al., 1984). In other studies, significant increases were not found when exposures were <25 ppm (Austin et al., 1988). Some investigators reported increases in the frequency of chromosomal damage and of sister chromatid exchange (SCE) in peripheral blood lymphocytes of workers exposed to concentrations as low as 1 ppm (Dean, 1985). Others failed to detect a statistically significant increase in SCE frequency in workers exposed to higher concentrations (Watanabe et al., 1980; Clarke et al., 1984; Sarto et al., 1984).

Exposure of CD-1 mice for 22 h/d, 7 d/w for 6 w at 0.04 ppm or 0.01 ppm increased the frequencies of spleen lymphocytes with mutations at the hypoxanthine-guanine phosphoribosyl transferase (hrpt) locus and the frequencies of chromosomal aberrations (chromatid breaks) (Au et al., 1991; Ward et al., 1992). Reduced effects at 1 ppm were attributed possibly to increased glutathione-S-transferase levels at higher doses, resulting in more detoxification metabolites and less putative toxic compounds (Ward et al., 1992). Female mice were found to be more sensitive than males in the Au et al. (1991) and the Ward et al. (1992) studies, but in studies using high doses, male mice were more sensitive than female mice (Uyeki et al., 1977; Choy et al., 1985). Exposure of mice (DBA/2, B6C3F$_1$, and C57BL mice) at 300 ppm for 6 h/d, 5 d/w (regimen 1) or 3 d/w (regimen 2) for 13 w induced a highly significant increase in the frequency of micronucleated polychromatic erythrocytes (Luke et al., 1988a). The magnitude of the increase was strain-specific, with DBA/2 greater than C57BL, and C57BL equal to B6C3F$_1$ mice, but independent of exposure regimen and, except for B6C3F$_1$ mice, independent of exposure duration. In DBA/2 mice, this genotoxic injury to the bone marrow was accompanied by a decreased percentage of polychromatic erythrocytes, indicating depression of erythropoiesis (Luke et al., 1988b).

Carcinogenicity

Epidemiological studies and case reports provide convincing evidence of the carcinogenic (leukemogenic) effects of benzene inhalation (Vigliani, 1976; Infante et al., 1977; Ott et al., 1978; Rinsky et al., 1981; Maltoni et al., 1989). The first epidemiological study of benzene, published in 1974, reported a leukemia incidence during 1967-1973 of 13/100,000 among 28,500 Turkish shoe workers exposed to benzene at concentrations of 150-650 ppm for 4 mo to 15 y (Aksoy et al., 1974). This incidence was significantly higher than the estimated 6/100,000 for the general population, and the incidence decreased after use of benzene was discontinued in 1969 (Aksoy, 1980). A mortality study and continued followup studies of rubber-industry workers exposed at 10-100 ppm for up to 10 y or more reported excessive mortality from myelogenous leukemia and reported a direct correlation between benzene exposure and other forms of leukemia (Infante et al., 1977; Infante, 1978; Rinsky et al., 1981, 1987). Epidemiological studies or case reports of chemical workers or other workers suggested a direct or possible correlation between exposure and excess mortalities from, or the development of, one or more forms of leukemia (Aksoy, 1978; Infante, 1978; Bond et al., 1986; Rinsky et al., 1987). Some of these studies, however, were criticized for inappropriate sampling techniques, exposure determinations, mortality standards, and experimental design (Van Raalte and Grasso, 1982).

IARC (1982), EPA (1989), NIOSH (1977), and others (ATSDR, 1989) concluded that benzene is carcinogenic to humans and is associated with an increased incidence of myelogenous leukemia. The data, however, do not establish an association with other types of leukemia. Several risk-assessment models were developed to estimate the probability of developing leukemia from a particular exposure level, utilizing data from the major epidemiological studies. A number of these assessments, including the Crump and Allen, the Rinsky et al. (and variations), and the White, Infante, and Chu assessments, were critically reviewed (Brett et al., 1989). The models yield widely varying risk estimates, depending on the particular model, the data selected for the model, and the exposure assumptions. In 1987, OSHA established an 8-h permissible exposure limit (PEL) of 1.0 ppm for benzene, relying on the Crump and Allen linear risk assessment, which was based on combined data from three high-quality epidemiological studies (Brett et al., 1989). The assessment projects a risk of 10 excess leukemia deaths per 1000 workers as a result of a 45-y occupational exposure to benzene at 1 ppm.

In 1990, the ACGIH proposed revision of its threshold limit value (TLV) for benzene from 10 ppm to 0.1 ppm, with a skin notation and designation as an A1 carcinogen (confirmed human carcinogen) (ACGIH, 1991). The ACGIH based its proposal on: (1) leukemia risk assessments emphasizing NIOSH case control data; (2) exposure levels in a cohort mortality study of leukemia in chemical workers; and (3) exposure levels associated with chromosomal breakage. Recently, the proposed lower limit was defended on the basis of the need to consider effects other than leukemia, chromosomal aberrations in workers at low exposure levels, and increased toxicity from dermal and intermittent exposures (Infante, 1992).

In animals, benzene by inhalation or gavage is a multipotent carcinogen, capable of producing a variety of neoplasms at several sites. Elevated incidences of tumors are reported in the Zymbal gland, oral cavity, preputial gland, harderian gland, liver, mammary gland, lungs, and ovaries, in addition to lymphomas and leukemias (Huff et al., 1988; Maltoni et al., 1989). Many attempts to induce leukemia in animals yielded negative or debatable results because of difficulties in establishing a suitable animal model (ATSDR, 1989). In an early study of 40 C57BL/6 mice, occurrences of leukemia (1), thymic lymphoma (6), plasmacytoma (1), and bone-marrow hyperplasia (13) were reported after exposures at 300 ppm for 6 h/d, 5 d/w, for life, as compared with the occurrence of two lymphomas (nonthymic) in controls (Snyder et al., 1980). A similar exposure of AKR mice caused bone-marrow hypoplasia, but did not alter the incidence or induction time of viral-induced lymphomas common in this strain.

In subsequent studies by the same laboratory, myelogenous leukemia occurred in 1 of 40 CD-1 mice exposed at 100 ppm and in 2 of 40 exposed at 300 ppm for 6 h/d, 5 d/w for life (Goldstein et al., 1982); it also occurred in 1 of 40 Sprague-Dawley rats similarly exposed at 100 ppm (Snyder et al., 1984). Liver tumors (4 in 40 rats) and Zymbal-gland carcinomas (2 in 40 rats) also were observed. The investigators suggested a causative role for benzene because spontaneous myelogenous leukemia had not been observed in these strains. However, the results might be of questionable significance because of the small numbers of animals and marginally increased incidences (Maltoni et al., 1989).

More convincing evidence of the leukemogenicity of benzene was provided by a major series of studies in which animals were exposed 6 h/d, 5 d/w, for 16 w to approximate the average exposure duration of workers (15% of lifetime). In C57BL/6 mice, exposure at 300 ppm resulted in a highly significant increase in lymphoma-leukemia, from 0% (0/90) in controls to 8% (8/90) (Cronkite et al., 1984). The exposure produced a definite pattern of lymphoma and mortality, with the first wave of deaths at 330-390 d of age primarily due to thymic lymphoma and the second wave beginning at 570 d of age due to nonthymic lymphoma and solid tumors. In a recent study from the same laboratory, exposure at 300 ppm 6 h/d, 5 d/w, for 16 w significantly increased the incidence of myelogenous leukemia from 0% in controls to 19.3% in male CBA/Ca mice and of nonhematopoietic nonhepatic neoplasms (Zymbal, harderian, mammary) from 21.7% to 52.6% in males and from 35.0% to 79.6% in females (Cronkite et al., 1989). A similar exposure at 100 ppm did not affect the incidence of myelogenous neoplasms, but did increase the incidence of nonhematopoietic nonhepatic tumors from 20.0% to 44.7% in males.

The Bologna Institute of Oncology conducted several chronic exposure studies in which benzene was administered by inhalation or gavage to various strains of rats and mice (Maltoni et al., 1989). Sprague-Dawley rats were exposed by inhalation either at 200 ppm for 15 w or at 200 ppm for 19 w followed by 300 ppm for 85 w (total of 104 w). Although extensive data from the studies were presented, the data do not appear to be complete for all tumor sites. Positive results were summarized as an increase or marginal increase in the incidence of the various tumor types or sites without statistical analysis. In the Sprague-Dawley rat, inhalation of benzene was reported to be "associated" with an increase in the incidence of total malignant tumors and carcinomas of the Zymbal glands and oral cavity and with a marginal increase in the incidence of hepatomas and carcinomas of the nasal cavities and mammary gland (Maltoni et al., 1989). At an exposure of 200 ppm, 4-7 h/d, 5 d/w, for 19 w, followed by an exposure at 300 ppm, 7 h/d, 5 d/w, for 85 w, with exposure started in embryonal life, the incidence of malignant tumors increased from 17.3% in male and female controls to 43.6%, Zymbal gland carcinomas increased from 0.7% to 10.0%, and hepatomas increased from 0.3% to 6.4%. In females, mammary gland tumors increased from 5.4% in controls to 13.8%. The investigators concluded that the carcinogenic effects of benzene increased with increasing doses (daily dose and length of treatment) and that the carcinogenic effect is enhanced when exposure is started during embryonal life (Maltoni et al., 1989).

Many mechanisms have been suggested for the carcinogenicity of benzene. One involves modification by benzene or its metabolites of "immune surveillance," thereby allowing development of unusual cellular species that cause leukemia and other neoplasms in humans (Leong, 1977). Another suggestion is that benzene or its metabolites might act as a promoter, rather than an initiator, by forcing compensatory hematopoiesis (regenerative hyperplasia), with the resultant appearance of preleukemic and leukemogenic clones from stem cells exposed to leukemogenic-initiating agents prior to benzene exposure (Harigaya et al., 1981).

One of the favored mechanisms involves the covalent binding of benzene metabolites to cellular macromolecules. Covalent binding to DNA was observed in the livers of rats exposed to benzene vapor (Lutz and Schlatter, 1977). In mice, radiolabeled metabolites were covalently bound to liver, bone marrow, kidney, spleen, blood, and fat; the label was bound to nucleic acids of hematopoietic cells and to nucleic acids and other macromolecules of mitochondria (Gill and Ahmed, 1981). Also, bioactivation of benzene by mitochrondia caused adduct formation with DNA and inhibited the ability of RNA polymerase to transcribe the genome (Kalf et al., 1982).

A recent review of the molecular pathology of benzene points out that its carcinogenic, hematotoxic, cytotoxic, and genotoxic effects are the consequences of highly complex, interactive biological processes (Yardley-Jones et al., 1991). Possible processes include increased production of hydroxyl radicals, generation of oxygen radicals, depletion of endogenous glutathione, activation of protein kinase c (an enzyme involved in cell transformation and tumor promotion), covalent binding to glutathione, protein and other cellular macromolecules, DNA damage, and induction of micronuclei.

Reproductive Toxicity

Studies of the effects of benzene on reproductive functions in humans are limited in number and scope but suggest a possible association between chronic exposure and adverse effects in females. In a study of 30 employed women with symptoms of benzene toxicity (indicative of higher exposure levels than currently allowed), 12 had menstrual disorders (Vara and Kinnunen, 1946). There were two spontaneous abortions and no births during employment, even though no contraceptive measures were taken by the 12 women, 10 of whom were married. Gynecological examinations revealed that the scanty menstruations in five of the women were due to hypoplasia of the ovaries.

Two European studies reported menstrual disturbances (heavy

bleeding) in workers exposed to benzene at concentrations of 31 ppm (Michon, 1965) and ovarian hypofunction in workers in another factory (Pushkina et al., 1968). In another European study of 360 gluing operators, all of whom were women who were exposed to petroleum (containing benzene) and chlorinated hydrocarbons both dermally and by inhalation, no significant difference in fertility between exposed workers and unexposed controls was found, but spontaneous abortion and premature birth increased (Mukhametova and Vozovaya, 1972).

Inhalation studies with animals have demonstrated adverse effects of benzene on the reproductive systems of both sexes, but particularly of males. In CD-1 mice exposed at 1, 10, 30, or 300 ppm, 6 h/d, 5 d/w, for 13 w, exposure at 300 ppm resulted in histopathological changes to the testes and ovaries (Ward et al., 1985). Changes to the testes included atrophy and degeneration, decreases in spermatozoa, and moderate increases in abnormal sperm forms. Pathological changes to the ovaries were less severe and consisted of bilateral cysts. In chronic exposure studies of rabbits and guinea pigs, slight increases in average testicular weight occurred in guinea pigs at 88 ppm 7-8 h/d, 5 d/w, for up to 6 mo and slight histopathological changes to the testes (degeneration of the germinal epithelium) occurred in rabbits at 80 ppm (Wolf et al., 1956).

Developmental Toxicity

There is little information on the developmental toxicity of benzene in humans. It is known, however, that benzene crosses the human placenta and is present in cord blood in amounts equal to those in maternal blood (Dowty et al., 1976).

A few case reports and epidemiological studies of benzene-exposed pregnant workers are available in the literature. The results generally are mixed or inconclusive and do not provide direct evidence of the developmental toxicity or the teratogenicity of benzene. One study reports on two infants with no evidence of chromosomal alterations delivered from a worker who had severe pancytopenia and increased chromosomal aberrations and who had been exposed to benzene during her entire pregnancy (Forni et al., 1971b). In another study, an increased frequency of chromatid and isochromatid breaks and SCE was found in lymphocytes from 14 children of female workers exposed to benzene during pregnancy; however, the workers were also exposed to other organic solvents (Funes-Cravioto et al., 1977). Other epidemiological studies of pregnant women occupationally exposed to undefined organic solvents or living near waste dumps contaminated with benzene and other carcinogens found no evidence of developmental toxicity or teratogenicity (ATSDR, 1989).

Numerous inhalation studies have shown that benzene is embryotoxic and fetotoxic in animals, as evidenced by increased incidences of resorption, reduced fetal weight, skeletal variation, and altered fetal hematopoiesis (ATSDR, 1989). However, no studies have shown benzene to be teratogenic or embryolethal in animals, even at concentrations that are toxic (reduced weight gain) to the mother. Exposure to benzene vapor adversely affected pregnant rabbits and rats at concentrations above 100 ppm (Green et al., 1978). Maternal toxicity was evidenced by a decrease in maternal-weight gain; the decrease was accompanied by retarded fetal growth. Some investigators found no increase in resorption in rats exposed at 100, 300, or 2200 ppm (Green et al., 1978), or at 10, 50, or 500 ppm (Murray et al., 1979), but others reported increased resorption in rodents, mostly at concentrations above 150 ppm (ATSDR, 1989). Exposures of mice at 500 ppm 7 h/d on gestation days 6-15 (Murray et al., 1979) and at 156 or 313 ppm 24 h/d or 4 h/d on gestation days 6-15 and exposures of rabbits at 313 ppm 24 h/d (Ungvary and Tatrai, 1985) resulted in growth retardation and increased skeletal variants in fetuses, but no malformations. In rats, concentrations of 50-2200 ppm caused decreased fetal weight, but numbers of skeletal variants increased significantly at 125 ppm and higher (Green et al., 1978). No pregnancies were reported in 10 female rats exposed to benzene at 210 ppm for 10-15 d and then joined by two unexposed males (Gofmekler, 1968). Changes in the weights of body organs were also reported at lower exposures, but the data are difficult to interpret because of a lack of any dose-response relationship. Hematopoietic alterations were reported in the fetuses and offspring of pregnant Swiss-Webster mice exposed by inhalation to benzene. At concentrations of 5, 10, or 20 ppm 6 h/d on gestation days 6-15, the number of erythroid colony-forming cells of progeny was markedly decreased, and at 10 and 20 ppm, granulocytic colony-forming cells also were reduced (Keller and Snyder, 1986).

Interaction with Other Chemicals

Ethanol has been shown to consistently increase the hematotoxicity of benzene in animals. In mice, decreased blood cell counts and bone-marrow cellularity induced by benzene were further reduced by ethanol administration (Seidel et al., 1990). The administration of ethanol increased the depression of hematopoietic progenitor cells, CFU-E, BFU-E, and CFU-C, induced by exposure to benzene at 300 or 900 ppm in BDF1 mice (Seidel et al., 1990).

TABLE 2-1 Toxicity Summary

Concentration, ppm	Exposure Duration	Species	Effects	Reference
<10	5 y	Human (workers) (n = 52)	Increased chromosomal aberrations	Picciano, 1979
10-100	Up to 10 y	Human (workers) (large n)	Excessive mortality from myelogenous leukemia in followup studies	Infante et al., 1977; Infante, 1978; Rinsky et al., 1981, 1987
11-1069	6-60 mo	Human (workers) (n = 332)	Severe cytopenia in 23 workers	Goldwater, 1941
13	11 y	Human (workers) (n = 22)	Increased chromosomal aberrations	Sarto et al., 1984
25-20,000	5 min to 8 h	Human (estimates)	No effects at 25 ppm, 8 h; headache, lassitude, weakness at 50-150 ppm, 5 h; effects tolerated at 3000 ppm, 0.5-1.0 h; dangerous to life at 7500 ppm, 0.5 h; can be fatal at 20,000 ppm, 5 or 10 min	Von Oettingen, 1940
30-210	3 mo to 17 y	Human (workers) (n = 217)	Pancytopenia in 6; 45 others with hematological abnormalities	Aksoy et al., 1971
3.4-49	N.S.	Human (workers) (n = 35)	Serum IgG and IgA decreased, IgM increased; leukocyte agglutinins increased in some; other solvent exposures were involved	Lange et al., 1973
Up to 500 (100 average)	N.S.	Human (workers) (n = 1104)	Severe pancytopenia in 25, hematological disorders in 83 others	Wilson, 1942
125-532	1-22 y	Human (workers)	Increased unstable and stable chromosomal aberrations in lymphocytes	Forni et al., 1971a

TABLE 2-1 (Continued)

Concentration, ppm	Exposure Duration	Species	Effects	Reference
150-650 (maximum)	4 mo-15 y	Human (workers) (n = 28,500)	Pancytopenia in 44, aplastic anemia in 32 with 8 deaths due to thrombocytopenic hemorrhage and infection; 6 developed leukemia within 6 y of followup	Aksoy et al., 1972; Aksoy and Erdem, 1978
0.04, 0.01, 1.0	22 h/d, 7 d/w, 6 w	Mouse (CD-1)	Increased no. of spleen lymphocytes with mutations at hrpt locus and of chromatid breaks (effects reduced at 1.0 ppm)	Au et al., 1991; Ward et al., 1992
1, 10, 30, 300	6 h/d, 5 d/w, 13 w	Rat (Sprague-Dawley)	No effects at 1, 10, 30 ppm; at 300 ppm, decrease in WBCs, percentage of lymphocytes, and decreased marrow cellularity	Ward et al., 1985
1, 10, 30	6 h/d, 5 d/w, 13 w	Mouse (CD-1)	No histopathology to testes or ovaries	Ward et al., 1985
300	6 h/d, 5 d/w, 13 w	Mouse (CD-1)	Atrophy and degeneration of testes, decrease in spermatozoa, moderate increase in abnormal sperm; bilateral cysts in ovaries	Ward et al., 1985
1, 10, 30	6 h/d, 5 d/w, 13 w	Mouse (CD-1)	No hematological changes	Ward et al., 1985
300	6 h/d, 5 d/w, 13 w	Mouse (CD-1)	Increase in mean cell volume, Hb; decrease in hematocrit, Hb, RBC, WBC, platelet counts, percentage of lymphocytes; histopathology to bone marrow, lymphoid tissue, and spleen	Ward et al., 1985
4, 20	5.5 mo	Rat	Conditioned reflex response time delayed at 20 ppm but not affected at 4 ppm	Novikov, 1956
5, 10, 20	6 h/d on d 6-16 of pregnancy	Mouse (Swiss Webster)	No. of CFU-E markedly reduced in progeny; at 10 and 20 ppm, granulocytic colony-forming cells also reduced	Keller and Snyder, 1986

Concentrations (ppm)	Duration	Species	Effects	Reference
10-28	4-6 h	Mouse	Induced SCE in peripheral blood lymphocytes, bone marrow, and micronuclei in bone-marrow polychromatic erythrocytes	Erexson et al., 1985
10, 25, 100, 300, 400	6 h/d, 5 d/w, up to 16 w	Mouse (C57BL)	Bone-marrow cellularity and pluripotential stem cells reduced at 100 ppm (2 w) but not at 10 and 25 ppm; peripheral blood lymphocytes not affected at 10 ppm but reduced at 25-400 ppm	Cronkite et al., 1985
10, 30	6 h/d, 6 d	Mouse (male C57BL)	Femoral LPS-induced B-lymphocyte colony-forming ability depressed at 10 ppm; splenic PHA-induced blastogenesis depressed at 30 ppm	Rozen et al., 1984
10, 30, 100, 300	6 h/d, 5 d before infection (pre-exposure regimen) or 5 d before and 7 d after infection (continuous regimen)	Mouse	Pre-exposure regimen had no effect at 10, 30, 100 ppm, but 300 ppm increased splenic bacterial counts (*Listeria monocytogenes*) 730% on d 4; continuous regimen had no effect at 10 ppm, but increased counts at 100 and 300 ppm	Rosenthal and Snyder, 1985
20, 50, 100, 300	8 h/d, 7 d	Rat (female Wistar)	Peripheral leukocytes depressed at all concentrations except 20 ppm; leukocyte AP levels increased at 100 and 300 ppm and unaffected at 20 and 50 ppm.	Li et al., 1986
21	4-10 d continuous or 8 h/d, 5 d/w, 2 w	Mouse (NMRI)	Continuous 4-10 d exposures decreased bone-marrow cellularity and no. of CFU-C; 2-w intermittent exposures decreased CFU-C; no such effects found in 8-w exposures at 10 ppm.	Toft et al., 1982
44, 47, 61, 65, 831	Variable	Rat	Leukopenia within 2-4 w at 61, 65, and 831 ppm and after 5-8 w at 44 and 47 ppm.	Deichmann et al., 1963

TABLE 2-1 (Continued)

Concentration, ppm	Exposure Duration	Species	Effects	Reference
50, 200, 400	6 h/d, 10-20 d	Mouse (Swiss albino)	Suppression of primary antibody response to fluid (FTT) and adsorbed (APTT) tetanus toxoid at 200 ppm (10-20 d), at 400 ppm (5, 12, or 22 d), but not at 50 ppm; secondary response unaffected at 50, 200, and 400 ppm	Stoner et al., 1981
50-2200	N.S.	Rat	Decreased fetal weight; significant increase in no. of skeletal variants at 125 ppm and above	Green et al., 1978
80-88	7-8 h/d, 5 d/w, 136, 175, 193 exposed	Rat, rabbit, guinea pig	Leukopenia, increased spleen weight, histopathology to bone marrow in rat (136 exposed), rabbit (175 exposed), and guinea pig (193 exposed)	Wolf et al., 1956
80-88	7-8 h/d, 5 d/w, 6 mo	Rabbit, guinea pig	Slight degeneration of germinal epithelium in rabbit; slight increase in testicular weight in guinea pig	Wolf et al., 1956
100	2 d, continuous	Mouse	Leukocytopenia	Gill et al., 1980
100	6 h/d, 5 d/w, 20 exposed	Mouse (male C57BL/6)	Injection of viral tumor cells induced 90% incidence of lethal tumors vs. 30% in controls	Rosenthal and Snyder, 1986
100	6 h/d, 5 d/w, lifetime	Rat (Sprague-Dawley)	Myelogenous leukemia (1/40), liver tumors (4/40), Zymbal gland carcinomas (2/40).	Snyder et al., 1984
100	6 h/d, 5 d/w, 16 w	Mouse (CBA/Ca)	Increased nonhematopoietic nonhepatic tumors from 20.0% in controls to 44.7% in males; no effect on myelogenous leukemia	Cronkite et al., 1989
300	6 h/d, 5 d/w, 16 w	Mouse (CBA/Ca)	Significantly increased myelogenous leukemia from 0% in controls to 19.3% in males and of nonhematopoietic nonhepatic tumors from 21.7% to 52.6% in males and from 35.0 to 79.6% in females	Cronkite et al., 1989
100, 300	6 h/d 5 d/w, lifetime	Mouse (CD-1)	Myelogenous leukemia in 1/40 at 100 ppm and in 2/40 at 300 ppm	Goldstein et al., 1982

Dose (ppm)	Exposure	Species	Effect	Reference
100, 300, 900	6 h/d, 5 d/w up to 16 w	Mouse (female BDF1)	Marked decrease in CFU-E at 100 ppm after 1 w; BFU-E, CFU-S, CFU-C decreased at 300 and 900 ppm	Seidel et al., 1990
100, 300, 1000, 3000	6 h/d, 5 d or Ct = 3000 ppm d	Mouse (C57BL)	Increased milk licking after first wk of exposure at 100 ppm or 300 ppm; food intake, hind-limb grip strength reduced at 1000 and 3000 ppm	Dempster et al., 1984
103, 302	6 h/d, 5 d/w, up to 26 w	Mouse (CD-1)	Marrow and spleen cellularity; GM-CFU-C in spleen reduced at 103 and 302 ppm for 5 d; at 302 ppm for 26 w; CFU-S and GM-CFU-S also reduced	Harigaya et al., 1981
156, 313	4 h/d, 3 or 24 h/d, on 6-15 of pregnancy	Mouse	Growth retardation and increased skeletal variants in fetuses but no malformations	Ungvary and Tatrai, 1985
200, 300	200 for 4-7 h/d, 5 d/w, 19 w; 300 for 7 h/d, 5 d/w, 85 w; exposure stated in embryonal life	Rat (Sprague-Dawley)	Increase in males and females in total malignant tumors from 17.3% (controls) to 43.6%; Zymbal gland carcinomas from 0.7% (controls) to 10.0%; hepatomas from 0.3% (controls) to 6.4%; in females; mammary gland tumors from 5.4% (controls) to 13.8%	Maltoni et al., 1989
210	continuous, 10-15 d before mating and 3 w after mating	Rat (female)	No litters	Gofmekler, 1968
300	6 h/d, 5 d	Mouse	Decreased peripheral blood erythrocytes, lymphocytes	Snyder et al., 1978
300	6 h/d, 6, 30, or 115 d	Mouse (male C57BL)	Reduced mitogen-induced proliferation of bone-marrow and splenic B lymphocytes and splenic T cells; reduced number of B lymphocytes in bone marrow and spleen and of T lymphocytes in thymus and spleen; thymic lymphoma after 115 exposures	Rozen and Snyder, 1985

TABLE 2-1 (Continued)

Concentration, ppm	Exposure Duration	Species	Effects	Reference
300	6 h/d, 3 d/w or 5 d/w, 13 w	Mouse (DBA/2, B6C3F$_1$, C57BL)	Increased frequency of MN-PCE in all strains; in DBA/2, decrease in percent of PCE.	Luke et al., 1988a,b
300	6 h/d, 5 d/w, 16 w	Mouse (C57BL/6)	Increased in lymphoma-leukemia from 0/90 (controls) to 8/90 (exposed)	Cronkite et al., 1984
300	6 h/d, 5 d/w, lifetime	Mouse (C57BL/6, AKR)	Leukemia (1/40), thymic lymphoma (6/40), plasmacytoma (1/40), bone-marrow hyperplasia (13/40 vs. nonthymic lymphoma (2/40 control) in C57BL/6; in AKR, bone-marrow hypoplasia, but lymphomas not affected	Snyder et al., 1980
300, 900	6 h/d, 5 d, followed by 2-w nonexposure before repeat exposure	Mouse (male CD-1, C57BL)	Behavior not affected on nonexposure days; on exposure days, eating, grooming more frequent, less inactivity; effects more marked at 300 than 900 ppm	Evans et al., 1981
313	24 h/d, d 6-15 of pregnancy	Rabbit	Growth retardation and increased skeletal variants in fetuses, but no malformations	Ungvary and Tatrai, 1985
400	6 h/d, intermittently for 9 d or consecutively for 11 d	Mouse (C57BL)	Depression of CFU-S	Harigaya et al., 1981
1500, 3000, 7000	15 min	Rat (Wistar)	Increased ectopic ventricular beats induced by coronary ligation or i.v. aconitine at 3000 and 7000 ppm, but not at 1500 ppm	Magos et al., 1990

Concentration (ppm)	Duration	Species	Effects	Reference
2200, 4600, 11,800, 24,000	Variable	Mouse	Narcosis at 2200 ppm; narcosis in 51 min at 4600 ppm, in 8 min at 11,800 ppm; deaths in 38-295 min at 11,800 ppm and in 50 min at 24,000 ppm	Von Oettingen, 1940
3100	4 h	Mouse (DBA/2)	Increased SCE frequency in bone-marrow cells; inhibited marrow cellular proliferation in males; no effect on chromosomal aberrations	Tice et al., 1980
4000, >10,000	N.S.	Rabbit	Narcosis at 4000 ppm; deaths at >10,000 ppm	Leong, 1977
4680	8 h	Mouse	Depletion of bone-marrow colony-forming cells	Uyeki et al., 1977
12,000-52,000	10 min/d repeated	Cat	Restless, rapid respiration, head nodding with hyper-synchronous, amygdaloid EEG at 12,000 ppm initially; ataxia, collapse with increase to 52,000 ppm; 3 Hz spike-wave EEG, seizures with repeated exposures.	Contreras et al., 1979
13,700	4 h	Rat	LC_{50}	Drew and Fouts, 1974
16,000	4 h	Rat	Deaths of 4/6	Smyth et al., 1962
35,000-45,000	Variable	Rabbit	Anesthesia in 4 min, with excitation, tremors at 5 min, loss of pupillary reflexes at 6.5 min, involuntary blinking at 15 min, deaths at 22-71 min	Carpenter et al., 1944

[a] Only the more important results are included.
[b] N.S. = not specified.

TABLE 2-2 Exposure Limits Set by Other Organizations

Organization	Concentration, ppm
ACGIH's TLV	0.1 (TWA, proposed; skin, A1 carcinogen)
OSHA's PEL	1 (TWA)
OSHA's STEL	5 (ceiling limit; carcinogen)
NIOSH's REL	0.1 (TWA)
NIOSH's STEL	1 (ceiling limit; carcinogen)
NIOSH's IDLH	3000
NRC's 1-h EEGL	50
NRC's 24-h EEGL	2

TLV = threshold limit value. TWA = time-weighted average. PEL = permissible exposure limit. STEL = short-term exposure limit. REL = recommended exposure limit. IDLH = immediately dangerous to life and health. EEGL = emergency exposure guidance level.

TABLE 2-3 Spacecraft Maximum Allowable Concentrations

Duration	ppm	mg/m^3	Target Toxicity
1 h	10	35	Immune system
24 h	3	10	Immune system
7 d	0.5	1.5	Immune system
30 d	0.1	0.3	Immune system
180 d	0.07	0.2	Immune system (leukemia)

RATIONALE FOR ACCEPTABLE CONCENTRATIONS

Because a large database exists on benzene's toxicity to animals and humans, multiple toxic effects must be considered in setting safe exposure concentrations. The immediate effects of benzene exposure are thought to be due to benzene itself, whereas the delayed effects are caused by toxic metabolites reaching target cells, particularly in the bone marrow. The most important threshold-type effect induced by benzene itself is CNS depression; however, mucosal irritation and cardiac effects also will be considered briefly. The clinical diseases associated with myelotoxicity are caused by benzene metabolites that injure cells in the bone marrow; the weight of evidence indicates that these effects are more severe with either increasing benzene concentration or increasing exposure time (cumulative-type effects).

Species Extrapolation

Based on work by Sabourin et al. (1992), it appears that mice are the best model for studying the effects induced by benzene metabolites. Analyses of urinary metabolites indicated that both humans and mice have a higher propensity to metabolize benzene to its toxic metabolites, muconic acid and hydroquinone, than do rats, monkeys, or chimpanzees. The ratio of hydroquinone or muconic acid to phenol in the urine of mice was 80% and 300%, respectively, as compared with concentrations in human urine (Henderson et al., 1992). Mice are also the most sensitive rodent species to metabolite-mediated effects, but do not seem to show leukemogenic changes similar to those found in industrial workers (see below). The usual species-extrapolation factor of 10 will not be used because that factor compensates for species differences in both metabolism and target-tissue susceptibility. Although the comparative tissue susceptibility of mice and humans is unknown, mice seem to produce more toxic metabolites; hence, a species factor of 3 will be used to extrapolate metabolite-induced effects in mice to human estimates. A factor of 10 will still be used for CNS effects thought to be caused directly by benzene itself. This factor is chosen because of potential differences in tissue susceptibility and in the reduced sensitivity of CNS tests in animals compared with the sensitivity of CNS tests in humans (e.g., performance decrements).

Benzene-Induced Toxicity

The acute irritancy and cardiac arrhythmogenic effects of benzene occur only at concentrations that induce significant CNS effects. In rodents, aconitine-induced cardiac arrhythmias were induced by preadministration of benzene at a few thousand parts per million for 15 min; however, narcotic effects were pronounced in the animals exposed to benzene (Magos et al., 1990). Likewise, in monkeys and cats, cardiac rhythm disturbances occurred primarily during induction of narcosis (Nahum and Hoff, 1934). Protection against CNS effects should protect against any sort of cardiac sensitization similar to that observed in the experimental models. Benzene apparently is not very irritating; humans have been reported to tolerate a concentration of 3000 ppm for up to 1 h (Flury, 1928), whereas CNS effects (headache, lassitude, and weariness) are estimated to occur in the range of 50-150 ppm (Gerarde, 1960). Concentrations low enough to protect against CNS effects will protect against irritation from benzene vapor.

Metabolite-Induced Toxicity

During longer exposures that are below the CNS threshold, the metabolites of benzene may induce several toxic effects that are mediated through damage to stem cells (and other cells) in the bone marrow. Each toxic effect will be analyzed separately; they include hematological, immunological, and neoplastic effects. The first two effects are particularly important because spaceflight causes loss of red-blood-cell mass and reduced immune function in some cells.

Benzene has also been reported to cause reproductive toxicity, developmental toxicity, and genotoxicity. High (but unknown) concentrations of benzene apparently caused menstrual disorders in 12 of 30 female workers exhibiting other signs of benzene toxicity (Vara and Kinnunen, 1946); however, other studies in humans have not confirmed this finding (Barlow and Sullivan, 1982). Animal studies have shown ovarian and testicular changes in mice after subchronic exposure at 300 ppm, but nonreproductive effects (e.g., hematological) occurred much earlier in the same study (Ward et al., 1985). Benzene concentrations low enough to protect against myelogenic toxicity also are not likely to produce any reproductive effects. No convincing human or animal data exist to show that inhalation exposures to benzene cause developmental abnormalities except at concentrations that are toxic to the mother. Benzene is clearly a genetic toxicant; however, the clinically recognized effects that it causes (hematotoxicity, immunotoxicity, and leukemia) are largely mediated through genetic mechanisms; hence, the genotoxic effects of benzene will not be analyzed separately.

Nervous System Effects

The early literature contains reports on the CNS effects induced by inhaling high concentrations of benzene for brief periods (Flury, 1928; Von Oettingen, 1940). Unfortunately, these human data are not sufficiently specific in terms of exposure concentrations, number of test subjects, and sensitivity of end points to be useful in setting acceptable concentrations for benzene. Animal studies typically focus on the myelogenic effects; however, Demster et al. (1984) reported that hind-limb grip strength was diminished 12% in mice exposed for 6 hr to benzene at 1000 ppm, but ten 6-hr exposures at 300 ppm produced no significant effect. Using 300 ppm as the no-observed-adverse-effect level (NOAEL) and applying a species extrapolation factor of 10, gave an acceptable concentration (AC) for benzene exposure of 30 ppm. This value seems to be below the estimated threshold (50-150 ppm) for induction of CNS effects in humans; therefore, it was adopted for all exposure times from 1 hr to 180 d. (See Table 2-5 at the end of this section.)

Hematological Effects

The hematological effects of benzene (e.g., hemorrhage of skin and mucous membranes and anemia) have been known for almost a century; however, quantitative human data on benzene's ability to induce these effects are not available. Most human studies have focused on the leukemogenic properties of benzene (considered below). Suitable data from three studies of mice were used to estimate safe human exposure limits to protect against hematological effects. For reasons cited above, the species extrapolation factor was 3; however, because a 10% loss in red-blood-cell mass is typical during spaceflight, a spaceflight factor of 3 was also applied. From the study by Dempster et al. (1984), two 6-h exposures at 300 ppm did not induce hematological effects. The human limits were calculated as follows:

AC (1 h) = 300 ppm × 1/3 × 1/3 = 33 ppm.
AC (24 h) = 300 ppm × 1/3 × 1/3 × 12/24 = 16 ppm.

Extrapolation to longer exposures is not reasonable from these data; for longer exposures, the data of Green et al. (1981b) were used. They found that 50 6-h exposures at 10 ppm did not induce hematological effects. This information was used to estimate human exposure limits as follows:

AC (7 d) = 10 ppm × 1/3 × 1/3 = 1.1 ppm.
AC (30 d) = 10 ppm × 1/3 × 1/3 × 50 × 6/720 = 0.5 ppm.

Extrapolation to 180 d was not necessary because data from Toft et al. (1982) show that an 8-w continuous exposure at 10 ppm caused no measurable changes in bone-marrow cellularity. From this 56-d exposure, the human limit was estimated as follows:

AC (180 d) = 10 ppm × 1/3 × 1/3 × 56/180 = 0.3 ppm.

This estimate was only slightly below the 30-d AC, suggesting that concentrations in the few tenths of a parts-per-million range approach a no-effect level even when exposures are very long.

Immunological Effects

Dempster et al. (1984) reported that five 6-h exposures to benzene at a concentration of 100 ppm induced a 30% reduction in circulating lymphocytes in mice. No significant change was detected after a single 6-h exposure; hence, the NOAEL was 100 ppm for 6 h. From this finding, the short-term ACs were derived as follows:

AC (1 h) = 100 ppm × 1/3 × 1/3 = 11 ppm.
AC (24 h) = 100 ppm × 1/3 × 1/3 × 6/24 = 3 ppm.

Since the immunological effects, which are similar (or greater) in mice and humans, were presumably induced by benzene's toxic metabolites, the species factor was only 3. Likewise, it was concluded that a spaceflight factor of 3 was appropriate because of the numerous reports of spaceflight effects on immune function in rats and, to a lesser extent, in astronauts (Lesnyak et al., 1993; Taylor, 1993). At landing, shuttle astronauts show depression in blast-cell transformation, changes in cytokine function, and depression of peripheral T-inducer, T-cytotoxic, and natural-killer cells (Taylor, 1993). The changes noted in animals include altered lymphocyte blastogenesis, cytokine function, killer-cell activity, and colony-stimulating factors (Lesnyak et al., 1993).

Rosenthal and Snyder (1985) showed that 12 6-h exposures (72 h total) to benzene at 10 ppm did not increase the susceptibility of mice to infection by *Listeria monocytogenes*. This observation was used to calculate ACs for 7 d and 30 d of exposure as follows:

AC (7 d) = 10 × 1/3 × 1/3 × 72/168 = 0.5 ppm.
AC (30 d) = 10 ppm × 1/3 × 1/3 × 72/720 = 0.1 ppm.

These were the lowest of the immunotoxicity ACs and were also the lowest for any toxic effect known to be caused by benzene.

Data were available to set 7-d and 30-d ACs from Green et al. (1981b). They found that mice exposed to benzene at 9.6 ppm for 6 h/d for 50 d (300 h = 12.5 d total) showed no changes in peripheral blood or bone marrow. Although splenic cellularity and weight changed, these changes are considered adaptive rather than adverse effects, because the direction of splenic changes in heavily exposed mice was opposite to the changes noted in mice exposed at 9.6 ppm. This information was used to calculate ACs as follows:

AC (7 d) = 9.6 ppm × 1/3 × 1/3 = 1.1 ppm.
AC (30 d) = 9.6 ppm × 1/3 × 1/3 × 300/720 = 0.4 ppm.

No long-term exposure data are available on the immunological effects of benzene exposure; however, Haber's rule can be used to extrapolate a value of 0.07 ppm for continuous exposures of 180 d.

Risk of Leukemia

The risk of death from leukemia induced by occupational and environmental exposure to benzene vapor has been estimated by many investigators. Most estimates have been based on positive epidemiological findings from relatively high industrial exposures; however, other estimates have been based on "detection limits" of negative epidemiological studies or extrapolation from animal data. Numerous controversial issues affect the outcome of the risk estimates. These may be categorized as follows:

1. The spectrum of neoplasms that can be ascribed to benzene exposure.
2. The concentrations found in industrial workers exposed at high concentrations.
3. The most appropriate model to extrapolate risk from high- to low-concentration exposures.
4. The most appropriate control population from which to judge excess leukemia risk.

The only point of agreement among investigators seems to be that cumulative benzene exposures above 300 ppm-y induce a significant increase in the occurrence of leukemia in workers. This observation is based on the results of a thorough epidemiological study of 1165 workers exposed at two industrial sites in Ohio (Infante et al., 1977). In that study, standard mortality ratios (SMRs) were not statistically above 100 except for the group with an exposure at the highest concentration (>400 ppm-y) and the group with a cumulative exposure at 200-400 ppm-y. In the group exposed at 40-200 ppm-y, the SMR was 322 (95% confidence interval 36-1165). Using such observations to predict a benzene concentration to which crew members could be continuously exposed for 180 d with no more than a 0.01% increase in leukemia is not simple. At the end of this section, there is a chronological summary in Table 2-4 of risk assessments that have been published since the late 1970s for benzene inhalation exposure. The concentrations calculated in this document and shown in Table 2-4 are for continuous 0.5-y exposures and are typically determined from the long-term intermittent-exposure predictions provided by the original risk assessors. The values are not the 95% confidence limits; for ease of comparison, only mean or maximum-likelihood values are calculated.

Since the mid-1970s, when the threshold limit value (TLV) for benzene was lowered from 25 ppm to 10 ppm (ACGIH, 1991), thought concerning the leukemogenic properties of benzene has evolved. In 1979, the U.S. Environmental Protection Agency (EPA) estimated that the risk of dying from leukemia as a result of a 70-y exposure to benzene at 1 ppm was 0.024 (EPA, 1979). This value was based on the geometric mean of three risk estimates that varied from 0.014 to 0.046 (Aksoy et al., 1974; Infante et al., 1977; Ott et al., 1978). However, these risk estimates were based on epidemiological studies that had substantial shortcomings. In 1986, the National Research Council's Committee on Toxicology used the EPA values to estimate that a 0.004-ppm lifetime exposure would impart a risk of 0.01% (NRC, 1986). Similarly, the EPA conclusions can be used to calculate a 180-d (0.5-y) exposure concentration (C) that would impart a 0.01% risk:

$$C\ (0.5\ y) = 1\ ppm \times 70/0.5 \times 0.0001/0.024 = 0.6\ ppm.$$

This value is not the upper confidence limit of a 0.01% risk; it is the most likely value estimated from the cited studies.

In the mid-1980s, the preliminary epidemiological investigations of the rubber workers in Ohio (Infante et al., 1977) were reexamined (Rinsky et al., 1981) and updated (Rinsky et al., 1987). These data have been interpreted in several ways, and several shortcomings have been indicated. The interpretations will be described below; however, no attempt will be made to refine the original author's conclusions.

One of the major shortcomings of the positive epidemiological findings is that worker exposures to benzene must be estimated; hence, the leukemia risk varies depending on variations in the exposure estimates. In 1985, the 5-y occupational-exposure (about 1 y of continuous exposure) risk of exposure to benzene at 1 ppm was estimated to range from 0.5/1000 to 2/1000. The range is due to uncertainty about the concentrations of benzene in the exposures of workers in the cohort (Infante and White, 1985). The Ohio cohort and a Michigan cohort (Ott et al., 1978) gave essentially the same risk estimate when the one-hit linear model was applied to the leukemia incidence. Using the geometric average of the range (1/1000), the concentration yielding a risk of 0.01% after a continuous exposure for 0.5 y was calculated as follows:

$$C\ (0.5\ y) = 1\ ppm \times 0.0001/0.001 \times 1/0.5 = 0.2\ ppm.$$

This value does not take into account statistical uncertainty from the incidence of leukemia or possibly inappropriate modeling.

The question of the link between benzene and cancer was subjected to cancer modeling in which the increase in cancers relative to the background incidence was estimated. One model, based on the concept that both initiation and promotion mechanisms must be considered (the linear nonthreshold model does not model epigenetic cancer mechanisms), predicted that 2 ppb would yield an incidence of 0.00001 for all cancers based on extrapolation from the most sensitive orally dosed rodents (Albert, 1988). This estimate assumes that a concentration that doubled malignant tumors in exposed animals would double the natural incidence of all cancers in humans (0.02). If this calculation is applied to a working lifetime of 45 y (10 y continuous exposure), then the half-year concentration can be calculated for a 0.0001 risk as follows:

$$C (0.5 \text{ y}) = 0.002 \text{ ppm} \times 10/0.5 \times 0.0001/0.00001 = 0.4 \text{ ppm}.$$

This is not an upper limit, and it applies, in principle, to all benzene-induced cancers.

In a somewhat similar relative-risk approach, epidemiological data were used to predict the odds ratio of benzene-exposed workers dying of leukemia relative to unexposed workers dying of leukemia (Rinsky et al., 1987). Using a linear low-dose extrapolation, the ratio was estimated to be 1.05 (range 1.01 to 1.09) for lifetime industrial exposure to benzene at 0.1 ppm. Because the odds of males dying of myelogenous leukemia is about 0.002, a 1/20 increase in this value is 0.0001. The half-year continuous-exposure concentration would be calculated as follows:

$$C (0.5 \text{ y}) = 0.1 \text{ ppm} \times 0.0001/0.0001 \times 10/0.5 = 2.0 \text{ ppm}.$$

This number is the most likely value; however, the concentration would be nearly halved if the upper 95% limit of the odds ratio (1.09) were used.

At about the same time as the above analyses were published, the use of a linear low-dose extrapolation was questioned (Grilli et al., 1987). The authors concluded from animal metabolism data that the tumor risk from exposures at 10 ppm seemed to be 2-3 times the risk that would be predicted from incidence data on exposures at 100 ppm. This increase stems from a flattening of the dose-response curve above about

40 ppm, where the high-affinity, low-capacity metabolic activation pathways start to become saturated.

Using a weighted average from four epidemiological studies, Austin et al. (1988) concluded that a 30-y lifetime occupational exposure (6.6 continuous years) to benzene at 1 ppm would impart an excess risk of death from leukemia of 0.005. Again the authors used a linear extrapolation to lower concentrations from conclusions reached at much higher concentrations. The prediction for the condition of 0.5 y of continuous exposure is as follows:

$$C\ (0.5\ y) = 1\ ppm \times 6.6/0.5 \times 0.0001/0.005 = 0.3\ ppm.$$

Again, this value is not based on the upper-confidence interval of the risk; if it were, the value would be about 0.1 ppm. It includes only deaths from leukemia, even though the analysis by Rinsky et al. (1987) suggested that multiple myeloma might also be an outcome of benzene exposure.

In an analysis of data on hematopoietic neoplasms in C57BL mice (Snyder et al., 1980), Beliles and Totman (1989) concluded that the maximum likelihood estimate for neoplasm risk in humans was 0.014 for lifetime occupational exposure at 10 ppm. This result was derived by noting that the mice were exposed experimentally to benzene for 15% of their lifetimes and humans working for 45 y (10 y continuous exposure) in a benzene-contaminated environment would be exposed for 14% of their lifetimes. Species extrapolation was done by allometric equations based on surface area or metabolic-rate analyses. It was found that body-weight ratios to the exponent 0.74 provided the best fit to mouse and human data. The half-year risk calculation from this estimate was as follows:

$$C\ (0.5\ y) = 10\ ppm \times 10/0.5 \times 0.0001/0.014 = 1.4\ ppm.$$

The actual production of myelogenous leukemia in mice by benzene inhalation has not been achieved by any investigator (Farris et al., 1993).

Using both positive and negative epidemiological data and a linear model, Swaen and Meijers (1989) estimated five to six excess deaths from nonlymphatic leukemia per 1000 deaths in workers exposed at 300 ppm-y of benzene (10 ppm × 30 y work exposure, or 6.6 y continuous exposure). The positive data were from Rinsky et al. (1987) and the negative data were from several massive studies of 34,000-38,000 workers presumably exposed to low concentrations of benzene (and many other chemicals). The authors of the latter studies used the unproven assumption that the low-concentration exposures averaged 50 ppm-y. From their analysis of the positive epidemiological data, the half-year concentration is as follows:

$$C (0.5 \text{ y}) = 10 \text{ ppm} \times 6.6/0.5 \times 0.0001/0.006 = 2.2 \text{ ppm}.$$

This result is similar to the value derived from the original work by Rinsky.

Recognizing the wide range of risk estimates for benzene, Byrd and Barfield (1989) asked whether this uncertainty was due to differences in data, methods, or concept. They concluded that methodological differences contributed most to the uncertainty. As an example, Parodi et al. (1989) argued that benzene lacks promoting potential below 10 ppm; hence, a sublinear response at low-concentration exposures might be predicted relative to exposures in the range of 10-30 ppm. However, they noted that a leveling-off appears in the range of 50-100 ppm, resulting in a sigmoidal dose-response curve.

Brett et al. (1989) examined various assumptions about the concentrations of benzene that the Ohio rubber workers were exposed to and the matching to control populations. Variations in control matching with Rinsky's exposure assumptions gave predicted increases in leukemia death rates in the range of 4.2-6.4/1000 after 45 ppm-y of exposure (equivalent to 10 y continuous exposure at 1 ppm), whereas the exposure assumptions of Crump and Allen (1984) gave a range of leukemia death rates of 0.5-1.6/1000 for the same variations in control matching. Clearly, the exposure assumptions have a larger effect on the estimates of leukemia risk than the control-matching choices. The authors' preference, using conditional logistic regression, gave leukemia death rates of 1.2-1.6/1000 workers (45 ppm-y of exposure) based on the three control-matching options (Brett et al., 1989). Using the average of this range, gives the following half-year continuous exposure limit:

$$C (0.5 \text{ y}) = 1 \text{ ppm} \times 10/0.5 \times 0.0001/0.0014 = 1.4 \text{ ppm}.$$

This value falls between the low-concentration exposure group (<0.6 ppm) and the high-concentration exposure group (>2 ppm).

In 1990, the ACGIH proposed a TLV of 0.1 ppm for benzene (ACGIH, 1991) and that value has been defended on scientific grounds (Infante, 1992). In particular, inhalation exposures at 1 ppm have been noted to induce cytogenic effects in animals and humans. In addition, Infante asserts that the Michigan cohort has provided evidence that low cumulative exposures of benzene can increase the risk of leukemia. Using the TLV (45 working y = 10 y continuous exposure) as a starting point without any quantitative specification of leukemia risk, the half-year concentration can be calculated as follows:

$$C (0.5 \text{ y}) = 0.1 \times 10/0.5 = 2 \text{ ppm}.$$

This value is at the high end of the allowable concentrations and suggests that the proposed TLV is not especially conservative.

Factors affecting the outcome of benzene risk estimates have been considered by a panel whose general conclusions have been published (Voytek and Thorslund, 1991). The panel preferred the epidemiological data over the animal data for risk assessment; however, the animal data were considered a valuable adjunct to the human exposure results. It was not clear whether updating (for new leukemia deaths) and combining epidemiological studies would provide a more powerful basis for risk assessment. The panel felt that it was important to focus on specific disease entities (e.g., myelogenous leukemia) rather than on more general categories (e.g., leukemia). The linear quadratic model was preferred over linear models based on leukemia and bone-cancer risks from radiation exposure, provided the former was found to be a "sensitive" model at low concentrations. A preference for absolute risk over relative risk was expressed by the panel.

Review of the risk assessments described above shows that they predict a half-year exposure concentration range of 0.2-2.2 ppm to cause a 0.01% increase in the risk of death due to leukemia. Yardley-Jones et al. (1991) concluded, not surprisingly, that without the three highest exposure cases in the Rinsky study, the model was statistically insignificant in predicting leukemia risks at low concentrations. Furthermore, in the parts-per-billion range, the largest degree of uncertainty is due to dose-response extrapolation rather than to uncertainties in worker exposures in the parts-per-billion range (Snyder et al., 1993a). Moreover, factoring in the four multiple myeloma cases (one expected) found by Rinsky should be considered. The problem is that three-fourths of the cases occurred in the lowest-concentration exposure group (<40 ppm-y) with a very long latency period. It is difficult to understand why the multiple myeloma cases occurred in the lowest-concentration exposure group with such a preference if the myelomas were caused by exposure to benzene. It was decided not to include these neoplasms in the present risk analysis.

The average of the nine estimates, which are not independent of each other, is 1.2 ppm. As a conservative measure, the lowest of the concentrations expected to impart a risk of no more than 0.01% was selected—that is 0.2 ppm. Higher radiation exposures during spaceflight are inevitable when compared with earth environments. The radiomimetic properties of benzene are well known; hence, it would be prudent to consider the effects as interacting, probably in an additive way. The degree of radiation protection that will be provided to crews is unknown; hence, a reduction by a factor of 3 in benzene's leukemogenic allowable concentration was selected somewhat arbitrarily. Therefore, the AC to protect against excess leukemia for a continuous exposure of 180 d is 0.07 ppm. Using linear extrapolation to shorter times, the ACs for 30, 7, and 1 d were found to be 0.4 ppm, 1.7 ppm, and 12 ppm, respectively. An estimate for 1 h was not considered suitable because the value would be well into the dangerous range based on other toxic effects.

TABLE 2-4 Concentrations Predicted to Increase Leukemia Risk 0.01% in 6 Months of Continuous Exposure

Year	Predicted Concentration, ppm	Basis of the Original Estimate and Comments	Reference
1979	0.6	EPA estimates from geometric mean of three epidemiology studies, each with definite shortcomings.	EPA, 1979
1985	0.2	Risk of 5-y occupational exposure at 1 ppm was 0.001, based on Ohio and Dow cohorts. Actual worker exposures were not measured.	Infante and White, 1989
1987	2.0	From an odds ratio (1.05) of 0.1-ppm-exposed workers dying of myelogenous leukemia compared with unexposed workers.	Rinsky et al., 1987
1988	0.4	Considers all tumors in orally dosed rodents, based on the assumption that exposure causing a doubling of tumors in rodents would double natural incidence in humans. Mechanisms are considered in model.	Albert, 1988
1988	0.3	Weighted average of four epidemiology studies with linear extrapolation to low doses.	Austin et al., 1988
1989, 1980	1.4	Allometric extrapolation of mouse hematopoietic neoplasm data to humans.	Beliles and Totman, 1989; Snyder et al., 1980
1989	2.2	Estimates from Cronkite et al. (1989) data. Compares well with risks suggested by negative epidemiology studies.	Swaen and Meijers, 1989
1989	1.4	Author's preference in exposure assumptions and control groups applied to Rinsky's epidemiology data using conditional logistic regression.	Brett et al., 1989
1989, 1992	2.0	Using TLV as a starting basis without specific statement of risk from exposure.	ACGIH, 1991; Infante, 1992

Summary of Toxic Effects

The analysis of toxic effects induced by benzene generally followed guidelines provided to NASA by the NRC Committee on Toxicology (NRC, 1992). A summary of the analysis has been provided in Table 2-5. Important deviations from past practices were the following: (1) A species extrapolation factor of 3 was used rather than 10 for effects caused by metabolites of benzene. (2) For the first time, a spaceflight factor was applied to an immunotoxicant because of the immune-modulating effects of spaceflight. (3) A radiation uncertainty factor was applied because of the leukemogenic properties of benzene and the relatively high radiation exposure of astronauts. (4) Finally, the present analysis of leukemogenic effects deviated from the NRC-recommended linearized multistage model because of uncertainty involving the human epidemiology database and variations in low-dose extrapolation methods used by investigators.

TABLE 2-5 End Points and Acceptable Concentrations

End Point	Exposure Data	Species and Reference	Uncertainty Factors				Acceptable Concentrations, ppm					
			Time	Species	Space-flight		1 h	24 h	7 d	30 d	180 d	
Nervous system toxicity, loss of hind-limb grip strength	NOAEL at 300 ppm, 10 × 6 h	MUS (Dempster et al., 1984)	1	10	1		30	30	30	30	30	
Hematotoxicity												
Anemia	NOAEL at 300 ppm, 2 × 6 h	MUS (Dempster et al., 1984)	1	3	3		33	16	—[a]	—	—	
Hemotoxic effects	NOAEL at 10 ppm, 50 × 6 h	MUS (Green et al., 1981b)	1 or HR[b]	3	3		—	—	1.1	0.5	—	
	NOAEL at 10 ppm, 8 w continuous	MUS (Toft et al., 1982)	HR	3	3		—	—	—	—	0.3	
Immuntoxicity												
Decrease in peripheral lymphocytes	NOAEL at 100 ppm, 6 h	MUS (Dempster et al., 1984)	1 or HR	3	3		11	3	—	—	—	
Resistance to bacterial infection and reduced spleenic lymphocyte count	NOAEL at 10 ppm, 12 × 6 h	MUS (Rosenthal and Snyder, 1986)	1 or HR	3	3		—	—	0.5	0.1	—	
Decrease in peripheral lymphocytes	NOAEL at 9.6 ppm, 50 × 6 h	MUS (Green et al., 1981b)	1 or HR	3	3		—	12	1.1	0.4	0.07	
Leukemia	Lowest of 0.01% risk estimates at 0.2 ppm, 180 d continuous	Varieties	—	—	3 rad		—	—	1.7	0.4	0.07	
SMAC							10	3	0.5	0.1	0.07	

[a]Extrapolation to these exposure durations produces unacceptable uncertainty in the values.
[b]HR = Haber's rule.

RECOMMENDATIONS

Benzene ranks among the most well-studied chemicals known to cause toxic effects in humans. Nonetheless, during our efforts to set exposure limits for astronauts, important limitations in the database became apparent; these limitations should be targets of further research on benzene's toxicity. The acceptable concentration for protection from neurotoxicity induced by short-term benzene exposures was based on rodent data. That was because the reported effects in humans appeared to be based on impressions from industrial experience rather than on specific human data. Well-controlled short-term human exposures to assess neurotoxicity (e.g., performance decrements) are needed to place the acceptable concentrations for such effects on a more reliable foundation. Ethical constraints could limit the scope of studies involving controlled human exposures to benzene.

Continuation of scientific investigations into the role of various enzymes and cofactors in the activation of benzene to its myelotoxic products should continue (Snyder et al., 1993b). Discovery of new potentially toxic metabolites, such as 6-hydroxy-*trans,trans*-2,4-hexadienoic acid, will further elucidate the mechanism of benzene toxicity (Kline et al., 1993). Refinement of toxicokinetic models will lead to better definition of research aims and facilitate comparative toxicity study (Woodruff and Bois, 1993). Taken together, such scientific investigations will result in an improved definition of time and concentration dynamics, particularly in the areas of continuous-vs.-intermittent exposure and low-concentration extrapolation.

The uncertainty factors selected to compensate for "target-organ" effects common to both benzene and spaceflight were arbitrarily assigned. Experiments designed to assess the interaction of chemical toxicity and spaceflight-induced changes would be valuable in evaluating the accuracy of such selections. For example, rodent experiments involving concomitant exposure to benzene and spaceflight (or a model of spaceflight) could be designed to show whether spaceflight modulates benzene's hematotoxicity or immunotoxicity. Findings in such experiments would improve the risk assessment and circumvent the need to make arbitrary choices in uncertainty factors.

REFERENCES

ACGIH. 1991. 1991-1992 Threshold Limit Values for Chemical Substances and Physical Agents and Biological Exposure Indices. American Conference of Governmental Industrial Hygienists, Cincinnati, Ohio.

Aksoy, M. 1978. Benzene and leukaemia. Lancet i:441.

Aksoy, M. 1980. Different types of malignancies due to occupational exposure to benzene: A review of recent observations in Turkey. Environ. Res. 23:181-190.

Aksoy, M., and S. Erdem. 1978. Followup study on the mortality and the development of leukemia in 44 pancytopenic patients with chronic benzene exposure. Blood 52:285-292.

Aksoy, M., K. Dincol, T. Akgun, S. Erdem, and G. Dincol. 1971. Haematological effects of chronic benzene poisoning in 217 workers. Br. J. Ind. Med. 28:296-302.

Aksoy, M., K. Dincol, S. Erdem, T. Akgun, and G. Dincol. 1972. Details of blood changes in 32 patients with pancytopenia associated with long-term exposure to benzene. Br. J. Ind. Med. 29:56-64.

Aksoy, M., S. Erdem, and G. Dincol. 1974. Leukemia in shoe-workers exposed chronically to benzene. Blood 44:837-841.

Aksoy, M., S. Erdem, G. Erdogan, and G. Dincol. 1976. Combination of genetic factors and chronic exposure to benzene in the aetology of leukemia. Hum. Hered. 25:149-153.

Albert, R.E. 1988. Quantitated carcinogen risks assessment and carcinogen exposure standards. In Risk Assessment and Risk Management of Industrial and Environmental Chemicals, C.R. Cothern, M.A. Mehlman, and W.L. Marcus, eds. Princeton, N.J.: Princeton Scientific.

ATSDR. 1989. Toxicological Profile for Benzene. ATSDR/TP-88/03. Agency for Toxic Substances and Disease Registry, Washington, D.C.

Au, W.W., V.M.S. Ramanujam, J.B. Ward Jr., and M.S. Legator. 1991. Chromosome aberrations in lymphocytes of mice after subacute low-level inhalation exposure to benzene. Mutat. Res. 260: 219-224.

Austin, H., E. Delzell, and P. Cole. 1988. Benzene and leukemia. A review of the literature and a risk assessment. Am. J. Epidemiol. 127:419-439.

Baarson, K., C.A. Snyder, and R.E. Albert. 1984. Repeated exposures of C57Bl mice to 10 ppm inhaled benzene markedly depressed erythropoietic colony formation. Toxicol. Lett. 20:337-342.

Barlow, S.M., and F.M. Sullivan. 1982. Pp. 83-103 in Reproductive Hazards of Industrial Chemicals. London: Academic.

Beliles, R.P., and L.C. Totman. 1989. Pharmacologically based risk assessment of workplace exposure to benzene. Regul. Toxicol. Pharmacol. 9:186-195.

Blank, I.H., and D.J. McAuliffe. 1985. Penetration of benzene through human skin. J. Invest. Dermatol. 85:522-526.

Bois, F.Y., T.J. Woodruff, and R.C. Spear. 1991. Comparison of three physiologically based pharmacokinetic models of benzene disposition. Toxicol. Appl. Pharmacol. 110:79-88.

Bond, G.G., E.A. McLaren, C.L. Baldwin, and R.R. Cook. 1986. An update of mortality among chemical workers exposed to benzene. Br. J. Ind. Med. 43:685-691.

Brett, S.M., J.V. Rodricks, and V.M. Chinchilli. 1989. Review and update of leukemia risk potentially associated with occupational exposure to benzene. Environ. Health Perspect. 82:267-281.

Byrd, D.M., and E.T. Barfield. 1989. Uncertainty in the estimation of benzene risks: Application of an uncertainty taxonomy to risk assessments based on an epidemiology study of rubber hydrochloride workers. Environ. Health Perspectives 82:282-287.

Carpenter, C., C. Shaffer, C. Weil, and H. Smyth. 1944. Studies on the inhalation of 1:3 butatiene; with comparison of its narcotic effect with benzol, toluol, and styrene and a note on the elimination of styrene by the human. J. Ind. Hyg. Toxicol. 26:69-78.

Chang, I.W. 1972. Study on the threshold limit value of benzene and early diagnosis of benzene poisoning. J. Cathol. Med. Coll. 23:429-434.

Choy, W.N., J.T. MacGregor, M.D. Shelby, and R.R. Maronpot. 1985. Induction of micronuclei by benzene in B6C3F$_1$ mice: Retrospective analysis of peripheral blood smears from the NTP carcinogenesis bioassay. Mutat. Res. 143:55-59.

Clarke, M.G., A. Yardley-Jones, A.C. MacLean, and B.J. Dean.

1984. Chromosome analysis from peripheral blood lymphocytes of workers after an acute exposure to benzene. Br. J. Ind. Med. 41: 249-253.

Contreras, C.M., T. Gonzalez-Estrada, D. Zarabozo, and A. Fernandez-Guardiola. 1979. Petit mal and grand mal seizures produced by toluene or benzene intoxication in the cat. Electroenceph. Clin. Neurophysiol. 46:290-301.

Cornish, H.H., and R.C. Ryan. 1965. Metabolism of benzene in nonfasted, fasted, and aryl-hydroxylase inhibited rats. Toxicol. Appl. Pharmacol. 7:767-771.

Cronkite, E.P., J.E. Bullis, T. Inoue, and R.T. Drew. 1984. Benzene inhalation produces leukemia in mice. Toxicol. Appl. Pharmacol. 75:358-361.

Cronkite, E.P., R.T. Drew, T. Inoue, and J.E. Bullis. 1985. Benzene hematotoxicity and leukemogenesis. Am. J. Ind. Med. 7:447-456.

Cronkite, E.P., R.T. Drew, T. Inoue, Y. Hirabayashi, and J.E. Bullis. 1989. Hematotoxicity and carcinogenicity of inhaled benzene. Environ. Health Perspect. 82:97-108.

Crump, K.C., and B.C. Allen. 1984. Quantitative Estimates of the Risk of Leukemia from Occupational Exposure to Benzene. Paper prepared for the Occupational Safety and Health Administration, Washington, D.C.

Dean, B.J. 1978. Genetic toxicology of benzene, toluene, xylenes and phenols. Mutat. Res. 47:75-97.

Dean, B.J. 1985. Recent findings on the genetic toxicology of benzene, toluene, xylenes and phenols. Mutat. Res. 154:153-181.

Deichmann, W.B., W.E. MacDonald, and E. Bernal. 1963. The hemopoietic toxicity of benzene vapors. Toxicol. Appl. Pharmacol. 5:210-224.

Dempster, A.M., H.L. Evans, and C.A. Snyder. 1984. The temporal relationship between behavioral and hematological effects of inhaled benzene. Toxicol. Appl. Pharmacol. 76:195-203.

Dowty, B.J., J.L. Laseter, and J. Storer. 1976. The transplacental migration and accumulation in blood of volatile organic constituents. Pediatr. Res. 10:696-701.

Drew, R.T., and J.R. Fouts. 1974. The lack of effects of pretreatment with phenobarbital and chlorpromazine on the acute toxicity of benzene in rats. Toxicol. Appl. Pharmacol. 27:183-193.

Driscoll, K.E., and C.A. Snyder. 1984. The effects of ethanol ingestion and repeated benzene exposures on benzene pharmacokinetics. Toxicol. Appl. Pharmacol. 73:525-532.

Duvoir, M.R., A. Fabre, and L. Derobert. 1946. [The significance of benzene in the bone marrow in the course of benzene blood diseases.] Arch. Mal. Prof. 7:77-79.

EPA. 1979. Carcinogenic Assessment Groups Final Report of Population Risk to Ambient Benzene Exposures. EPA/450/S-80-004. U.S. Environmental Protection Agency, Washington, D.C.

EPA. 1989. Health Effects Assessment for Benzene. EPA/600/8-89/086. U.S. Environmental Protection Agency, Washington, D.C.

Erexson, G.L., J.L. Wilmer, and A.D. Kligerman. 1985. Sister chromatid exchange induction in human lymphocytes exposed to benzene and its metabolites in vitro. Cancer Res. 45:2471-2477.

Evans, H.L., A.M. Dempster, and C.A. Snyder. 1981. Behavioral changes in mice following benzene inhalation. Neurobehav. Toxicol. Teratol. 3:481-485.

Farris, G.M., J. Everitt, R. Irons, and J. Popp. 1993. Carcinogenicity of inhaled benzene in CBA mice. Fundam. Appl. Toxicol. 20:503-507.

Flury, F. 1928. Modern industrial poisonings in pharmacological and toxicological respect. Arch. Exp. Pathol. Pharmakol. 138:65-82.

Forni, A., E. Pacifico, and A. Limonta. 1971a. Chromosome studies in workers exposed to benzene or toluene or both. Arch. Environ. Health 22:373-378.

Forni, A.M., A. Cappellini, E. Pacifico, and E.C. Vigliani. 1971b. Chromosome changes and their evolution in subjects with past exposure to benzene. Arch. Environ. Health 23:385-391.

Funes-Cravioto, F., C. Zapata-Gayon, and B. Kolmodin-Hedman. 1977. Chromosome aberrations and sister chromatid exchange in workers in chemical laboratories and a rotoprinting factory and in children of women laboratory workers. Lancet ii:322-325.

Gerarde, H.W. 1960. Pp. 97-108 in Toxicology and Biochemistry of Aromatic Hydrocarbons, E. Browning, ed. New York: Elsevier.

Gerarde, H.W. 1962. The aromatic hydrocarbons. Pp. 1219-1225 in Patty's Industrial Hygiene and Toxicology, Vol. 2, 2nd Rev. Ed., D.W. Fassett and D.D. Irish, eds. New York: Interscience.

Ghantous H., and B.R.G. Danielsson. 1986. Placental transfer and

distribution of toluene, xylene and benzene, and their metabolites during gestation in mice. Biol. Res. Pregnancy Perinatol. 7:98-105.
Gill, D.P., and A.E. Ahmed. 1981. Covalent binding of [^{14}C]benzene to cellular organelles and bone marrow nucleic acids. Biochem. Pharmacol. 30:1127-1131.
Gill, D.P., V.K. Jenkins, R.R. Kempen, and S. Ellis. 1980. The importance of pluripotential stem cells in benzene toxicity. Toxicology 16:163-171.
Gofmekler, V.A. 1968. Embryotropic action of benzene and formaldehyde inhalation. Hyg. Sanit. (USSR) 331(3):327-332.
Goldstein, B.D. 1977. Hematotoxicity in humans. J. Toxicol. Environ. Health Suppl. 2:69-105.
Goldstein, B.D., C.A. Snyder, S. Laskin, I. Bromberg, R.E. Albert, and N. Nelson. 1982. Myelogenous leukemia in rodents inhaling benzene. Toxicol. Lett. 13:169-173.
Goldwater, L.J. 1941. Disturbances in the blood following exposure to benzol. J. Lab. Clin. Med. 26:957-973.
Gonasun, L., C. Witmer, J. Kocsis, and R. Snyder. 1973. Benzene metabolism in mouse liver microsomes. Toxicol. Appl. Pharmacol. 26:398-406.
Green, J.D., B.K.J. Leong, and S. Laskin. 1978. Inhaled benzene fetotoxicity in rats. Toxicol. Appl. Pharmacol. 46:9-18.
Green, J.D., C.A. Snyder, J. LoBue, B.D. Goldstein, and R.E. Albert. 1981a. Acute and chronic dose/response effects of inhaled benzene on multipotential hematopoietic stem (CFU-S) and granulocyte/macrophage progenitor (GM-CFU-C) cells in CD-1 mice. Toxicol. Appl. Pharmacol. 58:492-503.
Green, J.D., C.A. Snyder, J. LoBue, B.D. Goldstein, and R.E. Albert. 1981b. Acute and chronic dose/response effect of benzene inhalation on the peripheral blood, bone marrow, and spleen cells of CD-1 male mice. Toxicol. Appl. Pharmacol. 59:204-214.
Greenlee, W.F., E.A. Gross, and R.D. Irons. 1981. Relationship between benzene toxicity and the disposition of ^{14}C-labelled benzene metabolites in the rat. Chem. Biol. Interact. 33:285-299.
Grilli, S., W. Lutz, and S. Parodi. 1987. Possible implications from results of animal studies in human risk estimations for benzene: Nonlinear dose-response relationship due to saturation of metabolism. J. Cancer Res. Clin Oncol. 113:349-358.

Guberan, E., and P. Kocher. 1971. Pronostic lointain de l'intoxication benzolique chronique: Controle d'une population 10 ans apres l'exposition. Schweiz Med. Worchenscher 101:1789-1790.

Harigaya, K., M.E. Miller, E.P. Cronkite, and R.T. Drew. 1981. The detection of in vivo hematotoxicity of benzene by in vitro liquid bone marrow cultures. Toxicol. Appl. Pharmacol. 60:346-353.

Hirokawa, T., and K. Nomiyama. 1962. Studies on the poisoning by benzene and its homologues: Oxidation rate of benzene in the rat liver homogenate. Med. J. Shinshu Univ. 7:29-39.

Henderson, R.F., P.J. Sabourin, M.A. Medinsky, L.S. Birnbaum, and G.L. Lucier. 1992. Benzene dosimetry in experimental animals: Relevance for risk assessment. Pp. 93-105 in Relevance of Animal Studies to the Evaluation of Human Cancer Risk. New York: Wiley-Liss.

Huff, J.E., W. Eastin, J. Roycroft, S.L. Eustis, and J.K. Haseman. 1988. Carcinogenesis studies of benzene, methyl benzene, and dimethyl benzenes. Ann. N.Y. Acad. Sci. 534:427-440.

IARC. 1982. International Agency for Research on Cancer. Monographs on the evaluation of the carcinogenic risk of chemicals to humans. Some industrial chemicals and dyestuffs. IARC Monogr. Eval. Carcinog. Risk Chem. Man. 29:93-148.

Ikeda, M., and H. Ohtsuji. 1971. Phenobarbital-induced protection against toxicity of toluene and benzene in the rat. Toxicol. Appl. Pharmacol. 20:30-43.

Infante, P.F. 1978. Leukemia among workers exposed to benzene. Tox. Rep. Biol. Med. 37:153-161.

Infante, P.F. 1992. Benzene and leukemia: The 0.1 ppm ACGIH proposed threshold limit value for benzene. Appl. Occup. Environ. Hyg. 7:253-262.

Infante, P.F., and M.C. White. 1985. Projections of leukemia risk associated with occupational exposure to benzene. Am. J. Ind. Med. 7:403-413.

Infante, P.F., R.A. Rinsky, J.K. Waggoner, and R.J. Young. 1977. Leukemia in benzene workers. Lancet. ii:76-78.

Inoue, O.K., K. Seiji, and M. Kasahira. 1986. Quantitative relation of urinary phenol levels to breath zone benzene concentrations: A factory survey. Br. J. Ind. Med. 43:692-697.

Irons, R.D., and D.A. Neptun. 1980. Effects of the principal hydroxy-metabolites of benzene on microtubule polymerization. Arch. Toxicol. 45:297-305.

Irons, R.D., D. Wierda, and R.W. Pfeifer. 1982. The immunotoxicity of benzene and its metabolites. Pp. 37-50 in Carcinogenicity and Toxicity of Benzene, M.A. Mehlman, ed. Princeton, N.J.: Princeton Scientific.

James, J.T., T.F. Limero, H.J. Leano, J.F. Boyd, and P.A. Covington. 1994. Volatile organic contaminants found in the habitable environment of the space shuttle: STS-26 to STS-55. Aviat. Space Environ. Med. 65:851-857.

Kalf, G.F., T. Rushmore, and R. Snyder. 1982. Benzene inhibits RNA synthesis in mitochondria from liver and bone marrow. Chem. Biol. Interact. 42:353-370.

Kalf, G.F., G.B. Post, and R. Snyder. 1987. Solvent toxicology: Recent advances in the toxicology of benzene, the glycol ethers, and carbon tetrachloride. Annu. Rev. Pharmacol. Toxicol. 27:399-427.

Keller, K.A., and C.A. Snyder. 1986. Mice exposed in utero to low concentrations of benzene exhibit enduring changes in their colony forming hematopoietic cells. Toxicology 42:171-181.

Kline, S., J. Forbes-Robertson, V. Lee-Grotz, B.D. Goldstein, and G. Witz. 1993. Identification of 6-hydroxy-*trans,trans*-2,4-hexadienoic acid, a novel ring-opened urinary metabolite of benzene. Environ. Health Perspect. 101:310-312.

Lange, A., R. Smolik, W. Zatonski, and J. Szymanska. 1973. Serum immunoglobulin levels in workers exposed to benzene, toluene and xylene. Int. Arch. Arbeitsmed. 31:37-44.

Leong, B.K. 1977. Experimental benzene intoxication. J. Toxicol. Environ. Health Suppl. 2:45-61.

Lesnyak, A.T., G. Sonnenfeld, M.P. Rykova, D.O. Meshkov, A. Mastro, and I. Konstantinova. 1993. Immune changes in test animals during space flight. J. Leuk. Biol. 54:214-226.

Li, G.L., N. Yin, and T. Watanabe. 1986. Benzene-specific increase in leukocyte alkaline phosphatase activity in rats exposed to vapors of various organic solvents. J. Toxicol. Environ. Health. 19:581-589.

Longacre, S., J. Kocsis, and R. Snyder. 1981. Influence of strain differences in mice on the metabolism and toxicity of benzene. Toxicol. Appl. Pharmacol. 60:398-409.

Luke, C.A., R.R. Tice, and R.T. Drew. 1988a. The effect of exposure regimen and duration on benzene-induced bone marrow damage in mice. II. Strain comparisons involving B6C3F$_1$, C57Bl/6 and DBA/2 male mice. Mutat. Res. 203:273-295.

Luke, C.A., R.R. Tice, and R.T. Drew. 1988b. The effect of exposure regimen and duration on benzene-induced bone marrow damage in mice. I. Sex comparison in DBA/2 mice. Mutat. Res. 203:251-271.

Lutz, W.K., and C.H. Schlatter. 1977. Mechanism of the carcinogenic action of benzene: Irreversible binding to rat liver DNA. Chem. Biol. Interact. 18:241-245.

Marcus, W.L. 1987. Chemical of current interest—Benzene. Toxicol. Ind. Health 3:205-266.

Magos, G.A., M. Lorenzana-Jimenez, and H. Vidrio. 1990. Toluene and benzene inhalation influences on ventricular arrhythmias in the rat. Neurotoxicol. Teratol. 12:119-124.

Maltoni, C., A. Ciliberti, G. Cotti, B. Conti, and F. Belpoggi. 1989. Benzene, an experimental multipotential carcinogen: Results of the long-term bioassays performed at the Bologna Institute of Oncology. Environ. Health Perspect. 82:109-124.

Medinsky, M.A., P.J. Sabourin, G. Lucier, L.S. Birnbaum, and R.F. Henderson. 1989. A physiological model for simulation of benzene metabolism by rats and mice. Toxicol. Appl. Pharmacol. 99:193-206.

Michon, S. 1965. Disturbances of menstruation in women working in an atmosphere polluted with aromatic hydrocarbons. Pol. Tig. Lek. 20:1648-1649.

Morimoto, K. 1983. Induction of sister chromatid exchanges and cell cycle division delays in human lymphocytes by microsomal activation of benzene. Cancer Res. 43:1330-1334.

Morimoto, K., and S. Wolff. 1980. Increase of sister chromatid exchanges and cell cycle perturbations of cell division kinetics in human lymphocytes by benzene metabolites. Cancer Res. 40:1189-1193.

Mukhametova, I.M., and M.A. Vozovaya. 1972. Reproductive power and the incidence of gynecological disorders in female workers exposed to the combined effect of benzene and chlorinated hydrocarbons. Gig. Tr. Prof. Zabol. 16:6-9.

Murray, F.J., J.A. John, L.W. Rampy, R.A. Kuna, and B.A. Schwetz. 1979. Embryotoxicity of inhaled benzene in mice and rabbits. Am. Ind. Hyg. Assoc. J. 40:993-998.

Nahum, L.H., and H.E. Hoff. 1934. The mechanism of sudden death in experimental acute benzol poisoning. J. Pharmacol. Exp. Ther. 50:336-345.

NIOSH. 1974. Criteria for a Recommended Standard . . . Occupational Exposure to Benzene. National Institute for Occupational Safety and Health, Department of Health, Education and Welfare. NIOSH-74-137. National Institute for Occupational Safety and Health, Cincinnati, Ohio.

NIOSH. 1977. Revised Recommendation for an Occupational Exposure Standard for Benzene. National Institute for Occupational Safety and Health, Cincinnati, Ohio.

Nomiyama, K., and H. Nomiyama. 1974a. Respiratory retention, uptake and excretion of organic solvents in man. Benzene, toluene, n-hexane, trichloroethylene, acetone, ethyl acetate and ethyl alcohol. Int. Arch. Arbeitsmed. 32:75-83.

Nomiyama, K., and H. Nomiyama. 1974b. Respiratory elimination of organic solvents in man. Benzene, toluene, n-hexane, trichloroethylene, acetone, ethyl acetate and ethyl alcohol. Int. Arch. Arbeitsmed. 32:85-91.

Novikov, Y.V. 1956. Effects of small benzene concentrations on higher nervous activity of animals in chronic experiments. Gig. Sanit. 21:20.

NRC. 1986. Emergency and Continuous Exposure Guidance Levels for Selected Airborne Contaminants. Benzene and Ethylene Oxide, Vol. 6. Washington, D.C.: National Academy Press.

NRC. 1992. Guidelines for Developing Spacecraft Maximum Allowable Concentrations for Space Station Contaminants. Washington, D.C.: National Academy Press.

Oehme, F. 1969. Ph.D. thesis. Comparative Study of the Biotransformation and Excretion of Phenol. Columbia: University of Missouri. 207 pp.

Ott, M.G., J.C. Townsend, W.A. Fishbeck, and R.A. Langner. 1978. Mortality among workers occupationally exposed to benzene. Arch. Environ. Health 33:3-10.

Parke, D.V., and R.T. Williams. 1953. Studies in detoxication. 49. The metabolism of benzene containing [^{14}C]benzene. J. Biochem. 54:231-238.

Parodi, S., W.K. Lutz, A. Colacci, M. Mazzullo, M. Taningher, and S. Grilli. 1989. Results of animal studies suggest a nonlinear dose-response relationships for benzene effects. Environ. Health Perspect. 82:171-176.

Pfeifer, R., and R. Irons. 1981. Inhibition of lecithin-stimulated lymphocyte agglutination and mitogenesis by hydroquinone: Reactivity with intracellular sulfhydryl groups. Exp. Mol. Pathol. 35:189-198.

Picciano, D. 1979. Cytogenetic study of workers exposed to benzene. Environ. Res. 19:33-38.

Pushkina, N.N., V.A. Gofmekler, and G.N. Klevtsova. 1968. Changes in content of ascorbic acid and nucleic acids produced by benzene and formaldehyde. Bull. Exp. Biol. Med. (USSR) 66:868-870.

Rickert, D.E., T.S. Baker, and J.S. Bus. 1979. Benzene disposition in the rat after exposure by inhalation. Toxicol. Appl. Pharmacol. 49:417-423.

Rinsky, R.A., R.J. Young, and A.B. Smith. 1981. Leukemia in benzene workers. Am. J. Ind. Med. 2:217-245.

Rinsky, R.A., B. Alexander, and M.D. Smith. 1987. Benzene and leukemia: An epidemiological risk assessment. N. Engl. J. Med. 316:1044-1050.

Rosenthal, G.J., and C.A. Snyder. 1985. Modulation of the immune response to *Listeria monocytogenes* by benzene inhalation. Toxicol. Appl. Pharmacol. 80:502-510.

Rosenthal, G.J., and C.A. Snyder. 1986. Altered T-cell responses in C57Bl mice following sub-chronic benzene inhalation. Toxicologist 61:68.

Rozen, M.G., and C.A. Snyder. 1985. Protracted exposure of C57Bl/6 mice to 300 ppm benzene depresses B- and T-lymphocyte numbers and mitogen responses. Evidence for thymic and bone marrow proliferation in response to the exposures. Toxicology 371(2): 13-26.

Rozen, M.G., C.A. Snyder, and R.E. Albert. 1984. Depressions in B- and T- lymphocyte mitogen-induced blastogenesis in mice exposed

to low concentrations of benzene. Toxicol. Lett. 20:343-349.
Sabourin, P.J., B.T. Chen, G. Lucier, L.S. Birnbaum, E. Fisher, and R.F. Henderson. 1987. Effect of dose on the absorption and excretion of [^{14}C]benzene administered orally or by inhalation in rats and mice. Toxicol. Appl. Pharmacol. 87:325-336.
Sabourin, P.J., J.D. Sun, J.T. MacGregor, C.M. Wehr, L.S. Birnbaum, G. Lucier, and R.F. Henderson. 1990. Effect of repeated benzene inhalation exposures on benzene metabolism, binding to hemoglobin, and induction of micronuclei. Toxicol. Appl. Pharmacol. 103:452-462.
Sabourin, P.J., B.A. Muggenburg, R.C. Couch, D. Lefler, G. Lucier, L.S. Birnbaum, and R.F. Henderson. 1992. Metabolism of [^{14}C]benzene by cynomolgus monkeys and chimpanzees. Toxicol. Appl. Pharmacol. 114:277-284.
Sammett, D., E.W. Lee, J.J. Kocsis, and R. Snyder. 1979. Partial hepatectomy reduces both metabolism and toxicity of benzene. J. Toxicol. Environ. Health 5:785-792.
Sarto, F.S., I. Cominato, A.M. Pinton, P.G. Brovedani, E. Merler, M. Peruzzi, V. Bianchi, and A.G. Levis. 1984. A cytogenetic study on workers exposed to low concentrations of benzene. Carcinogenesis 5:827-832.
Sandmeyer, E.E. 1981. Aromatic hydrocarbons. Pp. 3253-3283 in Patty's Industrial Hygiene and Toxicology, Vol. 2B, 3rd Rev. Ed., G.D. Clayton and F.E. Clayton, eds. New York: John Wiley & Sons.
Sato, A., and T. Nakajima. 1979. Partition coefficients of some aromatic hydrocarbons and ketones in water, blood, and oil. Toxicol. Appl. Pharmacol. 48:49.
Sato, A., and T. Nakajima. 1985. Enhanced metabolism of volatile hydrocarbons in rat liver following food deprivation, restricted carbohydrate intake, and administration of ethanol, phenobarbital, polychlorinated biphenyl, and 3-methylcholanthrene: A comparative study. Xenobiotica 15:67-75.
Sato, A., Y. Fujiwara, and T. Nakajima. 1974. Solubility of benzene, toluene, and *m*-xylene in various body fluids and tissues of rabbits. Sangyo Igaku 16:30.
Sato, A., T. Nakajima, and Y. Fujiwara. 1975. Kinetic studies on sex difference in susceptibility to chronic benzene intoxication—With

special reference to body fat content. Br. J. Ind. Med. 32:321-328.

Schrenk, H.H., W.P. Yant, and S.J. Pearce. 1941. Absorption, distribution and elimination of benzene by body tissues and fluids of dogs exposed to benzene vapor. J. Ind. Hyg. Toxicol. 23:20-34.

Seidel, H.J., E. Barthel, and D. Zinser. 1989a. The hematopoietic stem cell compartments in mice during and after long-term inhalation of three doses of benzene. Exp. Hematol. 17:300-303.

Seidel, H.J., G. Beyvers, M. Pape, and E. Barthel. 1989b. The influence of benzene on the erythroid cell system in mice. Exp. Hematol. 17:760-764.

Seidel, H.J., R. Bader, L. Weber, and E. Barthel. 1990. The influence of ethanol on the stem cell toxicity of benzene in mice. Toxicol. Appl. Pharmacol. 105:13-18.

Sherwood, R.J. 1972. Benzene: The interpretation of monitoring results. Ann. Occup. Hyg. 15:409-421.

Smyth, H.F., C.P. Carpenter, C.S. Weil, U.C. Pozzani, and J.A. Striegel. 1962. Range-finding toxicity data: List VI. J. Ind. Hyg. Assoc. 231:95-107.

Snyder, C.A. 1987. Benzene. In Ethyl Browning's Toxicity and Metabolism of Industrial Solvents. Vol. 1: Hydrocarbons, 2nd Ed. R. Snyder, ed. New York: Elsevier Science.

Snyder, C.A, B.D. Goldstein, and A.R. Sellakumar. 1978. Hematotoxicity of inhaled benzene to Sprague-Dawley rats and AKR mice at 300 ppm. J. Toxicol. Environ. Health 4:605-618.

Snyder, C.A., B.D. Goldstein, A.R. Sellakumar, I. Bromberg, S. Laskin, and R.E. Albert. 1980. The inhalation toxicology of benzene: Incidence of hematopoietic neoplasms and hematotoxicity in AKR/J and C57Bl/6J mice. Toxicol. Appl. Pharmacol. 54:323-331.

Snyder, C.A., J.D. Green, J. LoBue, B.D. Goldstein, C.D. Valle, and R.E. Albert. 1981. Protracted benzene exposure causes a proliferation of myeloblasts and/or promyelocytes in CD-1 mice. Bull. Environ. Contam. Toxicol. 27:17-22.

Snyder, C.A., B.D. Goldstein, A.R. Sellakumar, and R.E. Albert. 1984. Evidence for hematoxicity and tumorigenesis in rats exposed to 100 ppm benzene. Am. J. Ind. Med. 5:429-434.

Snyder, R., E.W. Lee, J.J. Kocsis, and C.M. Witmer. 1977. Bone marrow depressant and leukemogenic actions of benzene. Life Sci. 21:1709-1722.

Snyder, R., S.L. Longacre, C.M. Witmer, and J.J. Kocsis. 1981. Biochemical toxicology of benzene. Pp. 123-153 in Reviews in Biochemical Toxicology 3, E. Hodgson, J.R. Bend, and R.M. Philpot, eds. Amsterdam: Elsevier-North Holland.

Snyder, R., G. Witz, and B.D. Goldstein. 1993a. The toxicity of benzene. Environ. Health Perspect. 100:293-306.

Snyder, R., T. Chepiga, C.S. Yang, H. Thomas, K. Platt, and F. Oesch. 1993b. Benzene metabolism by reconstituted cytochromes P450 2B1 and 2E1 and its modulation by cytochrome b microsomal epoxide hydrolase, and glutathione transferases: Evidence for an important role of microsomal epoxide hydrolase in the formation of hydroquinone. Toxicol. Appl. Pharmacol. 122:172-181.

Srbova, J., J. Teisinger, and S. Skramovsky. 1950. Absorption and elimination of inhaled benzene in man. Arch. Ind. Hyg. Occup. Med. 2:1-8.

Stommel, P., G. Muller, W. Stucker, C. Verkoyen, S. Schobel, and K. Norpoth. 1989. Determination of S-phenylmercapturic acid in the urine—An improvement in the biological monitoring of benzene exposure. Carcinogenesis 10:279-282.

Stoner, R.D., R.T. Drew, and D.M. Bernstein. 1981. Benzene inhalation effect upon tetanus antitoxin responses and leukemogenesis in mice. Pp. 445-461 in Coal Conversion and the Environment, D.D Mahlum, R.H. Gray, and W.D. Felix, eds. Technical Information Center, U.S. Department of Energy, Oak Ridge, Tenn.

Sun, J.D., M.A. Medinsky, L.S. Birnbaum, G. Lucier, and R.F. Henderson. 1990. Benzene hemoglobin adducts in mice and rats: Characterization of formation and physiological modeling. Fundam. Appl. Toxicol. 15:468-475.

Swaen, G.W.H., and Meijers, J.M.M. 1989. Risk assessment of leukaemia and occupational exposure to benzene. Br. J. Ind. Med. 46:826-830.

Taylor, G.R. 1993. Immune changes during short-duration missions. J. Leukocyte Biol. 54:202-208.

Teisinger, J., V. Bergerova-Fiserova, and J. Kudrna. 1952. The metabolism of benzene in man. Pracov. Lek. 4:175-188.

Tice, R.R., D.L. Costa, and R.T. Drew. 1980. Cytogenetic effects of inhaled benzene in murine bone marrow: Induction of sister chromatid exchanges, chromosomal aberrations and cellular proliferation inhibition in DBA/2 mice. Proc. Natl. Acad. Sci. USA 77:2148-2152.

Toft, K., T. Olofsson, A. Tunek, and M. Berlin. 1982. Toxic effects on mouse bone marrow caused by inhalation of benzene. Arch. Toxicol. 51:295-302.

Tough, I.M., and W.M. Court Brown. 1965. Chromosome aberrations and exposure to ambient benzene. Lancet i:684.

Travis, C.C., J.L. Quillen, and A.D. Arms. 1990. Pharmacokinetics of benzene. Toxicol. Appl. Pharmacol. 102:400-420.

Ungvary, G., and E. Tatrai. 1985. On the embryotoxic effects of benzene and its alkyl derivatives in mice, rats and rabbits. Arch. Toxicol. Suppl. 8:425-430.

Uyeki, E.M., A.E. Ashkar, D.W. Shoeman, and T.U. Bisel. 1977. Acute toxicity of benzene inhalation to hemopoietic precursor cells. Toxicol. Appl. Pharmacol. 40:49-57.

Valle-Paul, C., and C.A. Snyder. 1986. Effect of benzene exposure on bone marrow precursor cells of splenectomized mice. Toxicologist 61:285.

Van Raalte, H.G.S., and P. Grasso. 1982. Hematological, myelotoxic, clastogenic, carcinogenic, and leukemogenic effects of benzene. Regul. Toxicol. Pharmacol. 2:153-176.

Vara, P.,and O. Kinnunen. 1946. Benzene toxicity as a gynecologic problem. Acta Obstet. Gynecol. Scand. 26:433-452.

Vigliani, E.C. 1976. Leukemia associated with benzene exposure. Ann. N.Y. Acad. Sci. 271:143-151.

Vigliani, E.C., and A. Forni. 1976. Benzene and leukemia. Environ. Res. 11:122-127.

Von Oettingen, W.F. 1940. Public Health Bulletin No. 255, U.S. Public Health Service, Washington, D.C.

Voytek, P.E., and T.W. Thorslund. 1991. Benzene risk assessment: Status of quantifying the leukemogenic risk associated with low dose inhalation of benzene. Risk Anal. 11:355-357.

Ward, C.O., R.A. Kuna, N.K. Snyder, R.D. Alsaker, W.B. Coate, and P.H. Craig. 1985. Subchronic inhalation toxicity of benzene in rats and mice. Am. J. Ind. Med. 7:457-473.

Ward, J.B., Jr., M.M. Ammenheuser, V.M.S. Ramanujam, D.L. Morris, E.B. Whorton, Jr., and M.S. Legator. 1992. The mutagenic effects of low level sub-acute inhalation exposure to benzene in CD-1 mice. Mutat. Res. 268:49-57.

Watanabe, T., A. Endo, Y. Kato, S. Shima, T. Watanabe, and M. Ikeda. 1980. Cytogenetics and cytokinetics of cultured lymphocytes from benzene-exposed workers. Int. Arch. Occup. Environ. Health 46:31-41.

White, M.C., P.F. Infante, and B. Walker. 1980. Occupational exposure to benzene: A review of the carcinogenic and related health effects following the United States Supreme Court decision. Am. J. Ind. Med. 1:233-243.

Wierda, D., and R. Irons. 1982. Hydroquinone and catechol reduce the frequency of progenitor B lymphocytes in mouse spleen and bone marrow. Immunopharmacology 4:41-54.

Wilson, R.H. 1942. Benzene poisoning in industry. J. Lab. Clin. Med. 27:1517-1521.

Winek, C.L., and W.D. Collom. 1971. Benzene and toluene fatalities. J. Occup. Med. 131:259-261.

Winek, C.L., W.D. Collom, and C.H. Wecht. 1967. Fatal benzene exposure by glue sniffing. Lancet i:683.

Wolf, M.A., V.K. Rowe, D.D. McCollister, R.L. Hollingsworth, and F. Oyen. 1956. Toxicological studies of certain alkylated benzenes and benzene. AMA Arch. Ind. Health 14:387-398.

Woodruff, T.J., and F.Y. Bois. 1993. Optimization issues in physiological toxicokinetic modeling: A case study with benzene. Toxicol. Lett. 69:181-196.

Woodruff, T., F.Y. Bois, J. Parker, D. Auslander, S. Selvin, M. Smith, and R. Spear. 1989. Design and analysis of a model of benzene toxicokinetics in mammals. Pp. 254-255 in Proceedings of the 11th Annual International Conference of the IEEE Engineering in Medicine and Biology Society, IEEE, Seattle.

Yardley-Jones, A., D. Anderson, and D.V. Parke. 1991. The toxicity of benzene and its metabolism and molecular pathology in human risk assessment. Br. J. Ind. Med. 48:437-444.

B3 Carbon Dioxide

King Lit Wong, Ph.D.
Johnson Space Center Toxicology Group
Biomedical Operations and Research Branch
Houston, Texas

PHYSICAL AND CHEMICAL PROPERTIES

Carbon dioxide is an odorless and colorless gas (Sax, 1984).

Synonym: Carbonic anhydride
Formula: CO_2
CAS number: 124389
Molecular weight: 44
Boiling point: Not applicable
Melting point: Sublime at -78°C
Vapor pressure: Not applicable
Conversion factors 1 ppm = 1.80 mg/m^3
at 25°C, 1 atm: 1 mg/m^3 = 0.56 ppm

OCCURRENCE AND USE

CO_2 normally exists in the atmosphere at 0.03% (Morey and Shattuck, 1989). In a Danish study, the maximal CO_2 concentrations inside 14 town-hall buildings (6 had natural and 8 had mechanical ventilation) were measured to be 0.05-0.13% (Skov et al., 1987). Wang (1975) reported that the CO_2 concentration inside a university auditorium built up to about 0.06-0.09% during a lecture. CO_2 is not used in space shuttles, but it will be used as a fire extinguishant in the space station.

Metabolism is a source of CO_2 in spacecraft, and thermodegradation of organic materials is a potential source of CO_2 (Coleman et al., 1968; Terrill et al., 1978; Wooley et al., 1979). Humans produce CO_2 via oxidative metabolism of carbohydrates, fatty acids, and amino acids; the production rate is dependent on the caloric expenditure of the individual (Baggott, 1982; Diamondstone, 1982; LeBaron, 1982; Olson, 1982). A young adult male produces about 22,000 meq of CO_2 per day (Baggott, 1982). For a 70-kg adult doing light work in spaceflight, the amount of CO_2 exhaled was estimated to be 500 L/d (Clamann, 1959). The amount of CO_2 exhaled by a group of normal male subjects, aged 18-45, inside a steel chamber was measured at 469 L/d per person (Consolazio et al., 1947). During a 7-d shuttle mission with seven crew members, the mean CO_2 concentration in the cabin was about 2 mm Hg, which was equivalent to 0.26% in an atmosphere of 760 mm Hg, with a 5-h peak of 9 mm Hg or 1.2% (NASA, 1984).

PHARMACOKINETICS AND METABOLISM

When inhaled, CO_2 freely penetrates cellular membranes (Baggott, 1982). The diffusion rate of CO_2 through the alveolar membrane into blood is about 20 times that of O_2 (West, 1979). CO_2 is carried in blood in three forms, the bicarbonate being the major form. Ninety percent of the CO_2 in blood reacts with water, under the catalysis of carbonic anhydrase inside the erythrocytes, to form carbonic acid, which in turn is ionized to bicarbonate (Baggott, 1982). This reaction also takes place in serum in the absence of carbonic anhydrase, but it proceeds much more slowly than with catalysis (Baggott, 1982).

The other two forms of CO_2 transport in blood are relatively minor. About 5% of the CO_2 in blood is dissolved in serum and cytoplasm (Baggott, 1982). The solubility of CO_2 in water is approximately 20 times that of O_2, so that CO_2 dissolved in plasma is a more important form of transport in blood than dissolved O_2 (West, 1979). CO_2 is present in blood in the third form as carbamino compounds, which are formed from the reaction of CO_2 with uncharged amino groups in hemoglobin (Baggott, 1982). The carbamino form accounts for about 5% of the CO_2 in blood (Baggott, 1982).

Normally, CO_2 is eliminated from the body via exhalation. A healthy man exhales CO_2 at about 220 mL/min at rest and 1,650 mL/min during moderate exercise (Cotes, 1979, pp. 266, 276, 384).

The CO_2–bicarbonate system functions as the major buffering system in blood (Baggott, 1982). In acidosis, an individual is exposed to a high concentration of CO_2. Hyperventilation increases the CO_2 exhalation, which raises the pH in blood (Baggott, 1982). In alkalosis, the individual will hypoventilate to reduce CO_2 exhalation and the kidney will excrete bicarbonate ions into the urine, both of which lower the pH in blood (Baggott, 1982).

TOXICITY SUMMARY

Acute and Short-Term Toxicity

Miscellaneous Signs and Symptoms

Both hearing and vision can be impaired by CO_2. A 6-min exposure to 6.1-6.3% CO_2 resulted in a 3-8% decrease in hearing threshold in six human subjects (Gellhorn and Spiesman, 1935). For CO_2 exposures of six human subjects lasting 5-22 min, 3-4% CO_2 was the threshold for causing slight hearing impairment and 2.5% was the no-observed-adverse-effect level (NOAEL) (Gellhorn and Spiesman, 1934, 1935). Because the amount of hearing impairment produced by about 6% CO_2 is very small and because the SMACs are expected to be much lower than 6%, hearing impairment is not considered in setting the SMACs for CO_2. Acute exposures to 6% CO_2 affected vision by reducing visual intensity discrimination in 1-2 min (Gellhorn, 1936) and by causing visual disturbances in several hours in an unspecified number of men (Schulte, 1964).

CO_2 exposures can cause other symptoms, such as tremor, discomfort, dyspnea, headache, and intercostal pain. Tremor was produced in human subjects exposed to 6% CO_2 for several hours (number of subjects unknown) (Schulte, 1964) or 7-14% CO_2 for 10-20 min (12 subjects) (Sechzer et al., 1960). Exposures of six volunteers to 6% CO_2 for 20.5-22 min led to discomfort (Gellhorn and Spiesman, 1935).

Dyspnea

Available data indicate that acute exposures to CO_2 at concentrations higher than 3% definitely could produce dyspnea. For instance, White et al. (1952) found that, in a 16-min exposure to 6% CO_2 in O_2, 19 of 24 volunteers had slight or moderate dyspnea, and the dyspneic sensation was severe in the remaining five subjects. A 17-32 min exposure of 16 human subjects to 4-5% CO_2 (Schneider and Truesdale, 1922) or a 2.5-10 min exposure to 7.6% CO_2 (Dripps and Comroe, 1947) resulted in dyspnea.

There were conflicting data on whether 2.8-3% CO_2 would cause dyspnea. On one hand, Menn et al. (1970) found that, in a 30-min exposure to 2.8% CO_2, dyspnea was detected in three of eight human subjects during maximal exercise, but not during half-maximal or two-thirds-maximal exercises. On the other hand, Sinclair et al. (1971) showed that a 1-h or 15- to 20-d exposure of four volunteers to 2.8% CO_2 failed to produce any dyspnea during steady strenuous exercise. However, Schulte (1964) reported that an exposure to CO_2 at concentrations as low as 2% for several hours resulted in dyspnea on exertion in an unknown number of human subjects. In the study conducted by Menn et al., 1.1% CO_2 failed to cause dyspnea in eight subjects even during maximal exercise in 30 min. There were also conflicting data on CO_2's dyspneic effect in resting subjects. Brown (1930a) showed that 3.2% CO_2 or 2.5-2.8% CO_2 did not produce dyspnea in five resting human subjects. In contrast, Schulte (1964) reported that an exposure to 3% CO_2 for several hours resulted in dyspnea even at rest, without specifying the number of human subjects on which he based his conclusion. The bulk of the data indicate that the NOAEL for CO_2 exposures based on dyspnea appears to be 2.8% because astronauts will engage in moderate, but not maximal, exercise.

Headaches

In addition to dyspnea, acute CO_2 exposures could produce headaches. Without specifying the size of population he based his conclusion on, Schulte (1964) reported that human subjects exposed to 2% or 3% CO_2 for several hours developed headaches on mild exertion; the headache was more severe at 3% CO_2 than 2%. Sinclair et al. (1971) showed that a 1-h exposure of four human subjects to 2.8% CO_2 resulted in occasional mild headaches during strenuous steady-state exercise. Menn et al. (1970) found that mild-to-moderate frontal headaches developed in six of eight human subjects exposed to 3.9% CO_2 for 30 min while doing two-thirds-maximal exercise. A similar exposure to 1.1% or 2.8% CO_2 failed to cause headaches (Menn et al., 1970). Therefore, there is conflicting evidence whether 2.8% CO_2 produces headaches during exertion.

In a comparison of the data on exercising subjects (Schulte, 1964; Menn et al., 1970; Sinclair et al., 1971) and on subjects at rest (Schneider and Truesdale, 1922; Brackett et al., 1965), CO_2 appears to cause more headaches at a lower concentration during exercise than at rest. White et al. (1952) showed that, soon after a 16-min exposure of 24 subjects to 6% CO_2, one developed a severe headache and nine developed mild headaches of very short durations. In a study of five or six resting human subjects conducted by Brown (1930a), an exposure to 3.2% CO_2 in 13.4% O_2 for several hours produced headache and giddiness, but an exposure to 2.5-2.8% CO_2 in 14.6-15% O_2 was devoid of any symptoms. Schneider and Truesdale (1922) showed that, in 16 resting volunteers exposed to 1-8% CO_2 for 17-32 min, headaches developed only at a CO_2 concentration of 5% or more and the headache could be intense. In a study by Brackett et al. (1965), 7% CO_2 caused mild headache in approximately seven resting volunteers in 40-90 min.

CO_2 exposures do not cause headaches immediately. Menn et al. (1970) reported that headaches mostly developed near the end of a 30-min exposure to 3.9% CO_2 while the subjects were performing two-thirds-maximal exercise. Glatte et al. (1967a) found that, in a 5-d exposure to 3% CO_2, mild-to-moderate throbbing frontal headaches were detected in four of seven human subjects in the first day. A similar response was found in human subjects exposed to 4% CO_2 (Glatte et al., 1967b; Menn et al., 1968). The headaches usually began in the first few hours of exposure.

The headaches produced by CO_2 are not long lasting. In a 30-min exposure to 3.9% CO_2, the headaches disappeared an hour after the exposure (Menn et al., 1970). In human subjects exposed to 3% or 4% CO_2 for 5 d, they recovered from the headaches in 3 d (Glatte et al., 1967b; Menn et al., 1968). Menn et al. (1970) postulated that the headaches are caused by CO_2-induced dilation of cerebral blood vessels (Patterson et al., 1955). The disappearance of the headaches soon after an acute exposure or disappearing beginning on the third day of a 5-d exposure suggests, as another possibility, that the headaches are due to CO_2-induced acidosis.

As discussed above, it is not certain whether 2.8% CO_2 could cause headaches. Similarly, there is conflicting evidence on 2% CO_2. Without specifying the size of the study population, Schulte (1964) reported that headaches were detected in human subjects exposed to 2% CO_2 for several hours on mild exertion. In contrast, Radziszewski et al. (1988) showed that a 30-d exposure of six human subjects to 2% CO_2 rarely produced headaches, even when they exercised.

Intercostal Pain

Acute CO_2 exposures can produce intercostal pain. Menn et al. (1970) reported that a 30-min exposure to 2.8% CO_2 caused intercostal muscle pain during maximal exercise in two of eight human subjects. They did not report any intercostal pain in the subjects during two-thirds- or half-maximal exercise. However, Sinclair et al. (1971) showed that a 1-h exposure to 2.8% CO_2 failed to produce intercostal muscle pain in four volunteers during steady strenuous exercise. It is possible that the test subjects in Sinclair's study did not exercise maximally during the exposure to 2.8% CO_2, so that they did not experience the intercostal pain that was reported by those in Menn's study. Menn et al. failed to detect intercostal muscle pain in eight human subjects exposed to 1.1% CO_2 for 30 min even during maximal exercise. Because astronauts will not be exercising maximally in the spacecraft, 2.8% is chosen as the NOAEL for intercostal muscle pain resulting from acute CO_2 exposures.

Acid–Base Balance

An exposure to CO_2 at concentrations much higher than the normal value of 0.03% increases the pCO_2 in blood (Mines, 1981). The increased pCO_2 in blood lowers the blood pH, although the lowering is reduced somewhat by the bicarbonate and protein buffers in blood (Mines, 1981). Acidosis is known to occur in humans after a 1-h exposure to 2.8% CO_2 (Sinclair et al., 1971). Both the CO_2 absorption and acidosis happen very rapidly. During a 1-h exposure of volunteers to 7% CO_2, the arterial pCO_2 and HCO_3 concentrations were raised, while the arterial plasma pH dropped from 7.40 to 7.30 as early as 10 min into the exposure (Brackett et al., 1965). These arterial parameters remained at a plateau from min 10-60 during the CO_2 exposure. The decreases in arterial plasma pH in humans resulting from acute CO_2 exposures are tabulated as follows.

TABLE 3-1 Arterial pH Decreases After Acute CO_2 Exposures

Concentration, %	Exposure Duration	Arterial pH Drop	Reference
1.5	1 d	0.05	Schaefer, 1963b
2	2 h	0	Guillerm and Radziszewski, 1979
2	2-3 d	0.01	Guillerm and Radziszewski, 1979
2.8	1 d	0.02	Glatte et al., 1967a
3.0	6-24 h	0.025	Sinclair et al., 1969
7	10-60 min	0.10	Brackett et al., 1965
10	10-60 min	0.22	Brackett et al., 1965

Electrolyte Levels

Messier et al. (1976) reported some electrolyte changes in 7-15 human subjects in 57-d submarine patrols, the atmosphere of which was maintained at 0.8-1.2% CO_2, 19-21% O_2, and CO at <25 ppm. On the first day of a patrol, the plasma levels of calcium decreased, with no change in plasma phosphorus levels, but the erythrocyte level of calcium increased.

Respiratory System

The most obvious effect of CO_2 exposures is increased alveolar ventilation, which is not a toxic effect per se, but it and other physiological changes inducible by CO_2 will be described in the Toxicity Summary. If O_2 is maintained at a constant concentration, alveolar ventilation of humans varies linearly with the CO_2 concentration at ventilation up to about 60 L/min (Cotes, 1979, pp. 149, 258, 363). The amounts of ventilatory increase during an acute exposure of normal human subjects to CO_2 at various concentrations are summarized in Table 3-2.

The hyperventilatory response is due mainly to a tidal volume increase, although the respiratory rate was found to increase in one study but not in another (Schaefer, 1963b; Glatte et al., 1967a; Guillerm and Radziszewski, 1979). The hyperventilatory response to inhaled CO_2 is triggered by CO_2's effect on chemoreceptors in the brain and the carotid chemoreceptors (Cotes, 1979, pp. 149, 258, 363; Phillipson et al., 1981). When the CO_2 exposure terminates, residual hyperventilation helps to lower the pCO_2 in blood, and thus the hyperventilation plays a role in restoring the normal blood pH.

Three studies show that human subjects acclimate somewhat to the hyperventilatory effect of CO_2 (Chapin et al., 1955; Schaefer, 1958; Radziszewski et al., 1988). The alveolar ventilation at rest was 15.1 L/min shortly after an exposure to 3% CO_2 began, but it was lowered to 12.9 L/min near the end of the 78-h exposure (Chapin et al., 1955). Schaefer (1958) also reported acclimation to CO_2's ventilatory effect. He presented evidence that diving instructors, who had held their breaths daily for long durations under water (resulting in CO_2 accumulation in their bodies), showed a smaller hyperventilatory response toward acute CO_2 challenges than other volunteers who were not accustomed to CO_2 retention. Some of the data from Radziszewski et al. (1988), summarized in Table 3-2 showed that the hyperventilatory response to CO_2 was diminished about one fifth at 24 h compared with 2 h in a continuous CO_2 exposure.

Some evidence indicates that CO_2 can stimulate or depress ventilation depending on the concentration. As mentioned above, CO_2 stimulates respiration at a concentration as low as 1%. CO_2 at concentrations higher than 8% has been reported to depress respiration in humans (Cotes, 1979, pp. 149, 258, 363). However, a 3.8-min exposure of human subjects to 10.4% CO_2 is known to stimulate respiration (Dripps and Comroe, 1947). So the exact CO_2 concentration required to consistently depress respiration is unknown and it might be much higher than 8%.

TABLE 3-2 Hyperventilatory Responses to Acute CO_2 Exposures

Concentration, %	No.	Exposure Duration	Increase in Minimum Volume, % (mean ± SD)	Reference
0.5-0.6	5	10 min	14 ±4	Campbell et al., 1913
0.5	6	24 h	—[a]	Radziszewski et al., 1988
1	16	17-32 min	32	Schneider and Truesdale, 1922
1	6	24 h	19	Radziszewski et al., 1988
2	16	17-32 min	80	Schneider and Truesdale, 1922
2	6	2 h	60	Radziszewski et al., 1988
2	6	24 h	45	Radziszewski et al., 1988
2.1-2.5	3	10 min	63 ±13	Campbell et al., 1913
2.2	3	10 min	36 ±21	Eldridge and Davis, 1959
2.5	3	10-20 min	30 ± 9	Brown et al., 1948
2.5	9	≈20 min	33 ± 21	Tashkin and Simmons, 1972
3	16	17-32 min	148	Schneider and Truesdale, 1922
3	5	2 h	70	Radziszewski et al., 1988
3	5	24 h	50	Radziszewski et al., 1988
3.8	5	2 h	160	Radziszewski et al., 1988
3.8	5	24 h	130	Radziszewski et al., 1988
4	16	17-32 min	208	Schneider and Truesdale, 1922
4.2	3	10 min	184 ± 110	Eldridge and Davis, 1959
4.3	5	2 h	240	Radziszewski et al., 1988
4.3	5	24 h	180	Radziszewski et al., 1988
5	3	10-20 min	130 ± 30	Brown et al., 1948
5	9	≈20 min	91 ± 60	Tashkin and Simmons, 1972
5	16	17-32 min	309	Schneider and Truesdale, 1922
5.7-6.1	5	10 min	413 ± 57	Campbell et al., 1913
5.9	7	5 min	184	Brown, 1930a
6	3	20.5-22 min	203	Brown, 1930a
6	23	16 min	200	White et al., 1952
6	16	17-32 min	419	Schneider and Truesdale, 1922
7	16	17-32 min	512	Schneider and Truesdale, 1922
7.5	3	10-20 min	474 ± 242	Brown et al., 1948
7.5	9	≈20 min	269 ± 123	Tashkin and Simmons, 1972
8	16	17-32 min	640	Schneider and Truesdale, 1922
8.8	5	7-10 min	228	Brown, 1930a
10	9	≈20 min	456 ± 189	Tashkin and Simmons, 1972
12.4	7	0.75-2 min	153	Brown, 1930a

[a]Statistically not significant.

Exposures to CO_2 are also known to affect lung functions. CO_2 inhalation for 2 h at 5% or 7.5% decreased specific airway conductance in volunteers, but 2.5% CO_2 did not change the conductance (Tashkin and Simmons, 1972). A 120% increase in the total lung resistance was detected in human subjects who inhaled 8% CO_2 in 19% O_2 for 3-6 min (Nadel and Widdicombe, 1962).

There are no data on the structural effect of CO_2 on the lungs of human beings. However, Schaefer and his colleagues reported that acute exposures to CO_2 injured the lungs of guinea pigs (Niemoeller and Schaefer, 1962; Schaefer et al., 1964a). In some of the guinea pigs exposed to 15% CO_2 in 21% O_2, Schaefer's group detected subpleural atelectasis, an increase of lamellar bodies in alveolar lining cells, congestion, edema, and hemorrhage in the lungs in 1 or 6 h (Schaefer et al., 1964a). When the exposure was extended to 1 or 2 d, they reported that hyaline membranes were seen in the lungs, in addition to the pulmonary injuries seen at 1 and 6 h. As the exposure was further extended to 7 or 14 d, they described a decline in incidences of atelectasis, edema, hemorrhages, and hyaline membranes in the lung. In that 1964 study, Schaefer's group looked at a total of six time points, with 4-14 guinea pigs exposed to CO_2 per time point. However, they used only 13 guinea pigs as controls, and they did not specify how many control guinea pigs were sacrificed per time point. That means, on the average, only two control guinea pigs were sacrificed at each time point and that is grossly inadequate. In another study, Niemoeller and Schaefer (1962) reported that CO_2 exposures at 1.5% or 3% could produce similar lung injuries as 15% CO_2. In this study, the same problem existed. They used only four control guinea pigs in the 1.5%-CO_2 experiment in which a group of exposed guinea pigs was examined at four time points. Similarly, in the 3%-CO_2 experiment, they used seven guinea pigs to control for examinations of exposed guinea pigs at five time points. Consequently, their findings that CO_2 exposures produced lung injuries in guinea pigs might not be reliable. Therefore, their findings in the lungs of guinea pigs (Niemoeller and Schaefer, 1962; Schaefer et al., 1964a) are disregarded in setting SMACs.

Cardiovascular System

CO_2 exposures are known to affect the heart and the circulatory system. A 17-32-min exposure of humans to 1% or 2% CO_2 is known to cause slight increases of systolic and diastolic pressure (Schneider and Truesdale, 1922). In another human study, a 15-30-min exposure to 5% or 7% CO_2 caused increases in blood pressure and cerebral blood flow and a decrease in cerebrovascular resistance (Kety and Schmidt, 1948). In the same study, no change in cardiac output was detected, but in another study, a 4-25-min exposure of volunteers to 7.5% CO_2 increased the cardiac output and blood pressure (Grollman, 1930). In addition to changing the cardiac output, CO_2 can increase the heart rate. A 10-15-min exposure to 5.4% CO_2 or a 4-25-min exposure to 7.5% CO_2 increased the pulse rate in humans (Grollman, 1930; Schaefer, 1958).

Acute CO_2 exposures can result in some EKG changes. Nodal and atrial premature systoles, premature ventricular contractions, inversion of P waves, low P waves, and increased T-wave voltage were observed in psychiatric patients exposed to 30% CO_2 in 70% O_2 for 38 s (MacDonald and Simonson, 1953). Similarly, McArdle (1959) exposed psychiatric patients to 30% CO_2 in 70% O_2 for 10-15 breaths, and he detected acidosis, marked increases in systolic and diastolic pressures, atrial extrasystoles, atrial tachycardia (but no ventricular extrasystole), increased P-wave voltage, low or inverted P waves, spiked T waves with a broad base, increased T-wave voltage, slight increases in PR intervals and QRS intervals, and a marked increase in the QT interval, which was the most consistent finding. The fact that it took only 35-45 breaths of the mixture of 30% CO_2 in 70% O_2 to produce narcosis in these patients suggests that the CO_2 concentration used was very high.

In CO_2 exposures at lower concentrations, lower incidences of abnormal cardiac rhythm result. For instance, in human subjects breathing 7-14% CO_2, balance O_2, for 10-20 min at rest, premature nodal contraction was detected in only 2 of 27 subjects (versus 0 of 27 before the exposure) and premature ventricular contraction was found in only 3 of 27 subjects (versus 1 of 27 before the exposure) (Sechzer et al., 1960).

At even lower CO_2 concentrations, only minor EKG changes were produced without any abnormal rhythm. In human subjects, a 6-8 min exposure to 6% CO_2 depressed the amplitude of the QRS complex and T wave, but there were no T-wave inversions or changes in the S-T segment (Okajima and Simonson, 1962). These EKG changes were more severe in men of about 60 years of age than in men in their twenties. In volunteers doing moderate or maximal exercise while exposed to 2.8% or 3.9% CO_2 for 30 min, Menn et al. (1970) found no significant increase in premature atrial or ventricular contractions over the incidences normally seen in exercising individuals in room air.

These data indicate that, in acute exposures, CO_2 can produce clinically unimportant abnormal cardiac rhythm at a fairly high concentration of 7-14% and requires a very high concentration of 30% to produce atrial tachycardia. Therefore, CO_2's EKG effects are not used in setting the SMACs for CO_2.

The mechanism of EKG changes produced by CO_2 is unknown. Altschule and Sulzbach (1947) postulated that the CO_2-induced EKG changes were due to CO_2-induced acidosis because the changes were seen with acidosis in a 45-90-min exposure of two patients at 5% CO_2 in 95% O_2 and the changes disappeared within 30 min of terminating the CO_2 exposure.

Nervous System

Exposures to CO_2 at the appropriate concentrations could cause CNS depression. Consolazio et al. (1947) discovered a decrease in hand-arm steadiness, but no change in the ability to compute, translate, check numbers, and discriminate pitch and loudness in four volunteers exposed to 5-6.75% CO_2 in 19.2% O_2 for 37 h. Schulte reported that an exposure of an unspecified number of human subjects to 5% CO_2 for several hours produced CNS depression (Schulte, 1964). An exposure of fighter pilots to 5% CO_2 for a unspecified duration degraded their performance in landing maneuvers, such as lengthened flight time between gear down and touch down and unacceptable increases in touch-down sink rates (Wamsley et al., 1969). Therefore, these studies indicate that 5% CO_2 is depressive to the CNS.

Brown (1930a) conducted a study with five human subjects in a static exposure chamber for 8 h with the CO_2 concentration measured at 4.1% and 5.3% at the end of the fourth and seventh hours, respectively. Brown showed that the number of numbers canceled in a cancellation test dropped 24% at the end of the seventh hour when the CO_2 concentration was 5.3% with 21% O_2. Using the data provided by Brown, the 24% reduction is found to be statistically significant from the pre-exposure number. However, Brown commented that the reduction was not serious deterioration. It should be noted that the same CO_2 exposure caused no changes on the scores in Army Alpha intelligence and arithmetic tests, attention, and muscular coordination (Brown, 1930a). Whether the 24% reduction in the cancellation test score was due to the concentration of CO_2 at 5.3% in the exposure chamber or due to boredom from confinement in the exposure chamber is unknown because there was no sham-exposed control group.

Data on CNS depression resulting from exposure to CO_2 at concentrations below 5% were inconsistent. In the study of five human volunteers conducted by Brown, the number of numbers canceled in the cancellation test decreased by 13% at the end of the fourth hour when the atmosphere contained 4.1% CO_2 in 21% O_2 (Brown, 1930a). Although the 13% reduction is statistically significant (based on a paired t test using Brown's data), Brown did not consider it a serious deterioration. There were no effects observed in the Army Alpha intelligence and arithmetic tests, attention, and muscular coordination (Brown, 1930a). Because Brown did not have a sham-exposed control group in assessing the CNS effects of CO_2, interpretation of Brown's CNS data is difficult.

Schaefer et al. (1958, 1959, 1963a) reported that some crew members on board a German submarine that contained 3-3.5% CO_2 in 15-17% O_2, suffered impaired attentiveness in a 2-mo underwater patrol in World War II. Nevertheless, as noted by Glatte et al. (1967b) and Menn et al. (1968), the submarine atmosphere was not tightly controlled, so that simultaneous exposure of the crew to other contaminants, such as carbon monoxide, could not be ruled out. It is, therefore, possible that the CNS depression suffered by the crew was due to the relatively low oxygen concentration, carbon monoxide, or certain organic solvents instead of the 3-3.5% CO_2. Brown's test produced giddiness and headache in four human subjects exposed to 3.2% CO_2 for several hours (Brown, 1930b). Unfortunately, the subjects were also exposed to a relatively low O_2 concentration of 13.4%, which makes interpretation of the finding of giddiness difficult.

In contrast to the findings of Schaefer et al. and Brown, other evidence shows that exposures to CO_2 in the range of 2-4% do not depress the CNS. For instance, Glatte et al. showed that a 5-d exposure of seven human subjects to 3% or 4% CO_2 failed to influence hand steadiness, vigilance, auditory monitoring, memory, and arithmetic and problem solving performance (Glatte et al., 1967a,b; Menn et al., 1968).

Storm and Giannetta (1974) showed that there were no changes in aiming ability, closure flexibility, visualization, perceptual speed, and number facility in 12 human volunteers who were tested everyday during a 2-w exposure to 4% CO_2. Schulte (1964) reported no mental depression in subjects exposed to 2% or 3% CO_2 for a few hours. The number of subjects is not known. From these data, the NOAEL for CNS depression is estimated to be 4%.

At a CO_2 concentration higher than 5%, CO_2's effect on the CNS is not purely depressive. Restlessness and dizziness have been detected in human subjects exposed to 7.5% CO_2 for 15 min, 10% CO_2 for 15-25 min, and 10.4% CO_2 for 3.8 min (Dripps and Comroe, 1947; Schaefer, 1963a; Brackett et al., 1965). Some studies have shown that acute exposures to CO_2 at high concentrations produced purely depressive signs and symptoms. For instance, unconsciousness was detected in human subjects exposed to 10% CO_2 for several hours (Schulte, 1964) and drowsiness and near stupor were found in individuals who had inhaled 12.4% CO_2 for 0.75-2 min (Brown, 1930b). In contrast, some investigators have presented evidence that CO_2 exposures excite the CNS. Psychomotor excitation, eye flickering, myoclonic twitches, increased muscle tone, and restlessness were produced by exposures to 10% CO_2 for 1.5 min and 15% CO_2 for 3 min in a study by Lambersten (1971).

At even higher CO_2 concentrations, the CNS effects of CO_2 are mostly depressive. Unconsciousness was the predominant finding in human subjects exposed to 17% CO_2 in 17.3% O_2 for 20-52 s (Aero Medical Association, 1953) or 18.6% CO_2 in 17% O_2 for <2 min (Dalgaard et al., 1972). Of course, at such high CO_2 concentrations, it is difficult to separate the CO_2 effect from the hypoxic effect. The CNS effects of acute CO_2 exposures are summarized in Table 3-3.

At CO_2 concentrations much higher than 17%, the CNS depression could result in death. Several workers in a ship carrying fish were found dead in the holding tank with a CO_2 concentration at 20-22% (Dalgaard et al., 1972). Similarly, CO_2 could be the cause of death in some fires. Gormsen et al. (1984) examined the causes of death in fire victims. They concluded that CO_2 poisoning or oxygen deficiency or both is the second most common cause of death, carbon monoxide poisoning being the most common.

TABLE 3-3 CNS Effects Resulting from Acute CO_2 Exposures

Concentration, %	CNS Effects	Reference
1.5	No effect	Schaefer, 1959
3-4.5	No effect or marginal depression	Glatte et al., 1967a; Deitrick et al., 1948; Brown, 1930a
5	Depression	Wamsley et al., 1969; Consolazio et al., 1947
6	Subjective feelings of speech and movement difficulties that did not exist when determined objectively	White et al., 1952
7.5-15	Mixture of depression and excitation	Brown, 1930a; Dripps and Comroe, 1947; Lambertsen, 1971
>17	Unconsciousness	Aero Medical Association, 1953; Dalgaard et al., 1972

Kidneys

CO_2 exposures might produce physiological changes in the kidney. A 30-min exposure of human subjects to 5% CO_2 produced increases in renal blood flow, glomerular filtration rate, and renal venous pressure, as well as decreased renal vascular resistance (Yonezawa, 1968). These physiological changes in the kidney probably represent renal compensation for the CO_2-induced acidosis because the plasma HCO_3^- level was increased. Due to their innocuous nature, the SMACs are not set to prevent these renal physiological changes.

Male Reproductive System

Acute CO_2 exposures might affect some of the mature cell types in the testis of laboratory animals. Vandemark et al. (1972) showed that a 4- or 8-h exposure to 2.5% CO_2 resulted in a disappearance of mature spermatids in rats. The disappearance was apparently due to sloughing of mature spermatids and Sertoli cells in the seminiferous tubules, resulting in cellular debris in the lumen. The degenerative change showed a concentration response, with the testis responding more to 5% CO_2 and even more at 10%. The testicular degenerative change was reversible because the testis appeared completely normal histologically 36 h after the CO_2 exposure. Although acute CO_2 exposures could affect the mature spermatids and Sertoli cells in the rat, they did not affect the weight of the testis and seminal vesicles. The response in the testis to CO_2 was somewhat affected by the exposure duration. An acute exposure to 2.5%, 5%, or 10% CO_2 did not affect the testis in 1 or 2 h, but it caused a similar degree of sloughing of mature spermatids in 4 or 8 h.

Even though Vandemark et al. (1972) found no testicular changes immediately after a 1-h exposure to 2.5% CO_2, it does not mean that the 1-h exposure absolutely would not cause any testicular change. It is possible that had the rats been sacrificed at a later time rather than immediately after exposure, some testicular degeneration could show up. However, the key point is that the testicular degeneration produced by acute CO_2 exposures is transient, with complete structural recovery in 36 h (Vandemark et al., 1972). Therefore, the 1-h and 24-h SMACs are not set according to CO_2's testicular toxicity.

Finally, Mukherjee and Singh (1967) observed spermatozoa with smaller head and midpiece in the vas deferens of mice exposed alternatively to 2 h of 36% CO_2 in 13.4% O_2 and 0.5 h of air for a total of 6 h. In another experiment, they exposed male mice to about 4 h of CO_2 at the same concentration per day (two CO_2 exposure periods of 2 h each separated by an air exposure of 0.5 h) for 6 d. The fertility was reduced in these male mice. However, the meaning of Mukherjee and Singh's findings is uncertain due to the very high CO_2 concentration and very low O_2 concentration used.

Intestine and Spleen

Other than injuring the lung and testes, there was a report that acute exposures of guinea pigs to high concentrations of CO_2 might also damage other tissues. Schaefer et al. (1971) present evidence that in 1-d exposure of guinea pigs to 15% CO_2 led to hemorrhages in the intestine and the spleen. Unfortunately, as discussed above, due to an inadequate number of control animals used in that study, it appeared that they did not adequately control their experiments. That casts doubt on the meaning of their positive findings. As a result, the SMACs are not set based on any intestinal or splenic end point.

Subchronic and Chronic Toxicity

Dyspnea and Intercostal Pain

Acute CO_2 exposures could cause headaches, dyspnea, and intercostal muscle pain, especially during exercise or exertion. However, Sinclair et al. (1971) showed that a 15-20 d exposure to 2.8% CO_2 failed to produce any dyspnea or intercostal muscle pain in four human subjects, who performed, twice daily, 45-min of continuous steady state exercises on a bicycle ergometer at a low, moderate, or heavy level. Radziszewski and his colleagues reported no dyspnea or intercostal pain in six human subjects who were exposed to 2% CO_2 for 30 d or 2.9% for 8 d and who performed, twice a week, a 10-min exercise with a bicycle ergometer at a 150-watt workload (Guillerm and Radziszewski, 1979; Radziszewski et al., 1988).

Headaches

Subchronic CO_2 exposures are known to produce headaches at a concentration of 3% or higher. In a 30-d exposure of six human volunteers to CO_2 conducted by Radziszewski et al. (1988), the subjects rarely developed headaches to 2% CO_2, but slight headaches were detected to 2.9% CO_2. Sinclair et al. (1969, 1971) showed that four subjects exposed to 2.8% CO_2 for 15-30 d or 3.9% CO_2 for 11 d occasionally developed mild headaches during heavy exertion, but the headaches disappeared after the first day of exposure. Glatte et al. (1967a,b) and Menn et al. (1968) reported that, in the 5-d exposure to 3% or 4% CO_2, mild-to-moderate throbbing frontal headaches started to appear in the first day in about 60% of seven human subjects and the headaches disappeared in the third day. They claimed that the headaches were not severe enough to interfere with normal activities. However, because 75% of the subjects who were inflicted with headaches felt that the headache was sufficiently prominent to request an analgesic, headaches are considered in setting the long-term SMACs.

The above data are summarized in Table 3-4 to provide a glimpse of the CO_2's concentration-response relationship based on headaches developed in repetitive CO_2 exposures.

TABLE 3-4 Data on CO_2-Induced Headaches

Concentration, %	Exposure Duration	Intensity of Headache	Reference
2	30 d	Rare headache even during exercise	Radziszewski et al., 1988
2.8	15-30 d	Occasional mild transient headache during heavy exertion	Sinclair et al., 1969, 1971
2.9	30 d	Slight headache	Radziszewski et al., 1988
3	5 d	Mild-to-moderate throbbing frontal headache that disappeared in 2 d	Glatte et al., 1967; Menn et al., 1968
3.8	30 d	Intense and annoying headache	Radziszewski et al., 1988
3.9	11 d	Occasional mild transient headache during heavy exertion	Sinclair et al., 1969, 1971
4	5 d	Mild-to-moderate throbbing frontal headache that disappeared in 2 d	Glatte et al., 1967a; Menn et al., 1968
4.3	30 d	Intense and annoying headache	Radziszewski et al., 1988

Nervous System

As discussed in the Acute Toxicity subsection, it is questionable whether acute CO_2 exposures at less than 5% cause CNS depression. Similarly, there are conflicting data on the CNS effects of subchronic CO_2 exposures at 3-5%. Schaefer (1949a,b) reported that, in an 8-d exposure of human subjects to 3% CO_2, mild excitement (euphoria, troubled sleep with frequent dreams and nightmares) was seen in d 1, followed by inattentiveness, erratic behavior, exhaustion, and confusion in d 2-8. However, no behavioral changes were found by Glatte et al.

in seven human volunteers exposed to 3% or 4% CO_2 for 5 d (Glatte et al., 1967a,b; Menn et al., 1968).

Although Schaefer (1949a,b) found that human subjects suffered motor skill impairment on d 2-8 of an 8-d exposure to 3% CO_2, Glatte et al. (1967a,b), Menn et al. (1968), and Storm and Giannetta (1974) did not find any psychomotor impairment in volunteers exposed to 3% or 4% CO_2. Glatte et al. exposed seven volunteers to 3% or 4% CO_2 for 5 d, and Storm and Giannetta exposed 12 volunteers to 4% CO_2 for 2 w. Both Glatte et al. and Storm and Giannetta used a battery of tests called the Repetitive Psychometric Measures, which tested hand steadiness, visualization, the arithmetical ability to add, the ability to find four-letter words in rows of letters, the speed of canceling letters in a row of letters, and the speed of perception. Glatte et al. also tested the subjects' ability to solve arithmetical problems involving multiplication and memory, compensatory tracking maneuvers, pitch, roll, and yaw maneuvers, simple visual vigilance (monitoring the on and off of a light), complex auditory monitoring, and memory (counting and remembering the number of flashes of a light for 1-min periods, as well as listening and remembering combinations of letters and numbers). Neither Glatte et al. nor Storm and Giannetta found any effect on these tests with exposures to 3% or 4% CO_2. Due to the extensive psychometric testing done by these two groups of investigators in the Air Force, an exposure of humans to 3% CO_2 is not likely to impair the CNS or the motor ability.

The data available indicate that a subchronic CO_2 exposure at less than 2% definitely has no CNS impairment effect. In a summary report of a 42-d study of 23 human volunteers exposed to 1.5% CO_2 in a submarine, Schaefer (1961) reported that there were no effects on immediate memory, problem-solving abilities, a letter-canceling test, the Minnesota Manual Dexterity test, a complex coordination test, the McQuarrie test of mechanical ability, strength, visual accommodation, visual acuity, depth perception, and pitch discrimination. There were, however, moderate increases in anxiety, apathy, increased uncooperativeness, a desire to leave, and increased sexual desire. In a study sponsored by NASA, Jackson et al. (1972) showed that there were no changes in psychomotor performances, as determined with a Langley Complex Coordinator, in four human volunteers continuously exposed to CO_2 for 90 d (0.6% CO_2 in the first 46 d and 0.8% in the remainder).

Acid-Base Balance

Similar to acute CO_2 exposures, subchronic CO_2 exposures also lower the blood pH. Table 3-5 summarizes the effect of CO_2 exposure on the acid-base balance of human subjects. The amount of plasma pH drop caused by CO_2 varied somewhat with the CO_2 exposure concentration. In a subchronic exposure of human subjects to CO_2 at 5 mm Hg (equivalent to 0.7% at sealevel) or 1.5% CO_2, a plasma pH drop of 0.05 unit was detected in the first 20 or so days (Schaefer, 1963b; Messier et al., 1971). However, Guillerm and Radziszewski (1979) showed that a 3-d exposure to 2% CO_2 lowered the plasma pH by 0.01 unit. In comparison, Glatte et al. (1967a) found that an exposure of seven humans to CO_2 at 21 mm Hg (equivalent to 2.8% at sealevel) lowered the plasma pH by 0.02 unit in 2-3 d and 0.01 unit in 4-5 d, but these pH drops were not statistically significant.

TABLE 3-5 Plasma pH Decreases During CO_2 Exposures

Concentration, %	No.	Exposure Duration	Plasma pH Drop	Reference
0.85	15	56 d	—[a]	Peck, 1971
0.7	12	3-24 d	0.05	Messier et al., 1971
0.7	12	31-38 d	—	Messier et al., 1971
1	15	44 d	0.02	Pingree, 1977
1.5	21	1-20 d	0.05	Schaefer, 1958
1.5	21	24-42 d	—	Schaefer, 1958
2	6	3 d	0.01	Guillerm and Radziszewski, 1979
2	6	8-30 d	—	Guillerm and Radziszewski, 1979
2.8	7	1-5 d	0.01-0.02	Glatte et al., 1967a
3.9	3-4	1-2 d	0.02	Sinclair et al., 1969
3.9	3-4	5 d	—	Sinclair et al., 1969

[a]Not significant.

During subchronic hypercapnia, the kidney compensates for the acidosis by increasing the secretion of H^+ in urine and conserving HCO_3^+ (Kryger, 1981). The renal compensation of the acidosis is, however, rather slow; it takes days before its effect is manifested (Kryger, 1981). The data summarized in Table 3-5 show that the body compensated for the respiratory acidosis in 5-8 d in two studies (Sinclair et al., 1969; Guillerm and Radziszewski, 1979) and in about 30 d in two other studies (Schaefer, 1963a; Messier et al., 1971).

The CO_2-induced acid-base change in animals is similar to that in humans. Schaefer et al. (1964a) reported that the arterial pCO_2 increased maximally as early as 1 h into a 14-d exposure of guinea pigs to 15% CO_2. The arterial pCO_2 remained higher than controls all through the 14-d CO_2 exposure period, but it gradually declined with time starting at 1 h, which was the first sampling point in that study (Schaefer et al., 1964). Schaefer et al. (1964) showed that the time course of arterial pH changes in guinea pigs exposed to 15% CO_2 followed that of pCO_2 quite closely. Barbour and Seevers (1943) showed that, in rats exposed to 11% CO_2 for 17 d, the arterial pH dropped below the pre-exposure level as early as 0.5 h (the first sampling point) into the CO_2 exposure. Starting at 0.5 h, the arterial pH gradually rose, but it remained lower than the pre-exposure level throughout the 17-d exposure.

Electrolyte Levels

Similar to acute CO_2 exposures, subchronic CO_2 exposures might also change the electrolyte levels in the body. The data on plasma total calcium levels and urinary total calcium excretion gathered in humans are summarized in Table 3-6.

In a 57-d submarine-patrol study, with an atmosphere of 0.8-1.2% CO_2 and less than 25-ppm CO, Messier et al. (1976) detected that calcium levels decreased in plasma, but increased in erythrocytes, in 7-15 human subjects. Because there were no changes in the parathyroid hormone and calcitonin plasma levels, the calcium changes detected by Messier et al. were not due to parathyroid hormone or calcitonin. A

TABLE 3-6 CO_2-Induced Calcium Changes

Concentration, %	No.	Exposure Duration	Plasma Calcium Level	Urinary Calcium Excretion	Reference
0.5	6	13 d	Not measured	No change	Davies et al., 1978a
0.6 0.8	4	90 d	d 1-53: No change d 54-90: 4% decrease	No higher than the range of normal	Jackson et al., 1972; Schaefer, 1979
0.7	15	49 d	d 5: 3% increase d 12, 19: No change d 26: 2% increase d 33: No change d 40: 5% increase d 47: 3% increase	w 1: No change w 2-7: 24-37% decrease	Gray et al., 1973
0.65	14	56 d	Not measured	d 2: No change d 9, 17: 35-42% decrease d 30: No decrease d 42, 56: 23-40% decrease	Peck, 1971
1	7	57 d	d 7-57: 10% decrease	d 1-57: 30-40% decrease	Messier et al., 1976
1.5	20	42 d	d 3-21: 6% decrease d 30-42: No decrease	d 1-42: 47% decrease	Schaefer et al., 1963b
3	7	5 d	No change	No change	Glatte et al., 1967a

group in the United Kingdom's Institute of Naval Medicine also discounted the role of a reduction in vitamin D, which promotes intestinal calcium absorption, because the reduction in urinary calcium excretion occurred fairly rapidly (Davies and Morris, 1979).

Similarly, Schaefer et al. (1963b) found that, in 20 human subjects exposed to 1.5% CO_2 for 42 d, the plasma calcium levels were lowered in the first 3 w of the exposure, but they returned to the pre-exposure levels in the last 3 w. Taking the data from a 90-d study performed by Jackson et al. (1972) under NASA's sponsorship, paired t-tests show that there was no significant change in the serum calcium levels in four human volunteers exposed to a median CO_2 concentration of 0.6% in the first 53 d. The serum calcium levels, however, dropped 4% in 54-90 d when the median CO_2 concentration was at 0.8%.

There were, however, other studies that tended to dispute that CO_2 has any effect on the plasma level of calcium. For instance, Glatte et al. (1967a) found no changes in the plasma and urinary calcium levels in humans exposed to 3% CO_2 for 5 d. Davies et al. (1978a) also found no changes in the urinary excretion of calcium, phosphorus, sodium, potassium, and magnesium in humans exposed to 0.5% CO_2 for 13 d. Davies et al. (1978b) showed that the reduction in urinary excretion of minerals found by other investigators might be artifacts of urine collection methodology.

To further complicate the picture, Schaefer et al. (1979a,b) found that subchronic exposures of guinea pigs to CO_2 also resulted in time-dependent changes in plasma levels of calcium, but in an opposite direction compared with the human data of Schaefer et al. and Messier et al. discussed above. In guinea pigs exposed to 1% CO_2 for 6 w, Schaefer et al. (1979a) detected increases in the plasma calcium levels in w 6, but the plasma calcium levels did not differ from the control in w 1-4. In another study conducted by Schaefer et al. (1979b) with guinea pigs exposed to 0.5% CO_2 for 8 w, no change in plasma calcium levels was detected in w 4 and an increase was detected in w 8. Because the data indicate that CO_2 increases the plasma calcium levels in guinea pigs (Schaefer et al., 1979a,b), but CO_2 either decreases or causes no change in the plasma calcium levels in humans (Schaefer et al., 1963b; Glatte et al., 1967a; Messier et al., 1976), the meaning of these data gathered from guinea pigs is doubtful.

Gray et al. (1973) reported that the serum levels of calcium, magnesium, and inorganic phosphorus were raised during a 7-w exposure of 15 submariners exposed to 0.7% CO_2. During the CO_2 exposure, the urinary excretions of these three electrolytes were reduced. Because the inverse relationship between the serum levels of calcium and urinary excretion of calcium also existed during the pre-exposure period, Gray et al. admitted that the renal handling of calcium in these subjects was unusual. Gray et al. speculated that, when the 15 submariners took part in the 7-w exposure, they had not completely recovered from a 3-w exposure to 1% CO_2 in a submarine patrol 3 m earlier. That cast doubt on whether the findings of Gray et al. in the 7-w study were representative of the responses in normal individuals exposed to CO_2.

The effect of CO_2 on urinary excretion of calcium has also been studied. In the 20 human subjects exposed to 1.5% CO_2 for 42 d, Schaefer et al. (1963b) found that the amount of calcium excreted in the urine per day was reduced by about 45-50% throughout the CO_2 exposure. Since the daily urine volume was reduced by only about one-third during the 42-d CO_2 exposure, that means the calcium concentration in the urine must have been reduced during the CO_2 exposure (Schaefer et al., 1963). Schaefer et al. reported that the urine pH dropped in the first 23 d of CO_2 exposure, but it returned to the pre-exposure value in the last 19 d of the exposure. Similarly, Messier et al. (1976) showed that the amount of calcium excreted in the urine per day was lower in the 7-15 human subjects in a 57-d submarine patrol exposed to about 1% CO_2. However, Messier et al. failed to find any changes in the daily urine volume during the CO_2 exposure. So one can infer that the concentration of calcium in the urine was reduced during the 57-d exposure to 1% CO_2. Unlike the finding of Schaefer et al., Messier et al. detected that the urine pH was elevated during the CO_2 exposure. Davies et al. (1978a) exposed six men to fresh air for 9 d followed by 0.5% CO_2 for 13 d with intensive physical training during the chamber stay. They found no changes in the daily urine volume and the urinary and fecal excretions of calcium.

There were two studies in which the amounts of urinary calcium excretion were reported, but the investigators were silent about the daily urine volume. The first study was a 90-d study sponsored by NASA in which four volunteers were exposed to a median CO_2 concentration of 0.7% (Jackson et al., 1972). Schaefer (1979) displayed urinary data gathered in that 90-d study. According to the data, the amounts of daily urinary calcium excretion, averaged among the four volunteers, ranged from 95 mg to 170 mg in the 90-d exposure. Unfortunately, Schaefer did not present the pre-exposure data. Nevertheless, the 90-d exposure to 0.7% CO_2 most likely did not increase the urinary calcium excretion because the values were all below the maximum limit of urinary calcium excretion of, 250 mg/d determined by Pak et al. (1985, 1989) in at least 77 normal volunteers without kidney stones. The second study was conducted by Peck on 15 healthy male Navy servicemen in a 56-d submarine patrol (Peck, 1971). These 15 men were exposed to a mean CO_2 concentration of 0.85%, with a range of 0.72% to 0.95%. Peck did not measure the pre-exposure daily urinary calcium excretion, but all values measured during the entire 56 d fell within normal limits.

The results of the studies conducted by Messier et al. (1976) and Schaefer et al. (1963b) did not agree on CO_2's effect on phosphorus levels. Messier et al. showed that a 57-d exposure to 1% CO_2 caused no change in the plasma phosphorus levels in seven human subjects during the exposure, but it reduced urinary excretion of phosphorus. In contrast, Schaefer et al. found that the plasma phosphorus levels were raised during a 42-d exposure of 20 human subjects to 1.5% CO_2, and that, while the urinary phosphorus excretion was increased in the first 2 d, it declined with time so that it was lower than the pre-exposure level in the last 3 w of the CO_2 exposure. In 15 men exposed to 0.7% CO_2 for 52 d, Gray et al. (1973) reported that the urinary excretion of phosphorus was raised in d 1-2, but it was below the control value from d 3-52. The serum phosphorus levels in these 15 subjects were raised from d 5-47. However, Glatte et al. (1967a) did not detect any change in the plasma and urinary levels of phosphorus in seven human subjects exposed to 3% CO_2 for 5 d. Similarly, Davies et al. (1978a) found no changes in urinary and fecal excretion of phosphorus in six men exposed to 0.5% CO_2 for 13 d. The sum findings of these studies seem to indicate that CO_2's effect on the body levels of phosphorus varies.

Messier et al. (1976) also found increased sodium in the plasma every week during the 57-d submarine-patrol study with exposure to 0.8-1.2% CO_2. A decrease in the plasma levels of potassium started to appear in the third week. Opposite changes in the erythrocyte levels of these electrolytes were detected.

Bones and Kidneys

The changes in plasma calcium levels discovered by Schaefer et al. (1979a) in guinea pigs seem to be related somewhat to renal calcification, which might be assessed either histologically or biochemically. Histological evidence of renal calcification, in the form of focal calcification primarily in tubules in the renal cortex, was presented by

Schaefer et al. in guinea pigs exposed to 1.5% CO_2 for 35-42 d and in rats exposed to 1.5% CO_2 for 35 or 91 d. Meessen (1948) also showed renal tubular necrosis with calcification in rabbits exposed to 4.5% CO_2 for 13 d. The presence of renal calcification can be determined biochemically by measuring the renal calcium concentration. Schaefer et al. (1979a,b) defined renal calcification as any rise in renal calcium concentration larger than 25%. An 8-w exposure of guinea pigs to 0.5% CO_2 resulted in an increase in the plasma calcium level and a larger than 25% rise in the renal calcium levels with no change in bone calcium levels in w 8 (but not in w 4-6) (Schaefer et al., 1979b). Similarly, in a 6-w exposure of guinea pigs to 1% CO_2, Schaefer et al. found an increase in plasma calcium levels and a decrease in bone calcium levels in w 1 and 6, as well as a larger than 25% rise in the renal calcium levels in w 2-6 (Schaefer et al., 1979a). These changes in the calcium levels in the bone, plasma, and kidneys support the theory that CO_2-induced renal calcification in these animals was due to the mobilization of calcium from the bone.

There is no evidence of subchronic CO_2 exposures causing renal calcification in humans. In guinea pigs, CO_2-induced renal calcification appeared to be associated with a rise in plasma calcium level (Schaefer et al., 1979a,b, 1980). In contrast, subchronic exposures to CO_2 at about 1-3% are known to decrease or cause no change in plasma calcium levels in humans (Schaefer et al., 1963b; Glatte et al., 1967a; Messier et al., 1976). It is, therefore, doubtful that subchronic exposures to CO_2 at low concentrations produce renal calcification in humans. As a result, the renal calcification data gathered in animals are not relied on in setting the SMACs of CO_2. The unchanged or lower plasma calcium levels in human subjects exposed to CO_2 (Schaefer et al., 1963b; Glatte et al., 1967a; Messier et al., 1976) also discount the possibility of CO_2 causing bone demineralization.

Tansey et al. (1979) compared the medical records of the crew in over 1000 Polaris submarine patrols in two periods: 1963-1967 and 1968-1973. Each patrol lasted about 60 d with a crew of 140. Data on submariners in these two periods are summarized in Table 3-7 as follows.

The CO_2 concentration onboard was higher in 1963-1967 than in

TABLE 3-7 Information on Submarine Patrols[a]

	1963–1967	1968–1973
CO_2 concentration	>1% 70-90% of the time (0.9-1.2% in 1966-1967)	>1% <20% of the time (0.8-0.9% in 1968-1971)
CO concentration	44 ppm (1961)	15-20 ppm (1969)
Submarine personnel, no.	347	225
Smokers, %	73	60
Man-days	3,240,000	4,410,000
Cases of ureteral calculi per 1000 man-days	0.007	0.004
Workdays lost to ureteral calculi per 1000 man-days	0.030	0.010

[a]Data from Tansey et al. (1979).

1968-1973, with the concentration higher than 1% CO_2 70-90% of the time in 1963-1967, but less than 20% of the time in 1968-1973. Tansey et al. did not report any pre-1963 CO_2 concentration data. Citing data from other studies, Tansey et al. reported that the CO_2 concentrations onboard submarines were 0.9% to 1.2% in 1966-1967 and that they were 0.8% to 0.9% in 1968-1971. The rate of crewmen taking sick leave on board due to ureteral calculi in the 1963-1967 period was almost twice that in the 1968-1973 period. The number of workdays lost to ureteral calculi, after normalization by the number of man-days, was three times higher in 1963-1967 than in 1968-1973.

The question is whether the larger number of workdays lost to ureteral calculi in 1963-1967 was due to the higher CO_2 concentrations onboard. In other words, could subchronic CO_2 exposures cause ureteral calculi in humans? Three lines of reasoning tend to cast doubt that the larger number of workdays lost to ureteral calculi was caused by subchronic CO_2 exposures. First, Tansey et al. admitted that the submarine atmosphere contained contaminants such as CO_2, CO, hydrocarbons, and aerosols in low concentrations. Even though the CO_2 concentrations in 1963-1967 appeared to be higher than those in 1968-

1973, there were no data on the concentrations of hydrocarbons and aerosols in these two periods. Although Tansey et al. did not measure the CO concentration, they presented data gathered by others showing that CO concentrations in the submarine declined about 50% from 1961 to 1969, and they declined another 50% from 1969 to 1972 (Tansey et al., 1979). These declines in CO concentrations illustrate the possibility that, other than reductions in CO_2 and CO concentrations in the two periods studied, there could be reductions in the concentrations of other air contaminants onboard the Polaris submarines. The difference in the number of workdays lost to genitourinary diseases between 1963-1967 and 1968-1973 might be due to an air contaminant other than CO_2.

The second reason is that kidney stone formation is affected by a number of risk factors, such as the oxalate contents of food (Schwille and Herrmann, 1992), hypocitraturia (Goldberg et al., 1989; Hofbauer et al., 1990), low urine volume (Thun and Schober, 1991), low testosterone concentration in urine (van Aswegen et al, 1989), and dietary protein intakes (Breslau et al., 1988; Trinchieri et al., 1991). There could be dietary differences in oxalate contents or protein intakes in these two periods and the dietary differences could play a role in causing the difference in the rate of workdays lost to ureteral calculi.

The third reason, which is the most convincing one, relates to the mechanism of nephrolithiasis formation in humans. According to Coe and Favus (1987), 75% to 85% of kidney stones are calcium oxalate and calcium phosphate stones. Calcium phosphate stones usually consist of hydroxyapatite. Urinary excretion of calcium is the major risk factor for calcium stone formation (Wasserstein et al., 1987; Goldberg et al., 1989). Urinary stones are usually formed when calcium salts become supersaturated in the urine (Coe and Favus, 1987). Since CO_2 exposures have been shown to either lower or not change urinary calcium excretion (Schaefer et al., 1963b; Glatte et al., 1967a; Davies et al., 1978a) and to reduce the urinary calcium concentration (Schaefer et al., 1963b; Messier et al., 1976), the chance of CO_2 exposures causing calcium supersaturation is low. In addition, the lower urinary pH detected in CO_2 exposures (Schaefer et al., 1963b; Radziszewski et al., 1976) disfavors kidney stone formation because the solubility of calcium oxalate is independent of pH and the deposition of apatite and octocalcium phosphate is disfavored in acidic urine (Coe and Favus, 1987). Therefore, it is highly unlikely that CO_2 exposures produce kidney stones in humans.

Tansey et al. (1979) stated that the rate of ureteral calculi aboard the 1968-1973 submarine patrols was "exactly the same as that for the general population." It stands to reason that even if *all* the ureteral calculi cases detected in the 1968-1973 submarine patrols were due to 0.8-0.9% CO_2 in the submarine, exposures to 0.8-0.9% CO_2 does not increase the risk of ureteral calculi in human subjects. It will be shown in the latter part of this document that the long-term SMACs for CO_2 are recommended at 0.7% based on other toxic end points. So the long-term SMACs will not be associated with any increased risk of ureteral calculi according to the submarine patrol data of Tansey et al.

Schaefer et al. (1980) showed that an 8-w exposure of guinea pigs to 1% CO_2 increased the CO_2 content in the bone, in the fourth to the eighth week, with the increases in the sixth and eighth weeks due mainly to an increase in the bicarbonate contents of the bone. Commensurate with the bicarbonate content increase in the bone, they saw an increase in plasma calcium levels in the sixth and eighth week of the CO_2 exposure. Schaefer et al. hypothesized that the increase in plasma calcium levels was due to CO_2 binding to the bone and releasing calcium from the bone in guinea pigs. As a result of this hypothesis, several scientists in NASA raised their concerns on the potential of CO_2 in releasing calcium from the bone of astronauts in the space station and causing kidney stones. However, according to an analysis of this potential given below, it is unlikely that their concerns would become reality.

First, CO_2's effects on calcium in human beings differ from those in guinea pigs. All available data showed that continuous CO_2 exposures lasting from 13 to 90 d at concentrations ranging from 0.5% to 1.5% either lowered or did not change urinary calcium excretion in human subjects (Schaefer et al., 1963b; Jackson et al., 1972; Gray et al., 1973; Messier et al., 1976; Davies et al., 1978a). Four studies showed that exposures of volunteers to 0.6-3% CO_2 lasting from 5 to 90 d either decreased or did not change the plasma calcium levels (Schaefer et al., 1963b; Glatte et al., 1967a; Jackson et al., 1972; Messier et al., 1976). Only one study by Gray et al. (1973) showed that a 7-w exposure of 15 submariners to 0.7% CO_2 increased the serum level of calcium. However, because these submariners excreted less calcium in the urine, Gray et al. admitted that the increase in serum calcium levels in these submariners was an anomaly. All in all, because plasma calcium levels are usually not raised in humans exposed to CO_2, it is unlikely that CO_2 would displace calcium from bones in humans.

Because bone demineralization is associated with calcium changes in astronauts in space (Whedon et al., 1977; Leach and Rambaut, 1977, pp. 204-216), it is of interest to examine the CO_2 effect, if any, on calcium in space. Calcium data, in means and standard deviations, gathered by Leach and Rambaut (1977), Whedon et al. (1977), and Whedon (1984) in three Skylab missions are plotted by the solid lines in Figure 3-1. According to Hopson et al. (1974), the CO_2 partial pressures in Skylab missions ranged from 4.8 to 5.5 mm Hg, with a mean of 5.3 mm Hg, time-weighted average (TWA). The Skylab data showed that, in three to nine astronauts exposed to microgravity and CO_2 at 5.3 mm Hg for up to 82 d, the plasma calcium levels increased 4-5% and the daily urinary calcium excretion increased 60-80% starting from d 12 (Figure 3-1). Vogel (1975) reported bone losses in three of the nine Skylab crew members. Because immobilization bed-rest studies performed by Donaldson et al. (1970) and Deitrick et al. (1948) showed increases in urinary calcium excretion of approximately the same level as that seen in the Skylab crew, the increase in urinary calcium excretion detected in Skylab missions was associated with bone demineralization in microgravity. The exposure to CO_2 at 5.3 mm Hg in these Skylab astronauts probably played no role in the calcium changes. These Skylab data also showed that both the plasma calcium levels and urinary calcium excretion in space missions lasting up to 84 d were quite stable once a plateau was reached. The plasma calcium levels reached a plateau at about 5 d, and the urinary calcium excretion reached a plateau in about 20 d.

To prove that CO_2 exposures in spacecraft play no important role in calcium changes in astronauts, calcium data from a space mission with CO_2 concentrations at much less than 5.3 mm Hg, but of a similar duration as Skylab missions, is needed as control data. Unfortunately, there is no such "control" mission. Inflight calcium plasma data are, nevertheless, available from the Spacelab 2 mission, which lasted for 8 d with a CO_2 partial pressure of 2.4 mm Hg, TWA (Shih, 1987). So the plasma calcium data of Spacelab 2, reported by Morey-Holton et al. (1988), reflect the plasma calcium concentrations in four astronauts who stayed for several days in microgravity with less CO_2 than those in Skylab. The inflight calcium plasma data from the Spacelab 2 mission are plotted by the dashed line in Figure 3-1. Compared with the plasma calcium data of Skylab missions in Figure 3-1 (the solid line), the plas-

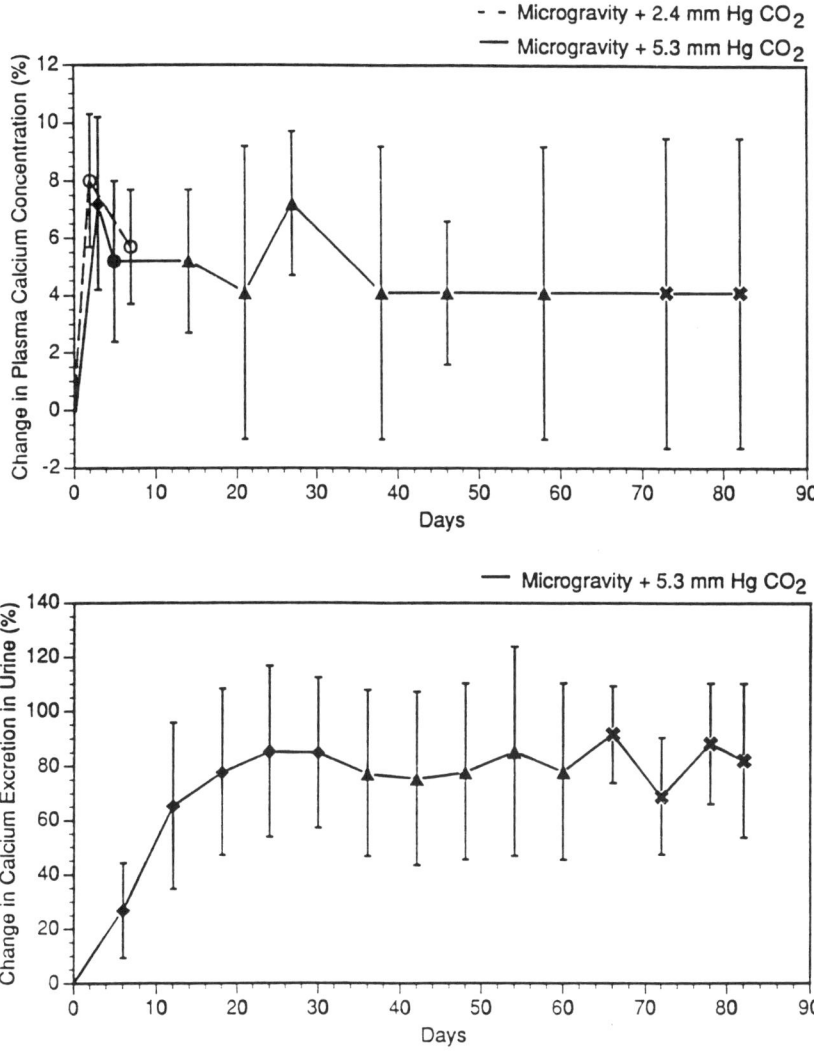

FIGURE 3-1 Inflight calcium data from Skylab and Spacelab 2 missions.

ma calcium concentration in microgravity appeared to be independent of CO_2 for at least 7 d. This lends credence to the belief that CO_2 exposures in spacecraft do not seem to play a major role in causing the calcium changes in microgravity. It should be noted that partial pressures of 5.3 mm Hg and 2.4 mm Hg are equivalent to concentrations of 0.7% and 0.3%, respectively, in an atmosphere of 760 mm Hg.

Respiratory System

Similar to acute CO_2 exposures, subchronic CO_2 exposures could also cause hyperventilation. Table 3-8 shows the amounts of ventilatory increase attained, after a plateau has been reached, in human subjects exposed to CO_2 for more than a day. The sum of the data shows that it takes at least 1% CO_2 to increase, with statistical significance, the minute volume after the hyperventilatory response reaches a plateau after the first few hours in a subchronic exposure. At 0.5% CO_2, the slight increase in minute volume at the plateau was masked by the physiological noise (Radziszewski et al., 1988).

In subchronic CO_2 exposure, the hyperventilatory response could diminish somewhat in human subjects after the first few days of exposure, indicative of a reduced sensitivity to CO_2's stimulation on respiration. Pingree (1977) showed that, in a 44-d exposure of 15 human subjects to 1% CO_2, the minute volume increased about 30% on the fourth day, but it returned to the control value starting on the eighth day. In contrast, Schaefer (1963b) reported that, at exposure to 1.5% CO_2, the respiratory ventilation was raised about 30% in normal volunteers throughout a 42-d exposure.

However, in a 30-d exposure of humans to 2% CO_2, the minute volume increase was diminished about one-third after 9 d of exposure and remained constant in the remaining 21 d of the CO_2 exposure (Guillerm and Radziszewski, 1979). Similarly, in a 30-d exposure to 2.7% CO_2, the minute volume increase was reduced after 4 d and the minute volume increase remained constant from d 5 to d 14 (Clark et al., 1971). On the thirtieth day of exposure to 2.7% CO_2, the hyperventilatory response recovered fully, so that the minute volume equaled that in the first day of CO_2 exposure (Clark et al., 1971).

TABLE 3-8 Hyperventilatory Responses to CO_2 Exposures

Concentration, %	No.	Exposure Duration	Minimum Volume Increase, %	Reference
1	15	4 d	30	Pingree, 1977
1	15	8-40 d	0	Pingree, 1977
1.5	21	1-42 d	30	Schaefer et al., 1963a; Schafer, 1963b
2	6	9-30 d	44	Guillerm and Radziszewski, 1979
2.8	7	5 d	25	Glatte et al., 1967a
3.9	3-4	3-11 d	130	Sinclair et al., 1969

The reduction in CO_2 hyperventilatory response during subchronic exposures appeared to occur sooner at higher CO_2 exposure concentrations. In an 11-d exposure of humans to 3.9% CO_2, the hyperventilatory response was diminished about one-third after two days of exposure (Sinclair et al., 1969). Another piece of evidence that humans developed reduced sensitivity toward CO_2's hyperventilatory effect was obtained by Schaefer (1963b). Schaefer showed that, after 35-40 d of continuous exposure of human subjects to 1.5% CO_2, the subjects did not increase their minute volume upon a 15-min challenge with 5% CO_2 as much as they did before the subchronic exposure to 1.5% CO_2.

An Air Force study showed that a 5-d exposure of seven human volunteers to 3% CO_2 resulted in no changes in maximum breathing capacity, vital capacity, and 1-s vital capacity (Glatte et al., 1967a). It is of interest that several studies done by the Navy indicate that subchronic CO_2 exposures might affect lung function. Schaefer et al. showed that a 42-d exposure to 1.5% CO_2 increased the anatomic dead space of the lung by about 40% and the physiologic dead space by 60% in 20-21 human subjects (Schaefer et al., 1963a; Schaefer, 1963b). An exposure of human subjects to 0.8-0.9% CO_2 raised the physiological dead space 50-60% in 20 d, which returned to normal soon after the exposure, indicating that the effect was reversible (Gude and Schaefer, 1969). The data on CO_2-induced increase in physiological dead space are not relied on in setting the SMACs. This is because the size of the decrease in physiological dead space caused by CO_2 exposures is similar to that caused by aging in a normal individual going from age 20 years to age 40 (Cotes, 1979, pp. 149, 358, 363).

There is no evidence of subchronic CO_2 exposures causing lung injuries in humans. However, based on electron microscopic studies of guinea pigs, subchronic CO_2 exposures are known to cause changes in type II pneumocytes. Schaefer et al. (1979b) reported, in guinea pigs exposed to 1% CO_2, increases in the size and number of type II pneumocytes, increases in the size and number of osmiophilic lamellar bodies inside type II pneumocytes, and clustering of 2-4 type II pneumocytes starting after 4 w of exposure (Douglas et al., 1979). These ultrastructural changes were also observed after 6 w of exposure. In comparison, an 8-w exposure to 0.5% CO_2 failed to cause any change in type II pneumocytes (Schaefer et al., 1979b). Schaefer et al. hypothesized that the proliferation of type II pneumocytes was a compensatory reaction to CO_2's impairment on type I pneumocytes (Douglas et al., 1979).

However, there was no evidence that type I pneumocytes were damaged by CO_2 (Schaefer et al., 1979b; Douglas et al., 1979). So there seems to be no support for the hypothesis of Schaefer et al. The type II pneumocyte changes probably represent a metabolic adaptation of the lung to CO_2 challenges because, among the alveolar lining cells, type II pneumocytes are the more metabolically active cell type (West, 1979). Since type II pneumocytes are thicker than type I pneumocytes (West, 1979), a potential adverse consequence of type II pneumocyte proliferation is impaired gas exchanges. Due to the fact that there was no difference between the arterial pO_2 in the guinea pigs with CO_2-induced type II pneumocyte changes and the control guinea pigs (Douglas et al., 1979), the proliferation of type II pneumocytes caused by 1% CO_2 in guinea pigs did not impair gas exchanges.

Another potential consequence of type II pneumocyte proliferation is the higher amount of lung surfactants that are synthesized by type II pneumocytes (Wright and Clements, 1987). Lung surfactants have been postulated to perform three functions: to help maintain a low lung compliance, to stabilize alveoli, and to reduce the chance of pulmonary edema (Notter and Finkelstein, 1984). Quite a bit is known about the biological effects of a lack of lung surfactants via studies of respiratory distress syndromes, but practically nothing is known about the biological effects of a higher than usual amount of lung surfactants. The only dose-response information gathered in a recent literature search is that, in the treatment of premature infants with respiratory distress syndrome, increasing the dose of surfactant given intratracheally by 300% up to 400 mg/kg body weight could improve the treatment (Gortner et al., 1990; Dunn et al., 1990). Since premature infants are deficient in lung surfactants to begin with, the dose-response data obtained from these infants probably do not reflect the biological effects of a higher than usual amount of lung surfactant in normal subjects. However, However, Douglas et al., 1979 showed that exposure to 1% CO_2 increased the number and size of lamellar bodies by only 30-50% in type II pneumocytes of guinea pigs. Assuming that the amount of lung surfactants secreted by type II pneumocytes in these guinea pigs was also increased by 30-50%, increases of such magnitude are not expected to have any harmful effect in the lung because any resultant decreases in surface tension would be of little clinical significance.

By considering the potential effects on gas exchanges and lung surfactants, it is safe to assume that the type II pneumocyte changes caused by subchronic exposures to 1% CO_2 are functionally insignificant. Therefore, type II pneumocyte changes are not a toxic end point used in setting SMACs for CO_2.

Finally, it should be noted that hyaline membranes and distended alveoli and alveolar ducts were seen in rabbits exposed to 4.5% CO_2 for 13 d by Meessen (1948). However, Meessen did not use a control group in the study, so the meaning of the findings is unclear.

Cardiovascular System

As discussed above, acute CO_2 exposures produced clinically significant arrhythmia in human subjects only at very high concentrations (30%). All the subchronic studies with EKG evaluations were performed with CO_2 concentrations of 4% or less and there are conflicting data on whether these concentrations of CO_2 cause arrhythmia. Glatte et al. (1967a) found no EKG problems in individuals exposed to 3% or 4% CO_2 for 5 d, in which they exercised an hour daily and were monitored with a 12-lead EKG. Sinclair et al. (1971) showed no increase in premature ventricular contractions in individuals exposed to 2.8% CO_2 for 15-20 d during near-maximal or maximal exercises. In another report, Sinclair et al. (1969) stated that a few individuals exposed to

3.9% CO_2 for 11 d or 2.7% CO_2 for 30 d developed "ectopic foci activities," presumably premature ventricular contractions (PVCs), during exercises. However, some of the ectopic foci were associated with exercises when breathing air (Sinclair et al., 1969). In addition, the ectopic foci activities during CO_2 breathing did not show a concentration-response relationship. The data of Glatte et al. and Sinclair et al. seem to suggest that subchronic exposures to 3-4% CO_2 are devoid of arrhythmia effects. In contrast, in two French studies, an exposure of human subjects to 2.9% or 3.8% CO_2 for 8 or 9 d resulted in extrasystoles (PVCs), but no extrasystoles were detected in a 30-d exposure to 1% or 1.9% CO_2 (Radziszewski et al., 1988; Guillerm and Radziszewski, 1979). Because extrasystoles are of little clinical significance (Massie and Sokolow, 1990), CO_2's SMACs are not set based on the EKG effects of CO_2.

Subchronic CO_2 exposures might affect heart morphology. In a 7-d exposure of guinea pigs to 15% CO_2, fat deposition in the myocardium was detected in d 7, but not at 1 h or d 1 (Schaefer et al., 1971). Other than fat deposition, there were no other changes in cardiac histology. According to the investigators, the experiment "failed to demonstrate any signs of myocardial damage in guinea pigs exposed for periods up to 7 days to 15% CO_2" (Schaefer et al., 1971). The fat deposition probably represents only a metabolic change in the heart and not any serious damage. For comparison, no cardiac histopathology was found in rats exposed to 8% CO_2 for 32 d (Pepelko, 1970). Due to the relatively minor nature of the myocardial changes, these findings are not relied on in setting the SMACs.

Structural Effects on Other Tissues

Other than affecting the kidney and lungs, subchronic CO_2 exposures might affect the liver. In rabbits exposed to 4.5% CO_2 in 21% O_2 for 13 d, necrosis was seen scattered throughout the liver lobules (Meessen, 1948). Unfortunately, no control group was used in this study, so its results are not relied on in setting SMACs. A 32-d exposure of rats to 8% CO_2 failed to cause any histological lesions in the liver, lungs, kidneys, adrenals, spleen, thyroid, and heart (Pepelko, 1970). Similarly, Schaefer et al. (1971) showed that exposures of guinea pigs to 3% CO_2 for 42 d or 15% CO_2 in 21% O_2 for 7 d failed to produce any histopathology in the liver. Schaefer et al. however, found a decrease in glycogen granules and an increase in fat granules in the liver of guinea pigs exposed to 3% CO_2 for 7 d. The granular changes recovered in 1 d after the end of the exposure. These changes in the granules were interpreted by the investigators to reflect functional changes in liver metabolism. Because these changes are not actual damages, they are not relied on in setting SMACs.

As mentioned above, 4- or 8-h exposures of CO_2 are known to produce injuries in the testis of rats (Vandemark et al., 1972). However, it is unclear whether subchronic CO_2 exposures could damage the testis. In a study without an adequate number of control animals, Schaefer et al. (1971) observed in the testis a marked reduction of mature spermatocytes with a concomitant increase in the precursor cells of spermatocytes in guinea pigs exposed to 15% CO_2 for 2 d. When the exposure was extended to 7 d, multinucleated giant cells were observed in the testis. Because on the average only about two control animals were examined per time point, it is not certain whether the testicular changes observed by Schaefer et al. in the 15% CO_2 group were due to CO_2 or whether they were artifacts. Nevertheless, some of the data gathered by Schaefer et al. in that subchronic study are of value in setting the SMACs. Schaefer et al. (1971) reported that the testes of the guinea pigs and rats exposed to 3% CO_2 for 42 d or to 1.5% CO_2 for 6 mo appeared normal histologically, it can be concluded that a subchronic exposure to 3% CO_2 is not toxic to the testis.

Hematological Changes

Guillerm and Radziszewski (1979) reported that a 10% reduction in hematocrit and a 9% reduction in red blood cell count were detected in six human subjects exposed to 2% CO_2 for 16-30 d. Because they failed to observe these reductions in humans exposed to 4% CO_2, they discounted hypercapnia as the cause of the hematological changes. Instead, they hypothesized that prolonged confinement might be the cause. Similarly, the Navy found that prolonged hypercapnia might not always produce hematological changes. Wilson and Schaefer (1979) showed that, in Polaris submarine patrols with CO_2 levels maintained between 0.7% and 1.2% and a CO level between 15 and 20 ppm, the hematological responses in smokers differed from that in nonsmokers.

In nine smokers in the patrol, the red blood cell count increased by 12% and the hematocrit increased by 4% on the sixth day, but not on the 32nd and 52nd day. However, these two hematological parameters did not change in 11 nonsmokers in the patrol on the sixth, 32nd, and 52nd day. Because astronauts will not be allowed to smoke cigarettes in spacecraft and most of them are nonsmokers, the hematological data are not relied on in setting CO_2 SMACs.

Carcinogenicity

No traditional carcinogenic bioassay has been known to be conducted with CO_2. However, Goldsmith et al. (1980) discovered that infusion of humidified 99.99% CO_2 into the peritoneal cavity of 4- to 6-mo-old BALB/c mice for 10-12 d led to lymphoma after a latent period of about 8 mo (the incidence in the air-exposed control group was 0% and that in the CO_2-exposed control group was about 60%) and a doubling of the incidence of pulmonary adenocarcinoma (from 15% in the control group to about 30% in the CO_2-exposed group). Due to the highly artificial nature of the CO_2 exposure, the practical meaning of the tumorigenic findings is uncertain.

Epidemiological Data

Only one epidemiological study involving CO_2 was found. In a criteria document for CO_2, NIOSH cited, an unpublished report submitted to NIOSH by the United States Brewers Association, Inc. (Riley and Barnea-Bromberger, 1976). The report concerned the acid-base effect of CO_2 exposures in brewery workers. In these workers, exposed to 1.1% CO_2 TWA with 3-min excursions up to 8%, the blood HCO_3^- levels did not differ from the control values (Riley and Barnea-Bromberger, 1976).

Genetic Toxicity

No genotoxic data of CO_2 have been found.

Developmental Toxicity

An exposure of rabbits to 10-13% CO_2 for 4-10 h on d 2 or 3 between d 7 and d 12 of pregnancy resulted in congenital hypoplasias in the vertebral column (Grote, 1965). The value of this teratogenic study was limited because only three pregnant rabbits were exposed to CO_2. In another teratogenic study, 71 pregnant rats were exposed to 6% CO_2 in 20% O_2 for 24 h between d 5 and d 21 of pregnancy (Haring, 1960). More increased incidences of cardiac and skeletal malformations were detected in the CO_2-exposed group than in the control group (Haring, 1960).

Interaction with Other Chemicals

Only one report of synergism involving CO_2 was found. Levin et al. (1987) reported that the amount of acidosis produced by a combined exposure of rats to 5% CO_2 and 2500 ppm CO was larger than that produced by either agent alone. The addition of CO_2 to an exposure atmosphere containing CO decreased the mean survival time of rats, compared with rats exposed to only CO (Rodkey and Collison, 1979). This potentiation of CO's lethal effect by CO_2 is thought to be due to CO_2's hyperventilatory effect (Rodkey and Collison, 1979). Indeed, co-exposures of rats to 5% CO_2 in 2500 ppm CO are known to increase the rate of COHb rise in blood compared with CO exposures alone (Levin et al., 1987). However, it should be noted that co-exposures to 5% CO_2 in CO do not always result in potentiation of CO's toxicity. For instance, exposure to CO_2 at 5% did not potentiate the incapacitation effect of 3500 ppm CO on the rat (Hartzell and Switzer, 1985). Finally, just as other substances might potentiate the effect of CO_2, the reverse is also true. A subcutaneous injection of naloxone at 5 mg/kg has been shown to increase the hyperventilatory response to CO_2 in rats because naloxone displaced endogenous endorphins from central opioid receptors (Isom and Elshowihy, 1982).

In addition to interacting with CO, CO_2 is known to interact with NO_2. Exposures of rats to 200 ppm NO_2 for 30 min resulted in an increase in methemoglobin level in blood (Levin et al., 1989). A co-exposure with 5% CO_2 led to a larger increase in methemoglobin than with 200 ppm NO_2 alone (Levin et al., 1989).

TABLE 3-9 Toxicity Summary

Concentration, %	Exposure Duration	Species	Effects	Reference
0.5	30 d	Human	No acidosis, hyperventilation, or symptoms; no changes in urinary excretion of potassium, sodium, or calcium	Radziszewski et al., 1988
0.6 in d 1-46, 0.8 in d 47-90	90 d	Human	No change in serum calcium level on d 1-53, but decreases accompanied with increases in serum phosphorus on d 54-90; no changes in hematological indices and psychomotor performance	Jackson et al., 1972
0.7	7 w	Human	Higher serum levels of calcium, magnesium, and inorganic phosphorus; lower urinary excretions of calcium, magnesium, and inorganic phosphorus; lower urinary excretion of acids except in w 3 and 4 when it was higher than pre-exposure level	Gray et al., 1973
0.8-0.9	20 d	Human	Physiological dead space increased by 50-60%, which returned to normal soon after the exposure	Gude and Schaefer, 1969
0.85-1.2	57 d	Human	In plasma, increase in sodium, decreases in K^+ and Ca^{++}, but no change in phosphorus; increased Mg^{++} only on d 51; decrease in Cl^- in w 5-7; decreases in pH, increases in pCO_2 and bicarbonate in w 4 with complete recovery by d 51; in urine, phosphorus and hydroxyproline decreased in w 1-3; Ca^{++} decreased in w 1-3, increased in w 4-5, then decreased in w 6-9; no change in parathyroid or calcitonin activity	Messier et al., 1976
1	17-32 min	Human	Alveolar ventilation increased by 24%; slight increases of systolic and diastolic blood pressure	Schneider and Truesdale, 1922
1	30 d	Human (n = 1)	Acidosis; increases in blood pCO_2 and respiratory ventilation; no change in performance in physical exercise	Zharov et al., 1963
1	30 d	Human	Hyperventilation; no symptoms; no changes in urinary excretion of potassium, sodium, or calcium; increased arterial pO_2	Radziszewski et al., 1988
1 or 2	30 min	Human	No symptoms during exercises at two-thirds maximum or maximum oxygen consumption	Menn et al., 1970

Conc.	Duration	Subject	Effects	Reference
1.1 TWA with excursions up to 8% for 3 min	8 h/d, 5 d/w	Human (workers)	No change in blood HCO_3^{1-} levels	Riley and Barnea-Bromberger, 1976
1.5	10-15 min	Human	Increases in respiratory rate and tidal volume; lower respiratory rate increase and higher tidal volume increase in instructors than in other individuals	Schaefer, 1958
1.5	15 h/d, 6 d	Human (n = 1)	Impairment in night vision sensitivity and green color sensitivity; all other visual functions were normal	Weitzman and Kinney, 1969
1.5	42 d	Human	Alveolar ventilation increased by 8%; ventilatory response to 5% CO_2 challenges decreased at end of w 6; anatomic dead space of lung increased; O_2 consumption increased in first 2 w; plasma Ca^+ and phosphorus followed changes in plasma pH; uncompensated respiratory acidosis in first 3 w (decrease in blood pH, urine pH, urinary HCO_3^- excretion, and CO_2 exhalation); compensated respiratory acidosis in last 3 w (normal blood pH, increases in urine pH, urinary HCO_3^- excretion, and CO_2 exhalation); no effects on weight, pulse rate, blood pressure, oral temperature, adaptation to darkness, visual acuity, visual accommodation, depth perception, pitch discrimination, manual dexterity, strength, coordination, immediate memory, and letter-canceling, problem-solving, and mechanical abilities; apathy, increased sexual desire, desire to leave, and uncooperativeness	Schaefer, 1961a,b, 1963b,c; Schaefer et al., 1963a,b, 1964b
1.8 or 3.5	11-40 min	Human	No changes in oxygen consumption, pulse rate, and cardiac output; increase in respiratory ventilation	Grollman, 1930
1.9	N.S.[b] (until subjects could not exercise further)	Human	Compared with exercising in normal air, 45% higher ventilation when doing submaximal exercise, but exposure did not increase ventilation further when doing maximal exercise (O_2 consumption was even lower)	Luft et al., 1974
2	30 d	Human	No acidosis, headache, or change in psychomotor performance; hyperventilation (more at 2 h than 24 h); good ability to exercise	Radziszewski et al., 1988

TABLE 3-9 (Continued)

Concentration, %	Exposure Duration	Species	Effects	Reference
2	17-32 min	Human	Alveolar ventilation increased by 50%; slight increases of systolic and diastolic blood pressure	Schneider and Truesdale, 1922
2	Several h	Human	Headache and dyspnea on mild exertion	Schulte, 1964
2	30 d	Human (n = 1)	Acidosis; increases in blood pCO_2 and respiratory ventilation; deterioration in performance in physical exercise	Zharov et al., 1963
2.5	2 h	Human	No changes in specific airway conductance	Tashkin and Simmons, 1972
2.5-2.8% CO_2 in 14.6-15% O_2	Several h	Human	No giddiness, headache, dyspnea, or drop in body temperature	Brown, 1930b
2.7	30 d	Human	Mild headache only on d 1; hyperventilation diminished after d 1	Sinclair et al., 1969
2.8	1 h or 15-20 d	Human	Acidosis; abilities to exercise moderately or heavily did not change; during exercise, occasional mild headaches, but no dyspnea, intercostal muscle pain, or EKG changes; no difference between acute and subchronic CO_2 exposures	Sinclair et al., 1971
2.8 or 3.9	30 min	Human	Intercostal muscle pain and respiratory difficulties during exercises at two-thirds maximum or maximum oxygen consumption; ability to do heavy exercise impaired; mild-to-moderate frontal headache at 3.9% CO_2 occurred near end of exercise; no significant increases in premature atrial or ventricular contractions usually seen with exercise in normal atmosphere	Menn et al., 1970
2.9	8 d	Human	Acidosis and hyperventilation at 2 and 24 h; slight headache; extrasystoles during exercise; no change in psychomotor performance	Radziszewski et al., 1988
3	Several h	Human	Dyspnea even at rest, headache (more severe than at 2% CO_2), and diffuse sweating	Schulte, 1964

Concentration	Duration	Subject	Effects	Reference
3	78 h	Human (n = 2)	Acclimation in the ventilatory effect of CO_2; minute volume 15.1 L/min at the start and 12.9 L/min near end of exposure	Chapin et al., 1955
3	5 d	Human	Very slight acidosis on d 1-3; raised arterial pCO_2 and serum HCO_3^- levels on d 3-5; respiratory ventilation increased by 15-60%, which was easily tolerated; mild-to-moderate headaches in 4/7 subjects on d 1-2; no changes in vital capacity and 1-s vital capacity, psychomotor functions (hand steadiness, vigilance, auditory monitoring, memory, arithmetic, and problem solving); no changes in urinary levels of Ca^{++}, phosphorus, K^+, Na^+, NH_3, and titratable acidity or in serum levels of Ca^{++}, phosphorus, K^+, Na^+, alkaline phosphatase, SGOT, SGPT, direct bilirubin, and indirect bilirubin; no change in the ability to exercise moderately for 1 h daily; no EKG problems	Glatte et al., 1967a
3	8 d	Human	A slight state of excitement on d 1 (euphoria, troubled sleep with frequent dreams and nightmares), followed by slight depression of the nervous systems in the remainder of the exposure (inattentiveness, erratic behavior, exhaustion, confusion, and decreased manual skills); uncompensated respiratory acidosis in the first 3 d; acidosis then compensated by increases in plasma HCO_3^{1-} level, urinary excretion of acid, and alkali retention by kidneys; when the subjects performed moderate work, the tidal volume decreased and the respiratory rate increased, leading to higher O_2 update and CO_2 excretion	Schaefer, 1949a,b
3.0-3.5	1.2 min	Human	Small loss in hearing threshold	Gellhorn and Spiesman, 1934, 1935
3.2% CO_2 in 13.4% O_2	Several h	Human	Giddiness and headache; no dyspnea or drop in body temperature	Brown, 1930b

TABLE 3-9 (Continued)

Concentration, %	Exposure Duration	Species	Effects	Reference
3.3	10-15 min	Human	Increases in respiratory rate and tidal volume (diving instructors and other subjects responded similarly); CNS-depression decrease in flicker-fusion threshold and increase in latent time of alpha blocking after light stimulus; increases in blood sugar and oxygen consumption; decrease in the eosinophil count	Schaefer, 1958, 1963a
3.8	9	Human	Acidosis and hyperventilation at 2 and 24 h; intense and headaches and gastralgia; extrasystoles during exercise; limited exercise capacity; no change in psychomotor performance	Radziszewski et al., 1988
3.9	5 or 11 d	Human	Acidosis on d 1-4; arterial and CSF pH returned to normal on d 5; mild headaches on d 1 only; hyperventilation (decreased in magnitude on d 2)	Sinclair et al., 1969
4	5 d	Human	Acidosis; tidal volume almost doubled but no change in respiratory rate; mild-to-moderate throbbing frontal headaches beginning in the first few hours but none by d 3	Menn et al., 1968
4	11 d	Human	Alveolar ventilation increased by 200% on day 1 but dropped to 150% after day 1; increased pCO_2 in arterial blood and cerebral spinal fluid	Clark et al., 1971
4	2 w	Human	No change in hand-eye coordination, complex tracking performance, and problem-solving ability	Storm and Giannetta, 1974
4.3	1 d	Human	Acidosis and hyperventilation at 2 and 24 h; intense and annoying headaches and gastralgia; not able to exercise; no change in psychomotor performance	Radziszewski et al., 1988
4-5	17-32 min	Human	Dyspnea	Schneider and Truesdale, 1922

4-5	≈4 h	Human	Body temperature dropped 1°F; deterioration in performance in cancellation test; no effects on Army alpha intelligence test, arithmetic test, muscular coordination, and attention	Brown, 1930b
4.7	Several h	Human	Headache and dyspnea	Brown, 1930b
4.7% CO_2, balance O_2	15 min	Human (emphysema patients)	Alveolar ventilation increased by < 100% (increased by 150-300% in normal subjects)	Tenney, 1954
5	N.S.	Human (fighter pilots, n = 2)	Significant degradation in pilot performance during landing; lengthened flight time between gear down and touch down, and unacceptable increase in touch-down sink rates	Wamsley et al., 1969
5	30 min	Human	Increased renal blood flow, glomerular filtration rate, and renal vascular resistance; increase in plasma HCO_3^- level but no increase in NA^+, K^+, and Cl^- levels	Yonezawa, 1968
5	17-32 min	Human	Headache, dizziness, hiccoughing	Schneider and Truesdale, 1922
5 or 7	15-30 min	Human	Increases in blood pressure and cerebral blood flow; decrease in cerebrovascular resistance; no changes in cardiac output or cerebral oxygen consumption	Kety and Schmidt, 1948
5% in 95% O_2	45-90 min	Human (n = 2 psychotic patients)	Arterial pH dropped to 6.9; A-V nodal beats; increases in amplitude of R and T waves; raise or depression of S-T segment; inverted T waves; EKG changes disappeared 30 min after exposure	Altschule and Sulzbach, 1947
5, 7.5 or 10	2 h	Human	Decreased specific airway conductance	Tashkin and Simmons, 1972

TABLE 3-9 (Continued)

Concentration, %	Exposure Duration	Species	Effects	Reference
5-6.75	37 h	Human	Headache; increased respiratory ventilation; soreness of respiratory musculature; 10-bpm increase in heart rate; slight decrease in hand-arm steadiness; no changes in blood pressure, auditory discrimination, hand-eye coordination, or abilities to stand still, walk 1-in rail, or compute, translate, and check numbers	Consolazio et al., 1947
5	Several h	Human	CNS depression	Schulte, 1964
5.4 or 7.5	10-15 min	Human	Increases in respiratory rate and tidal volume; diving instructors responded less than other subjects; increases in blood sugar and oxygen consumption; decrease in eosinophil count; increase in pulse rate at 7.5% CO_2	Schaefer, 1958
6	1-2 min	Human	Decreased visual intensity discrimination	Gellhorn, 1936
6	6-8 min	Human	More decreases in amplitude of QRS complex and T wave in older men (aged 61 y) than in younger men (aged 23 y); no change in S-T segments; no T inversion	Okajima and Simonson, 1962
6	16 min	Human	Dyspnea; headache; sweating; hyperventilation; subjective feeling of speech difficulty (speech understandable) and of movement difficulty; slightly slower rate of card sorting but no change in card-sorting error rate	White et al., 1952
6	20.5-22 min	Human	Considerable discomfort but tolerable; 9% rise in systolic pressure; 7% rises in diastolic pressure and pulse rate	Brown, 1930b
6	Several h	Human	Visual disturbances and tremors	Schulte, 1964
7% CO_2, 93% O_2	3 min	Human	Tidal volume, respiratory rate, and ventilation increased by 140, 50, and 250%, respectively	Sullivan and Yu, 1983

Condition	Subject	Duration	Effects	Reference
7% CO_2, 93% O_2	Human	5 min	Tidal volume, respiratory rate, and ventilation increased by 150, 60, and 290%, respectively	Sullivan and Yu, 1983
7	Human	60 min	Arterial pCO_2, H^+, and HCO_3^- levels increased in 10 min of exposure and remained at a plateau at min 10-60; arterial Na^+ level increased by <1%; no changes in arterial K^+, Cl^-, and phosphate levels; mild headache and burning of eyes	Brackett et al., 1965
7-14% CO_2, balance O_2	Human	10-20 min	Headache, moaning, belligerent complaining, coughing, restlessness, sweating, twitching, tremor, amnesia, unconsciousness; increased respiratory ventilation, arterial pressure, heart rate, and plasma concentrations of epinephrine, norepinephrine, and corticosteroids; premature nodal contraction (2/27 subjects vs 0/27 before exposure) and premature ventricular contraction (3/27 subjects vs 1/27 before exposure) on EKG	Sechzer et al., 1960
7.5% CO_2 in 16% O_2	Human	3.25-6 min	Considerable discomfort, but tolerable; 24% and 20% rises in systolic and diastolic pressures; 10% rise in pulse rate	Brown, 1930b
7.5	Human	4-25 min	Increases in pulse rate, cardiac output, blood pressure, and respiratory ventilation	Grollman, 1930
7.5	Human	15 min	Headache, dizziness, restlessness, and dyspnea	Schaefer, 1963a
7.6	Human	2.5-10	Dyspnea, dizziness, headache, head fullness, sweating, and increases in respiratory ventilation and systolic and diastolic pressures	Dripps and Comroe, 1947
8% CO_2, 19% O_2	Human	3-6 min	Total lung resistance increased by 120%; no change in static lung compliance	Nadel and Widdicombe, 1962
8	Human	17-32 min	Tolerance limit	Schneider and Truesdale, 1922
8.8% CO_2 in 39% O_2	Human	7-10 min	Approaching tolerance limit; 22% and 13% rises in systolic and diastolic pressures, respectively, and 13% rise in pulse rate	Brown, 1930b
10	Human	1.5 min	Eye flickering, myoclonic twitches, and psychomotor excitation	Lambertsen, 1971
10	Human	15-25 min	Restlessness, confusion, and listlessness	Brackett et al., 1965

TABLE 3-9 (Continued)

Concentration, %	Exposure Duration	Species	Effects	Reference
10	N.S.	Human	Unconsciousness	Schulte, 1964
10.4	3.8 min	Human	Dizziness, dyspnea, headache, head fullness, restlessness, hyperventilation, unconsciousness, and rises in systolic and diastolic pressures	Dripps and Comroe, 1947
10.4% CO_2 in 14.4% O_2	1-2.25 min	Human	33% and 38% rises in systolic and diastolic pressure, respectively, and 19% rise in pulse rate	Brown, 1930b
12.4%	0.75-2 min	Human	Dizziness, drowsiness, near stupor, dyspnea, head fullness, sweating, flushing sensation, sense of impending collapse, throat irritation, and slight choking sensation; 1/7 subjects collapsed; no nausea or throbbing of temples; 55% and 26% rises in systolic and diastolic pressures, respectively, and 13% rise in pulse rate	Brown, 1930b
15	3 min	Human	Eye flickering, myoclonic twitches, psychomotor excitation, increased muscle tone, sweating, flushing, dilated pupils, leg flexion, torsion spasms, and restlessness	Lambertsen, 1971
17% CO_2, 17.3% O_2	20-52 s	Human	Unconsciousness	Aero Medical Association, 1953
18.6, 17 O_2	<2 min	Human	Dullness, unconsciousness, cyanosis, and throbbing headache	Haldane and Smith, 1892
20-22% CO_2, ca. 16% O_2	N.S.	Human (workers)	Death; survivors experienced unconsciousness, cyanosis, sluggish reflexes, rattling respiration, and motor unrest	Dalgaard et al., 1972
20% CO_2, 80% O_2	3 min	Human	Eye flickering, myoclonic twitches, psychomotor excitation, increased muscle tone, sweating, flushing, dilated pupils, leg flexion, torsion spasms, restlessness, tonic and tonic-clonic seizures	Lambertsen, 1971

Gas	Duration	Species	Effects	Reference
30% CO$_2$, 70% O$_2$	38 s	Human (patients in psychiatry)	Narcosis; extrasystoles, premature atrial and nodal beats, atrial tachycardia, and supraventricular tachycardia on EKG	MacDonald and Simonson, 1953
30% CO$_2$, 70% O$_2$	50-52 s	Human (patients in psychiatry)	Unconsciousness and extrasystoles on EKG; consciousness regained 110 s after exposure	Friedlander and Hill, 1954
30% CO$_2$, 70% O$_2$	3 min	Human	Eye flickering, myoclonic twitches, psychomotor excitation, increased muscle tone, sweating, flushing, dilated pupils, leg flexion, torsion spasms, restlessness, tonic and tonic-clonic seizures; unconsciousness within 2 min	Lambertsen, 1971
30% CO$_2$, 70% O$_2$	N.S. (10-15 breaths)	Human (patients in psychiatry)	Auricular extrasystoles, auricular tachycardia, increased P-wave voltage, low or inverted P waves, spiked T waves with a broad base, increased T-wave voltage, slight increases in PR intervals and QRS intervals, and marked increase in QT interval; marked increases in systolic and diastolic pressure; acidosis; no ventricular extrasystole	McArdle, 1959
0.5	4 w	Guinea pig	No effects on body weight; calcium levels in kidney, bone, or plasma; type II pneumocyte cell size; and number of lamellar bodies in type II pneumocytes	Schaefer et al., 1979b
0.5	8 w	Guinea pig	Increased calcium levels in kidneys and plasma; no significant effects on bone calcium level, body-weight gain, type II pneumocyte cell size, and the number of lamellar bodies in type II pneumocytes	Schaefer et al., 1979b
1	1 w	Guinea pig	Acidosis; Ca^{++} increased in kidney; Ca^{++} and phosphorus increased in plasma; Ca^{++} and phosphorus decreased in bone; no change in body weight gain	Schaefer et al., 1979a
1	1 or 2 w	Guinea pig	Acidosis; no change in arterial pO$_2$, pCO$_2$, or HCO$_3^-$ level; no change in appearance of pneumocytes, alveolar macrophages, ciliated epithelial cells and Clara cells of terminal bronchioles, and endothelial cells in the lung under the electron microscope	Douglas et al., 1979

TABLE 3-9 (Continued)

Concentration, %	Exposure Duration	Species	Effects	Reference
1	2 or 4 w	Guinea pig	Acidosis; Ca^{++} increased in kidney; Ca^{++} and phosphorus levels in plasma did not differ from the control levels; Ca^{++} and phosphorus levels in bone did not differ from the control levels; no changes in body weight gain	Schaefer et al., 1979a
1	3 w	Guinea pig	No change in arterial pH and pO$_2$; increased arterial pCO$_2$; decreased arterial HCO$_3^-$; no change in the appearance of pneumocytes, alveolar macrophages, ciliated epithelial cells and Clara cells of terminal bronchioles, and endothelial cells in the lung under the electron microscope	Douglas et al., 1979
1	4 w	Guinea pig	Acidosis; increased arterial pCO$_2$; no change in arterial pO$_2$ and HCO$_3^-$; marked increases in the size and number of type II pneumocytes; clustering of 2-4 type II pneumocytes together; the changes in type II pneumocytes remained 2 or 4 w after exposure; no changes in the other cell types in lung under electron microscopy	Douglas et al., 1979
1	6 w	Guinea pig	No changes in arterial pH, pO$_2$, pCO$_2$, and HCO$_3^-$; marked increases in the size and number of type II pneumocytes; clustering of 2-4 type II pneumocytes; no changes in other cell types in lung under electron microscopy	Douglas et al., 1979
1	6 w	Guinea pig	No acidosis; Ca^{++} increased in kidney; Ca^{++} and phosphorus increased in plasma; Ca^{++} and phosphorus decreased in bone; no changes in body weight gain	Schaefer et al., 1979a
1-2	30 min	Rat	Respiratory frequency and ventilation increased by 30-40%	Lai et al., 1978
1-2	5 h	Mouse	All animals died	Zink and Reinhardt, 1975

1-5	24-42 d	Guinea pig	No lowering in blood pH; subpleural atelectasis in lung; no hyaline membranes or edema; atelectasis disappeared 2 w after the end of exposure	Niemoeller and Schaefer, 1962
1.5	1-14 w	Guinea pig	Depressed body weight gain	Schaefer et al., 1971
1.5	1 w	Guinea pig	No histopathology in liver, heart, testes, spleen, and pancreas	Schaefer et al., 1971
1.5	35-42 d	Guinea pig	Histological signs of focal renal tubular calcification in the cortex; the incidence increased with exposure duration; the calcification remained 27 d after exposure; no histological signs of renal calcification during the first 2 w of exposure	Schaefer et al., 1979a
1.5	6 mo	Guinea pig	No lowering in blood pH; subpleural atelectasis in lung; no hyaline membranes or edema	Niemoeller and Schaefer, 1962
2.5	1 or 2 h	Rat	No changes in testicular histology and weight	Vandemark et al., 1972
2.5	4 or 8 h	Rat	Sloughing of spermatid and Sertoli cells into seminiferous tubule lumen; no mature spermatids in the tubule; no change in testicular weight; return to normal 36 h after exposure	Vandemark et al., 1972
3	1 h	Guinea pig	Lowered blood pH; plasma GOT level doubled; plasma GPT level unchanged	Schaefer et al., 1971
3	1 d	Guinea pig	Lowered blood pH; no changes in plasma GOT and GPT levels	Schaefer et al., 1971
3	1 d	Guinea pig	Reduced body weight; increased kidney and testicular weight; no changes in liver, lung, thyroid, spleen, thymus, and adrenal weight	Schaefer et al., 1971
3	1-7 d	Guinea pig	Blood pH dropped more in guinea pigs than in rats; blood bicarbonate levels increased in rats, not in guinea pigs	Schaefer et al., 1971
3	2 d	Guinea pig	Blood pH lowered to 7.27; subpleural atelectasis and edema in lung; no hyaline membranes in lung	Niemoeller and Schaefer, 1962
3	3 d	Guinea pig	Reduced body and thymus weight; increased lung and testicular weight; no changes in kidney, liver, spleen, thyroid, and adrenal weight	Schaefer et al., 1971

155

TABLE 3-9 (Continued)

Concentration, %	Exposure Duration	Species	Effects	Reference
3	4 d	Guinea pig	Blood pH lowered to 7.31; subpleural atelectasis and hyaline membranes in lung; no edema in lung	Niemoeller and Schaefer, 1962
3	4 d	Guinea pig	Blood pH lowered to 7.32; increase in lung/body-weight ratio; no change in surface tension of lung extracts; increase in abnormal lamellar bodies in alveolar lining cells, subpleural atelectasis, edema, hyaline membranes, and phagocytic pneumocytes in lung	Schaefer et al., 1964a
3	7 d	Guinea pig	No changes in blood pH and plasma levels of GOT and GPT; depletion of glycogen vacuoles and increase in fat vacuoles in liver; no significant liver histopathology; a small incidence of hyaline membranes in lung; increase in zymogen granules in pancreas; no testicular histopathology	Schaefer et al., 1971
3	21 d	Guinea pig	Lower body-weight gain; increased adrenal weight and decreased spleen weight; no changes in liver, kidney, lung, thymus, thyroid, and testicular weight. Glycogen granules depleted in liver on 7 d were restored on 21 d; no testicular histopathology	Schaefer et al., 1971
3	42 d	Guinea pig	No lowering of blood pH; subpleural atelectasis and hyaline membranes in lung; no edema in lung	Niemoeller and Schaefer, 1962
3	42 d with 2 w recovery	Guinea pig	No lowering of blood pH; no subpleural atelectasis, hyaline membranes, or edema in lung	Niemoeller and Schaefer, 1962
3	93 d	Monkey	No changes in body weight, blood glucose, hemoglobin, hematocrit, total leukocyte count, serum levels of Ca^{++}, Cl^-, phosphate, blood urea nitrogen, serum bilirubin and cholesterol levels, and RCB sedimentation rate	Stein et al., 1959

45	13 d	Rabbit	Distended alveoli and alveolar ducts with hyaline membranes and leukocytic infiltration in lung; Necrosis scattered throughout the liver lobules. Necrotic renal tubular epithelium with calcium incrustation in the cortical-medullary zone; tubular lumen were obstructed with calcium casts	Meessen, 1948
5	1 or 2 h	Rat	No changes in testicular histology and weight	Vandemark et al., 1972
5	4 or 8 h	Rat	Sloughing of spermatid and Sertoli cells into seminiferous tubule lumen; no mature spermatids in the tubule; no change in testicular weight; return to normal 36 h after exposure	Vandemark et al., 1972
8	32 d	Rat	Reduced body-weight gain; increased heart/body-weight and kidney/body-weight ratios; no change in liver/body-weight and lung/body-weight ratios; increase in eosinophils, decreased hematocrit, and no change in reticulocyte percent in blood; no histological changes in spleen, thyroid, liver, kidney, adrenals, heart, and lungs	Pepelko, 1970
10	1 or 2 h	Rat	No change in testicular histology and weight	Vandemark et al., 1972
10	4 or 8 h	Rat	Sloughing of spermatic and Sertoli cells into seminiferous tubule lumen; no mature spermatids in the tubule; no change in testicular weight; return to normal 36 h after exposure	Vandemark et al., 1972
10	4 d	Rat	No death or CNS depression	Barbour and Seevers, 1943
10	30 d	Rat	Hyperventilation; body-weight loss and reduced food intake; marked reticulocytosis, but no changes in hemoglobin level, RBC count, and white-blood-cell count	Barbour and Seevers, 1943
11	2.5 to 5 h	Rat	O_2 consumption reduced	Barbour and Seevers, 1943

TABLE 3-9 (Continued)

Concentration, %	Exposure Duration	Species	Effects	Reference
15	1 h	Guinea pig	Blood pH decreased to 7.00; increases in blood corticosteroids and free fatty acids; decreases in adrenal epinephrine and adrenal cholesterol	Schaefer et al., 1968
15	1 or 6 h	Guinea pig	Blood pH decreased to 7.01-7.09; increase in lung/body-weight ratio; no change in surface tension of lung extracts; increase of lamellar bodies in alveolar lining cells, subpleural atelectasis, congestion, edema, and hemorrhage in lung; no hyaline membranes; increase in phagocytic pneumocytes at 6 h but not at 1 h	Schaefer et al., 1964a
15	6 h	Guinea pig	Blood pH decreased to 7.10; increases in blood corticosteroids and free fatty acids; decrease in adrenal epinephrine	Schaefer et al., 1968
15	6 h or 1 d	Guinea pig	Reduction in oxygen partial pressure required to half saturate the blood; decreases in blood pH and 2,3-diphosphoglycerate	Messier and Schaefer, 1971
15	1 d	Guinea pig	Blood pH decreased to 7.10; body weight decreased by ≈ 10%; increase in adrenal/body-weight ratio; decreases in thymus/body-weight and spleen/body-weight ratios; reduction in lymphocyte count; no change in total white-blood-cell count; increases in blood corticosteroids and free fatty acids; decreases in adrenal epinephrine and adrenal cholesterol	Schaefer et al., 1968
15	1 d	Guinea pig	Blood pH decreased to 7.10; increases in lung/body-weight ratio and surface tension of lung extracts; increase in abnormal lamellar bodies in alveolar lining cells, subpleural atelectasis, congestion, edema, hemorrhage, hyaline membranes, and phagocytic pneumocytes in lung	Schaefer et al., 1964a
15	1 d	Guinea pig	Intestinal hemorrhage that disappeared after 3 or 4 d of exposure; congestion and hemorrhages of spleen	Schaefer et al., 1971

15	1-7 d	Guinea pig	Blood pH decreased more in guinea pigs than in rats; blood bicarbonate level increased on 1 d in rats but not in guinea pigs	Schaefer et al., 1971
15	2 d	Guinea pig	A marked reduction of mature spermatocytes in testes	Schaefer et al., 1971
15	3 d	Guinea pig	Blood pH decreased to 7.3; body weight decreased by ≈ 10%; increase in adrenal/body-weight ratio; decrease in thymus/body-weight ratio; no change in spleen/body-weight ratio; reduced lymphocyte count; no changes in total white-blood-cell count and adrenal cholesterol; increase in free fatty acids in blood	Schaefer et al., 1968
15	8 h/d, 7 d	Guinea pig	Blood pH decreased to 7.11; increase in blood corticosteroids; decrease in adrenal epinephrine	Schaefer et al., 1968
15	7 d	Guinea pig	No significant changes in blood pH and body weight; increase in adrenal/body-weight and decrease in thymus/body-weight ratios; no change in spleen/body-weight ratio; no change in lymphocyte and total white-blood-cell counts; no change in corticosteroids and free fatty acids in blood or epinephrine in adrenal	Schaefer et al., 1968
15	7 d	Guinea pig	No decrease in blood pH; no changes in lung/body-weight ratio and surface tension of lung extracts; subpleural atelectasis, congestion, increases in phagocytic pneumocytes in lung; no edema, hemorrhage, hyaline membranes or abnormal lamellar bodies in alveolar lining cells in lung	Schaefer et al., 1964a
15	7 d	Guinea pig	No change in oxygen partial pressure required to half saturate the blood decreases and 2,3-diphosphoglycerate in RBC; lowered blood pH	Messier and Schaefer, 1971
15	7 d	Guinea pig	Multinucleated giants cells in testes; fat deposition in myocardium (not seen on 1 d); increase in zymogen granules in pancreas; congestion and hemorrhage in spleen; decreased blood pH; no changes in SGOT and SGPT levels	Schaefer et al., 1971

TABLE 3-9 (Continued)

Concentration, %	Exposure Duration	Species	Effects	Reference
15	14 d	Guinea pig	No decrease in blood pH; no changes in lung/body-weight ratio and surface tension of lung extracts; congestion and increase in phagocytic pneumocytes in lung; no subpleural atelectasis, edema, hemorrhage, or abnormal lamellar bodies in alveolar lining cells in lung	Schaefer et al., 1964a
15	42 d	Guinea pig	No change in adrenal cholesterol content, lymphocyte and total white-blood-cell counts	Schaefer et al., 1968
15.8	3 h	Dog	Gastrointestinal bleeding, multiple ulcerations, marked dilatation of arterioles and capillaries in the GI tract, markly reduced platelet counts, and increased clotting time	DeBellis et al., 1968
16	2 d	Guinea pig	Distended alveoli and alveolar ducts with hyaline membranes and leukocytic infiltration in lung; necrosis scattered throughout the liver lobules; Necrotic renal tubular epithelium with calcium incrustation in the cortical-medullary zone; tubular lumen were obstructed with calcium casts; irreversible degenerative changes of ganglion cells in cerebral cortex, basal ganglia, and brain stem	Meessen, 1948
20	4 d	Rat	80% mortality associated with pulmonary edema and sanguinous exudate; CNS depression	Barbour and Seevers, 1943
25	36 h	Rat	100% mortality associated with pulmonary edema and sanguinous exudate; CNS depression	Barbour and Seevers, 1943
25	N.S.	Rabbit	Transient convulsions followed by marked CNS depression	Barbour and Seevers, 1943
30	N.S.	Rat	Narcosis developed immediately	Barbour and Seevers, 1943

36% in 13.4% O_2	6 h (2 h CO_2 alternated with 0.5 h air)	Mouse	Decreases in the area and breadth of the head and midpiece of spermatozoa in vas deferens	Mukherjee and Singh, 1967
36% in 13.4% O_2	4 h/d, 6 d (2 h CO_2 alternated with 0.5 h air)	Mouse	Reduced fertility	Mukherjee and Singh, 1967
40	3 h	Rat	21% mortality; decreased body temperature, respiratory rate, heart rate, blood pH, urine pH, and testicular weight; increased hematocrit, SGOT, SGPT, serum Ca, P, and K levels, and lung and kidney weights	Mitsuda et al., 1982

[a]Only results of inhalation exposures are included. The O_2 concentrations of all the CO_2 exposures were maintained at 20-22%, unless otherwise noted.
[b]N.S. = not specified.

TABLE 3-10 Exposure Limits Set by Other Organizations

Organization	Concentration, ppm
ACGIH's TLV	5000 (TWA)
ACGIH's STEL	30,000
OSHA's PEL	5000 (TWA)
NIOSH's REL	10,000 (TWA)
	30,000 (ceiling)
NIOSH's IDLH	50,000
Navy's 90-d limit	5000[a]
Navy's 24-h limit	40,000
Navy's 1-h limit	40,000

[a]According to the Navy (1988), long-term exposures at 5000-8000 ppm probably have no significan health effect.
TLV = threshold limit value. TWA = time-weighted average. STEL = short-term exposure limit. PEL = permissible exposure limit. REL = recommended exposure limit. IDLH = immediately dangerous to life and health.

TABLE 3-11 Spacecraft Maximum Allowable Concentrations

Duration	ppm	mg/m^3	Target Toxicity
1 h	13,000	23,400	CNS depression, visual disturbance
24 h	13,000	23,400	CNS depression, visual disturbance
7 d[a]	7000	12,600	Hyperventilation
30 d	7000	12,600	Hyperventilation
180 d	7000	12,600	Hyperventilation

[a]There was no 7-d SMAC. Space-shuttle flight rules require mission termination at 2% or above and flight surgeon's evaluation at 1-2% (NASA, 1988).

RATIONALE FOR ACCEPTABLE CONCENTRATIONS

To set the SMACs, guidelines developed by a subcommittee of the Committee of Toxicology are consulted (NRC, 1992). First, an acceptable concentration (AC) is estimated for each relevant toxic end point based on data gathered from an exposure of the appropriate duration.

The lowest AC is then selected as the SMAC for that exposure duration.

The finding by Zink and Reinhardt (1975) that an exposure to 1-2% CO_2 for 5 h killed all the mice was not used in setting the SMACs. The reason is that no mortality was found in human subjects exposed to 2% or 2.7% CO_2 for 30 d (Zharov et al., 1963; Sinclair et al., 1969). The mortality finding of Zink and Reinhardt on the mouse is obviously of no value in setting exposure limits for humans.

In humans, subchronic CO_2 exposures are known to either decrease the plasma levels of calcium and phosphorus (Schaefer, 1961a, 1963a,b; Schaefer et al., 1963a,b, 1964a; Messier et al., 1976) or cause no change (Glatte et al., 1967a). One group of investigators even suggested that the decrease in urinary excretion of calcium and phosphorus during subchronic CO_2 exposures could be an artifact (Davies et al., 1978a,b). Therefore, subchronically CO_2 has no or only weak effects on calcium and phosphorus levels. Even if CO_2 does affect calcium and phosphorus plasma levels, the effects are opposite to those seen in space missions (Schaefer et al., 1963b), so the long-term SMACs are not set to prevent the calcium and phosphorus changes.

Visual Impairments, Tremors, and CNS Depression

An exposure of human subjects to 6% CO_2 for several hours produced visual disturbances and tremors (Schulte, 1964). Similarly, visual intensity discrimination was also found to be reduced in a 1-2 min exposure of human subjects to 6% CO_2 (Gellhorn, 1936). Both visual impairments and tremors are toxic end points that should be prevented because they might interfere with the astronauts in dealing with a contingency.

Several studies showed that the NOAEL for visual impairments and tremors is about 3-4%. Both Storm and Giannetta (1974) and Glatte et al. (1967a) used a battery of tests, called Repetitive Psychometric Measures, to evaluate the effects of CO_2 on vision, hand steadiness, and CNS functions. Storm and Giannetta exposed 12 volunteers to 4% CO_2 for 2 w and they detected no visual or tremor problems in the volunteers. Glatte et al. found that a 5-d exposure of seven subjects to 3% CO_2 had no effects on vision and hand steadiness.

There are other investigators who used methods other than the Repetitive Psychometric Measures to evaluate the effects of CO_2. For instance, Radziszewski et al. (1988) did not find any visual problems or tremors in six human subjects exposed to 2.9% CO_2 for 30 d. Similarly, Sinclair et al. (1971) found that 2.8% CO_2 did not produce visual problems or tremors in four human subjects after an exposure of 1 h or 15-20 d. Menn et al. (1970) also failed to detect visual problems or tremors in eight human subjects exposed to 2.8% CO_2 for 30 min.

The mechanisms of CO_2-induced visual disturbances and tremors are unknown. It is possible that they are related to the effect of CO_2-induced acidosis on the eye or the nervous system. Because there are indications that acidosis develops rapidly during CO_2 exposures, the visual impairments and tremors are assumed not to increase in severity with exposure duration. In rats, an exposure to 11% CO_2 resulted in acidosis as soon as 0.5 h into the exposure (the earliest blood pH determination in that study) and the blood pH gradually rose afterward (Barbour and Seevers, 1943). During a 1-h exposure of human subjects to 7% CO_2, the arterial plasma pH decreased as early as 10 min into the exposure and stayed constant from min 10-60 (Brackett et al., 1965). Due to the rapid development of acidosis and the fact that there is no evidence that the visual problems and tremors are exposure-duration dependent, the same AC is derived for an exposure lasting 1 h, 24 h, 7 d, 30 d, or 180 d.

As concluded in the Toxicity Summary section, 5% CO_2 causes mild CNS depression in acute exposures. Based on a 5-d exposure of seven human subjects conducted by Glatte et al. (1967a) and a 2-w exposure of 12 human subjects done by Storm and Giannetta (1974), 3% CO_2 is the NOAEL for the CNS effect of CO_2. Because the same method, Repetitive Psychometric Measures, was used to evaluate CO_2's effects on vision, hand steadiness, and CNS functions, the data of Glatte et al. and Storm and Giannetta are combined in deriving the ACs for the prevention of visual disturbances, tremors, and CNS impairment.

1-h, 24-h, 7-d, 30-d, and 180-d AC based on visual disturbances, tremors, and CNS depression
= NOAEL × 1/small n factor
= NOAEL × (square root of n)/10
= 3% × (square root of (7 + 12))/10
= 3% × 0.44
= 1.3%.

Headaches

Acute CO_2 exposures might also produce other symptoms, such as headaches, dyspnea, and intercostal muscle pain, especially during exercise or exertion (Schulte, 1964; Glatte et al., 1967b; Menn et al., 1968, 1970). Because CO_2-induced headaches are usually transient (Glatte et al., 1967a; Sinclair et al., 1969; Glatte et al., 1967b; Menn et al., 1968; Mines, 1981) and because Radziszewski et al. showed that an exposure to 2% CO_2 rarely produced headaches in 30 d in six human subjects who exercised twice weekly for 10 min each at a 150-watt workload (Radziszewski et al., 1988; Guillerm and Radziszewski, 1979), 2% CO_2 is chosen to be the ACs for headaches.

Dyspnea and Intercostal Pain

CO_2 has been shown to produce dyspnea and intercostal pain during exercise or exertion (Schulte, 1964; Menn et al., 1970). Menn et al. showed that exposure to 2.8% CO_2 did not produce dyspnea and intercostal pain in eight human subjects in 30 min. Similarly, Sinclair et al. (1971) did not detect dyspnea and intercostal pain in four human subjects who were exposed to 2.8% CO_2 for 1 h or 15-20 d. These individuals performed a 45-min continuous steady state exercise at low, moderate, or heavy load once during the 1-h exposure and twice daily during the 15-20 d of CO_2 exposure. Radziszewski and his colleagues reported no dyspnea or intercostal pain in six human subjects exposed to 2.9% CO_2 and who exercised at a 150-watt workload for 10 min twice a week (Radziszewski et al., 1988). Because there is no evidence that the production of dyspnea and intercostal pain by hypercapnia is time-dependent, the data of Menn et al. and Sinclair et al. are combined in deriving the 1-h and 24-h ACs for dyspnea and intercostal pain. The 1-h and 24-h ACs are derived without any adjustment for the small number of human subjects used since a large safety margin is not needed in short-term contingencies, in which astronauts can tolerate a little bit of dyspnea or intercostal pain on exertion for a short time.

1-h and 24-h ACs for dyspnea and intercostal pain
= 30-min or 1-h NOAEL
= 2.8%.

In deriving the 7-d, 30-d, and 180-d ACs for dyspnea and intercostal pain, the data of Sinclair et al. (1971) and Radziszewski et al. (1988) are combined. Because the six subjects in the study of Radziszewski et al. did not develop dyspnea and intercostal pain when exposed to 2.9% CO_2, the NOAEL of 2.8% obtained by Sinclair et al. should also be a NOAEL for these six subjects.

7-d, 30-d, and 180-d ACs for dyspnea and intercostal pain
= 15-20 d NOAEL × 1/small n factor
= 15-20 d NOAEL × (square root of n)/10
= 2.8% × (square root of (4 + 6))/10
= 2.8% × 0.32
= 0.9%.

Hyperventilation

The NRC subcommittee on SMACs advised NASA to consider the hyperventilatory effect in setting CO_2's SMACs. Because the 1-h and 24-h SMACs are designed for contingencies, it is acceptable for the astronauts to tolerate some hyperventilation. Therefore, the 1-h and 24-h SMACs are not set based on CO_2-induced hyperventilation.

The hyperventilatory effect is considered in establishing the 7-d, 30-d, and 180-d SMACs. Unlike other end points, the acceptable concentrations for a 7-d, 30-d, or 180-d exposure are not set at levels that will prevent any hyperventilation. Doing otherwise will be too conservative because CO_2-induced hyperventilation is not harmful per se. Under most situations, CO_2-induced hyperventilation is a protective response for the body when oxygen is displaced by CO_2 at an abnormally high level. The situation is different, however, in spacecraft. Since oxygen is artificially maintained in spacecraft at a level sufficiently high for metabolic needs, CO_2-induced hyperventilation is of lesser importance in spacecraft than on earth. In setting the acceptable CO_2 concentrations, toxic effects secondary to CO_2-induced hyperventilation should be taken into consideration. There are three potential secondary effects of CO_2-induced hyperventilation:

1. Discomfort associated with extreme hyperventilation.

2. Impairment of the ability to exercise or take on heavy workload.
3. Increased inhalation of airborne toxicants.

Consideration of the discomfort associated with extreme hyperventilation overlaps somewhat with that of the miscellaneous symptoms, such as dyspnea and intercostal muscle pain, discussed above. The levels set to prevent dyspnea and intercostal muscle pain will not be repeated here. Sinclair et al. in an Air force study stated that, "with the exception of occasional mild headaches and awareness of increased ventilation during the first 24 hours" of a 30-d exposure of four subjects to 2.8% CO_2 or a 11-d exposure of another four subjects to 3.9% CO_2, the humans subjects tolerated the hypercapnia "without apparent difficulty" (Sinclair et al., 1969). In a French study, there was no report of any symptoms or complaints in six human volunteers exposed for 30 d to 2% CO_2, which increased the minute volume by 45% in d 8-30 (Radziszewski et al., 1988; Guillerm and Radziszewski; 1979). So the Air Force and French studies indicate that humans can tolerate 2-3.9% CO_2; the low end of the range, 2%, appears to be a prudent choice as the NOAEL based on tolerability. The number of test subjects used in the Air Force and French studies are pooled, $4 + 4 + 6 = 14$, in calculating acceptable concentrations based on tolerability.

7-d, 30-d, and 180-d ACs based on tolerability
= NOAEL × 1/small n factor
= NOAEL × (square root of n)/10
= 2% × (square root of 14)/10
= 2% × 0.37
= 0.7%.

Another factor to be considered is whether a subchronic exposure to CO_2 will impair the ability of astronauts to exercise daily, which is very important in conditioning the muscle in microgravity. The French group showed that, in a 1-d exposure of five human subjects to 4.3% CO_2, the subjects were unable to exercise (Radziszewski et al., 1988). In a 9-d exposure to 3.8% CO_2, five subjects could exercise only at a limited capacity (Radziszewski et al., 1988). In contrast, in a 30-d exposure of six volunteers to 2% CO_2, the volunteers had good ability to exercise (Radziszewski et al., 1988), and their oxygen consumption rate, respiratory rate, and heart rate during the 10-min exercise at a 150-watt workload did not differ from performance of the same exercise in the normocapnic control period (Guillerm and Radziszewski, 1979). The only small change was that their minute volumes were raised about 17% when they were breathing 2% CO_2 than air (Guillerm and Radziszewski, 1979). Similarly, the Air Force group showed that there was no reduction in the ability of four human subjects to perform 45-min of steady state exercise at low, moderate, or heavy workload during a 15-20 d exposure to 2.8% CO_2 (Sinclair et al., 1971). The oxygen consumption rate and heart rate of the subjects during these exercises stayed the same in hypercapnic or normocapnic condition. The minute volumes during these exercises when they were breathing 2.8% CO_2 were about 20% higher than that when breathing air. The Glatte et al. (1967a) Air Force study showed that six out of seven human volunteers easily tolerated daily 1-h moderate exercises at a 100-watt workload during a 5-d exposure to 3% CO_2. Glatte et al. reported that the lone human volunteer who could not tolerate the exercises was of so small a statute that the exercise bike could not be lowered sufficiently to fit him. Therefore, the exercise data of this man can be ignored. The data from the French and Air Force studies (Glatte et al., 1967a; Sinclair et al., 1971; Radziszewski et al., 1988) showed that a subchronic exposure to 2-3% CO_2 should have no adverse impact on the ability of astronauts to exercise or work, so the number of test subjects used in these studies are pooled, 6 + 4 + 6 = 16, in estimating the acceptable concentrations based on the ability to exercise.

7-d, 30-d, and 180-d ACs based on exercise ability
= NOAEL × 1/small n factor
= NOAEL × (square root of n)/10
= 2% × (square root of 16)/10
= 2% × 0.40
= 0.8%.

The final factor worth considering is the increased inhalation of airborne toxicants caused by CO_2-induced hyperventilation. Based on the consideration of the subjective feeling of tolerability and the exercise ability, the acceptable concentration is about 0.7%. Radziszewski et al. (1988) exposed six volunteers to 0.5% and 1% CO_2 and they reported that 0.5% CO_2 failed to produce significant increase in the minute volume, while 1% CO_2 produced a 19% hyperventilation in the first day of exposure. In 15 human subjects exposed to 1% CO_2 for 44 d, Pingree (1977) showed that the minute volume rose 30% in d 4, but it returned to the control value on the eighth day. Judging from these data of Radziszewski et al. and Pingree, the AC of 0.7% will increase inhalation of airborne toxicants by less than 30% in the first few days of CO_2 exposure. A less than 30% increase is small compared with the potential respiratory ventilatory increase caused by moderate or heavy workload. According to the NRC Committee on Toxicology, a man inhales 7.5 L/min at rest and 20 L/min at light activity (NRC, 1992). NRC assumes that a man inhales 20 m^3/d, which is approximately equivalent to the amount of air inhaled in 10 h of rest and 14 h of light activity per day (NRC, 1992). The minute volume at light activity is about 170% higher than that at rest. According to the data of Sinclair et al. (1971), the minute volumes at moderate and heavy workload are 590% and 940% higher than that at rest. Therefore, the less-than-30% increase in minute volume in the first few days of an exposure to 0.7% CO_2 would be less than the exercise-induced minute volume increase if the astronauts were to engage in any moderate or heavy exercises. Especially because the less-than-30% increase in CO_2-induced minute volume will disappear in a few days, in the long run there will not be any significant increase in the inhalation of airborne toxicants.

From the analysis with these three secondary effects of hyperventilation, 0.7% CO_2 is selected to be the acceptable concentration for a 7-d, 30-d, or 180-d exposure based on hyperventilation.

Airway Resistance Increases

In humans, acute CO_2 exposures to 5% or 7.5% CO_2 in 2 h might result in increased airway resistance, and increased total lung resistance, without any lung compliance change, in exposure to 8% CO_2 in 3-6 min (Nadel and Widdicombe, 1962; Tashkin and Simmons, 1972). Theoretically, increased total lung resistance could be caused by an increase in tissue resistance in the lung or airway resistance. Although an increased tissue resistance could be produced by CO_2 because CO_2 is known to raise the surface tension of the alveolar extract (Schaefer et al., 1964a), the increased total lung resistance detected in the human subjects exposed to 8% CO_2 for 3-6 min (Nadel and Widdicombe, 1962) was most probably not due to any raised tissue resistance. This is because tissue resistance normally contributes only 20% of the total lung resistance, with airway resistance contributing the remaining 80% in young adults (West, 1979). As a result, the CO_2-induced increase in total lung resistance was probably due mostly to increased airway resistance. The finding by Nadel and Widdicombe (1962) that an acute exposure to 8% CO_2 increased the total lung resistance without changing the lung compliance supports the notion that CO_2 increases the total lung resistance by increasing the airway resistance. Therefore, a CO_2 concentration that does not increase airway resistance should also prevent any increases in total lung resistance. According to Tashkin and Simmons (1972), a 2-h exposure of nine human subjects to 2.5% CO_2 did not change the airway resistance. Glatte et al. (1967a) found that the forced expiratory volume in 1 s failed to change in seven volunteers during a 5-d exposure to 3% CO_2. Therefore, 3% CO_2 is considered the NOAEL for increases in airway resistance. Because a large safety margin is not needed in short-term contingencies, no adjustment is made for using a NOAEL based on only seven human subjects in deriving the short-term ACs.

1-h and 24-h ACs based on airway resistance increases
= 5-d NOAEL
= 3%.

The increase in airway resistance has been postulated to be due to the local action of hypercapnia in the larynx (Cotes, 1979, pp. 149, 258, 363). Being a local reaction on the upper airway, it is not expected to increase in severity with time of exposure. Therefore, the 5-d NOAEL of 3% is used to derive the ACs for 7-d, 30-d, or 180-d exposures without any time adjustment.

7-d, 30-d, and 180-d ACs based on lung mechanics changes
= 5-d NOAEL × 1/small n factor
= 5-d NOAEL × (square root of n)/10
= 3% × (square root of 7)/10
= 3% × 0.26
= 0.8%.

Testes

Although an acute exposure of rats to CO_2 at a concentration as low as 2.5% was found to cause the sloughing of mature spermatids and Sertoli cells in the testis at 4 or 8 h (but not at 1 or 2 h) (Vandemark et al., 1972), both the 1-h and 24-h SMACs are not set based on testicular toxicity. This is because the testicular morphology recovered completely in these rats in 36 h after the 8-h CO_2 exposure (Vandemark et al., 1972). So the only potential functional deficit is a 1- or 2-d period of temporarily reduced fertility within a week of the acute CO_2 exposure at 2.5%. Such a temporary reduced fertility is acceptable, considering that the 1-h and 24-h SMACs are aimed at emergency situations.

Although there is no solid proof that subchronic CO_2 exposures cause testicular injuries, for prudence' sake, it is assumed that CO_2 is toxic to the testis subchronically. This is because acute CO_2 exposures could injure the testis, albeit only temporarily (Vandemark et al., 1972). If the CO_2 exposure duration is extended from acute to subchronic, it is possible that the testicular injury will persist. Even though the study of Schaefer et al. (1971) was not adequately controlled, the absence of testicular damage in guinea pigs and rats exposed to 3% CO_2 for 42 d suggests that 3% could be treated as a NOAEL for subchronic exposures. The NRC Subcommittee on SMACs advised NASA not to apply the traditional interspecies factor of 10 with CO_2's testicular toxicity because they felt that the toxicity is due to the acidosis and they did not think that the testes in humans will be more sensitive than the testes in rodents toward CO_2-induced acidosis.

7-d, 30-d, and 180-d AC based on testicular toxicity
 = 42-d NOAEL in the rodent studies
 = 3%.

Establishment of SMACs

The ACs based on these toxic end points are summarized in Table 3-12. By comparing the ACs, the 1-h and 24-h SMACs are set at 1.3% (9.9 torr), while the 7-d, 30-d, and 180-d SMACs are all established at 0.7% (5.3 torr). For comparison purpose, it should be noted that it has been the Navy's position that "[h]uman exposures have been safely conducted in atmospheres containing up to 5 torr CO_2, for up to 90 days. Such exposures are therefore considered safe at this time" (Naval Submarine Medical Research Laboratory, 1982).

Finally, it should be pointed out that potential influences of microgravity-induced physiological changes on the acceptable concentrations are not needed in setting the SMACs for CO_2. Even though microgravity-induced hypercalciuria is a risk factor for kidney stone formation (Pak et al., 1989), CO_2's SMACs need not be adjusted for hypercalciuria because CO_2 exposures are not known to increase urinary calcium excretion in human subjects (in fact CO_2 exposures decreased urinary calcium excretion in human subjects) (Messier et al., 1976; Schaefer et al., 1963b).

TABLE 3-12 End Points and Acceptable Concentrations

End Point	Exposure Data	Species and Reference	Uncertainty factors Species	Uncertainty factors Small n	Acceptable Concentrations, % 1 h	24 h	7 d	30 d	180 d
Visual impairment, tremor, CNS depression	NOAEL at 3%, 24 h/d, 5 d or 2 w	Human (n = 7, 12) (Glatte et al., 1967; Storm and Giannetta, 1974)	—	10/(sq. rt. of 19)	1.3	1.3	1.3	1.3	1.3
Headache	NOAEL at 2%, 24 h/d, 30 d	Human (n = 6) (Radziszewski et al., 1988; Guillerm and Radziszewski, 1979)	—	—	2	2	2	2	2
Dyspnea, intercostal pain	NOAEL at 2.8%, 0.5 or 1 h	Human (n = 8, 4) (Menn et al., 1970; Sinclair et al., 1971)	—	—	2.8	2.8	—	—	—
	NOAEL at 2.8%, 15 or 20 d	Human (n = 4, 6) (Sinclair et al., 1971; Radziszewski et al., 1988)	—	10/(sq. rt. of 10)	—	—	0.9	0.9	0.9
Airway resistance increases	NOAEL at 3%, 24 h/d, 5 d	Human (n = 7) (Glatte et al., 1967)	—	—	3	3	—	—	—
	NOAEL at 3%, 24 h/d, 5 d	Human (n = 7) (Glatte et al., 1967)	—	10/(sq. rt. of 10)	—	—	0.8	0.8	0.8
Hyperventilation Tolerability	NOAEL at 2%, 24 h/d, 11 or 30 d	Human (n = 4, 4, 6) (Sinclair et al., 1969; Radziszewski et al., 1988 Guillerm and Radziszewski, 1979)	—	10/(sq. rt. of 14)	—	—	0.7	0.7	0.7
Exercise impairment	NOAEL at 2%, 24 h/d, 5, 15, or 30 d	Human (n = 6, 4, 6) (Glatte et al., 1967; Sinclair et al., 1971; Radziszewski et al., 1988)	—	10/(sq. rt. of 16)	—	—	0.8	0.8	0.8
Testicular injury	NOAEL at 3%, 24 h/d, 42 d	Rat and guinea pig (Schaefer et al., 1971)	1	—	—	—	3	3	3
SMAC					1.3	1.3	0.7	0.7	0.7

REFERENCES

Aero Medical Association, Committee on Aviation Toxicology. 1953. Pp. 6-9, 31-39, 52-55, 74-79, and 110-115 in Aviation Toxicology: An Introduction to the Subject and a Handbook of Data. New York: The Blakiston Co.

Altschule, M.D., and W.M. Sulzbach. 1947. Tolerance of the human heart to acidosis: Reversible changes in RS-T interval during severe acidosis caused by administration of carbon dioxide. Am. Heart J. 33:458-463.

Baggott, J. 1982. Gas transport and pH regulation. Pp. 1098-1101, 1114, 1120-1123 in Textbook of Biochemistry with Clinical Correlations, T.M. Devlin, ed. New York: John Wiley & Sons.

Barbour, J.H., and M.H. Seevers. 1943. A comparison of the acute and chronic toxicity of carbon dioxide with especial reference to its narcotic action. J. Pharmacol. Exp. Ther. 78:11-21.

Brackett, N.C., Jr., J.J. Cohen, and W.B. Schwartz. 1965. Carbon dioxide titration curve of normal man: Effect of increasing degrees of acute hypercapnia on acid-base equilibrium. New Engl. J. Med. 272:6-12.

Breslau, N.A., L. Brinkley, K.D. Hill, and C.Y. Pak. 1988. Relationship of animal protein-rich diet to kidney stone formation and calcium metabolism. J. Clin. Endocrinol. Metab. 66:140-146.

Brown, E.W. 1930a. The physiological effects of high concentrations of carbon dioxide. U.S. Naval Med. Bull. 28:721-934.

Brown, E.W. 1930b. The value of high oxygen in preventing the physiological effects of noxious concentrations of carbon dioxide. U.S. Naval Med. Bull. 32:523-553.

Brown, E.B., G.S. Campbell, M.N. Johnson, A. Hemingway, and M.B. Visscher. 1948. Changes in response to inhalation of CO_2 before and after 24 hours of hyperventilation in man. J. Appl. Physiol. 1:333-338.

Bungo, M.W. 1989. The cardiopulmonary system. Pp. 197-199 in Space Physiology and Medicine, A.E. Nicogossian, ed. Philadelphia: Lea & Febiger

Campbell, J.M.H., C.G. Douglas, J.S. Haldane, and F.G. Hodgson. 1913. The response of the respiratory centre to carbonic acid, oxygen, and hydrogen ion concentration. J. Physiol. 46:301-318.

Chapin, J.L., A.B. Otis, and H. Rahn. 1955. Changes in the Sensitivity of the Respiratory Center in Man After Prolonged Exposure to 3% CO_2. Pp. 250-254 in WADC Tech. Rep. No. 55-357, Wright-Patterson Air Force Base, Dayton, Ohio.

Clamann, H.G. 1959. Some metabolic problems of space flight. Fed. Proc. 18:1249-1255.

Clark, J.M., R.D. Sinclair, and B.E. Welch. 1971. Rate of acclimatization to chronic hypercapnia in man. Pp 399-408 in Underwater Physiology, C.J. Lambertsen, ed. New York: Academic Press.

Coe, F.L., and M.J. Favus. 1987. Nephrolithiasis. Pp. 1211-1215 in Harrison's Principles of Internal Medicine, E. Braunwald, K.J. Isselbacher, R.G. Petersdorf, J.D. Wilson, J.B. Martin, and A.S. Fauci, eds. New York: McGraw-Hill.

Coleman, W.E., L.D. Scheel, R.E. Kupel, and R.L. Larkin. 1968. The identification of toxic compounds in the pyrolysis products of polytetrafluoroethylene (PTFE). Am. Ind. Hyg. Assoc. J. 29:33-40.

Consolazio, W.B., M.B. Fisher, N. Pace, L.J. Pecora, and A.R. Behnke. 1947. Effects on man of high concentrations of carbon dioxide in relation to various oxygen pressures during exposures as long as 72 hours. Am. J. Physiol. 151:479-503.

Cotes, J.E. 1979. Lung function. In Assessment and Application in Medicine. Oxford, U.K.: Blackwell Scientific Publications.

Dalgaard, J.B., F. Dencker, B. Fallentin, P. Hansen, B. Kaempe, J. Steensberg, and P. Wilhardt. 1972. Fatal poisoning and other health hazards connected with industrial fishing. Br. J. Ind. Med. 29:307-316.

Davies, D.M., J.E.W. Morris, D.J. Smith, and R.J. Pethybridge. 1978a. Mineral metabolism and circadian patterns of renal excretion in man in a closed environment study involving hypercapnia and abnormal physical activity. INM Rep. No. 25/78. Institute of Naval Medicine, Gosport, Alverstoke, U.K.

Davies, D.M., W.M. Edmondstone, A. Bishop, and J.E.W. Morris. 1978b. Urinary mineral excretion in men on a long submarine patrol and effects upon it of the oral administration of Vitamin D. INM Rep. No. 26/78. Institute of Naval Medicine, Gosport, Alverstoke, U.K.

Davies, D.M. and J.E.W. Morris. 1979. Carbon dioxide and vitamin D effects on calcium metabolism in nuclear submariners: A review. Undersea Biomed. Res. 6(Suppl.):S71-S80.

DeBellis, J., H. Constantine, and M. Stein. 1968. Effect of acute hypercapnia on gastrointestinal blood loss. Fed. Proc. 28:509.

Deitrick, J.E., G.D. Whedon, and E. Shorr. 1948. Effects of immobilization upon various metabolic and physiologic functions of normal men. Am. J. Med. 4:3-36.

Diamondstone, T.I. 1982. Amino acid metabolism I. P. 545 in Textbook of Biochemistry with Clinical Correlations, T.M. Devlin, ed. New York: John Wiley & Sons.

Donaldson, C.L., S.B. Hulley, J.M. Vogel, R.S. Hattner, J.H. Bayes, and D.E. McMillan. 1970. Effect of prolonged bed rest on bone mineral. Metabolism 19:1071-1084.

Douglas, W.H.J., K.E. Schaefer, A.A. Messier, and S.M. Pasquale. 1979. Proliferation of pneumocyte II cells in prolonged exposure to 1% CO_2. Undersea Biomed. Res. (Submarine Suppl.):S135-S142.

Dripps, R.D., and J.H. Comroe, Jr. 1947. The respiratory and circulatory response of normal man to inhalation of 7.6 and 10.4 percent CO_2 in a comparison of the maximal ventilation produced by severe muscular exercise, inhalation of CO_2 and maximal voluntary hyperventilation. Am. J. Physiol. 149:43-51.

Dunn, M.S., A.T. Shennan, and F. Possmayer. 1990. Single-versus multiple-dose surfactant replacement therapy in neonates of 30 to 36 weeks' gestation with respiratory distress syndrome. Pediatrics 86: 564-571.

Eldridge, F., and J.M. Davis. 1959. Effect of mechanical factors on respiratory work and ventilatory responses to CO_2. J. Appl. Physiol. 14:721-726.

Fisch, C. 1988. Electrocardiography and vectorcardiography. Pp. 180-219 in Heart Disease: A Textbook of Cardiovascular Medicine, E. Braunwald, ed. Philadelphia: W.B. Saunders.

Friedlander, W.J., and T. Hill. 1954. EEG changes during administration of carbon dioxide. Dis. Nerv. Syst. 15:71-75.

Gellhorn, E. 1936. Effect of O_2 lack, variations in CO_2 content of inspired air, and hyperpnea on visual intensity discrimination. Am. J. Physiol. 115:679-684.

Gellhorn, E., and I. Spiesman. 1934. Influence of variations of O_2 and CO_2 tension in inspired air upon hearing. Proc. Soc. Exp. Biol. Med. 32:46-47.

Gellhorn, E., and I. Spiesman. 1935. Influence of hyperpnea and of variations of O_2- and CO_2-tension in inspired air upon hearing. Am. J. Physiol. 112:519-528.

Glatte, H.A., G.J. Motsay, and B.E. Welch. 1967a. Carbon Dioxide Tolerance Studies. Rep. No. SAM-TR-67-77. U.S. Air Force School of Aerospace Medicine, Brooks Air Force Base, San Antonio, Tex.

Glatte, H.A., B.O. Hartman, and B.E. Welch. 1967b. Nonpathologic hypercapnia in man. Pp. 110-129 in Lectures in Aerospace Medicine. 6th Series. Rep. No. SAM-TR-68-116. U.S. Air Force School of Aerospace Medicine, Brooks Air Force Base, San Antonio, Tex.

Goldberg, H., L. Grass, R. Vogl, A. Rapoport, and D.G. Oreopoulos. 1989. Urine citrate and renal stone disease. Can. Med. Assoc. J. 141:217-221.

Goldsmith, A.E., G.F. Ryan, and A.B. Joseph. 1980. Metabolic carcinogenesis: Induction of murine lymphoma by CO_2-treatment in vivo and in vitro. Jpn. J. Med. Sci. Biol. 33:7-18.

Gortner, L., F. Pohlandt, P. Bartmann, and B. Disse. 1990. Treatment of respiratory distress syndrome in very small premature infants with bovine surfactant. Monatsschr. Kinderheilkd. 138:8-12.

Gray, S.P., J.E.W. Morris, and C.J. Brooks. 1973. Renal handling of calcium, magnesium, inorganic phosphate and hydrogen ions during prolonged exposure to elevated carbon dioxide concentrations. Clin. Sci. Mol. Med. 45:751-764.

Grote, W. 1965. [Disturbances of embryonic development at elevated CO_2 and O_2 partial pressure and at reduced atmospheric pressure.] Z. Morphol. Anthropol. 56:165-194.

Gormsen, H., N. Jeppesen, and A. Lund. 1984. The causes of death in fire victims. Forensic Sci. Int. 24:107-111.

Grollman, A. 1930. Physiological variations in the cardiac output of man. IX. The effect of breathing carbon dioxide, and of voluntary forced ventilation on the cardiac output of man. Am. J. Physiol. 94:287-299.

Gude, J.K., and K.E. Schaefer. 1969. The Effects on Respiratory Dead Space of Prolonged Exposure to a Submarine Environment. Rep. No. 587. Submarine Medical Research Laboratory, Naval Submarine Medical Center, Groton, Connecticut.

Guillerm, R., and E. Radziszewski. 1979. Effects on man of 30-day exposure to a PI_{CO2} of 14 torr (2%): Application to exposure limits.

Undersea Biomed. Res. (Submarine Suppl.):S91-S114.

Haldane, J., and J.L. Smith. 1892. The physiological effects of air vitiated by respiration. J. Pathol. Bacteriol. 1:168-186.

Haring, O.M. 1960. Cardiac malformations in rats induced by exposure of the mother to carbon dioxide during pregnancy. Circ. Res. 8:1218-1227.

Hartzell, G., and W.G. Switzer. 1985. On the toxicities of atmospheres containing both carbon monoxide and carbon dioxide. J. Fire Sci. 3:307-309.

Hofbauer, J., K. Hobarth, and O. Zechner. 1990. The significance of citrate excretion and calcium/citrate quotients in urine in patients with calcium calculi. Z. Urol. Nephrol. 83:597-602.

Hopson, G.D., J.W. Littles, and W.C. Patterson. 1974. MSFC Skylab Thermal and Environmental Control System Mission Evaluation. Rep. No. NASA TM X-64822. National Aeronautics and Space Administration, Washington, D.C.

Huntoon, C.L., P.C. Johnson, and N.M. Cintron. 1989. Hematology, immunology, endocrinology, and biochemistry. Pp. 222-239 in Space Physiology and Medicine, A.E. Nicogossian, ed. Philadelphia: Lea & Febiger.

Isom, G.E., and R.M. Elshowihy. 1982. Naloxone-induced enhancement of carbon dioxide stimulated respiration. Life Sci. 31:113-118.

Jackson, J.K., J.R. Wamsley, M.S. Bonura, and J.S. Seeman. 1972. Pp. 34, 48-50 in Program Operational Summary: Operational 90 Day Manned Test of a Regenerative Life Support. Rep. No. NASA CR-1835. National Aeronautics and Space Administration, Washington, D.C.

Johnston, R.F. 1959. The syndrome of carbon dioxide intoxication: Its etiology, diagnosis, and treatment. Univ. Mich. Med. Bull. 25:280-292.

Kety, S.S., and C.F. Schmidt. 1948. The effects of altered arterial tensions of carbon dioxide and oxygen on cerebral blood flow and cerebral oxygen consumption of normal young men. J. Clin. Invest. 27:484-492.

Kryger, M.H. 1981. Respiratory failure 2: Carbon dioxide. Pp. 205-219 in Pathophysiology of Respiration, M.H. Kryger, ed. New York: John Wiley & Sons.

Lai, Y.-L., Y. Tsuya, and J. Hildebrandt. 1978. Ventilatory responses to acute CO_2 exposure in the rat. J. Appl. Physiol. 45:611-618.

Lambertsen, C.J. 1971. Therapeutic gases: Oxygen, carbon dioxide, and helium. Ch. 55 in Drill's Pharmacology in Medicine, J.R. DiPalma, ed. New York: McGraw-Hill.
Leach, C.S., and P.C. Rambaut. 1977. Biochemical responses of the Skylab crewmen: An overview. In Biomedical Results from Skylab, R.S. Johnston and L.F. Dietlein, eds. National Aeronautics and Space Administration, Washington, D.C.
LeBaron, F.N. 1982. Lipid metabolism I. P. 473 in Textbook of Biochemistry with Clinical Correlations, T.M. Devlin, ed. New York: John Wiley & Sons.
Levin, B.C., M. Paabo, J.L. Gurman, S.E. Harris, and E. Braun. 1987. Toxicological interactions between carbon monoxide and carbon dioxide. Toxicology 47:135-164.
Levin, B.C., M. Paabo, L. Highbarger, and N. Eller. 1989. Synergistic Effects of Nitrogen Dioxide and Carbon Dioxide Following Acute Inhalation Exposures in Rats. Society of the Plastics Industry, Inc. Available from the National Technical Information Services, Springfield, Va., Doc. No. PB89-214779.
Luft, U.C., S. Finkelstein, and J.C. Elliott. 1974. Respiratory gas exchange, acid-base balance, and electrolytes during and after maximal work breathing 15 mm Hg PI_{CO2}. Pp. 282-293 in Topics in Environmental Physiology and Medicine: Carbon Dioxide and Metabolic Regulations, G. Nahas and K.E. Schaefer, eds. New York: Springer-Verlag.
MacDonald, F.M., and E. Simonson. 1953. Human electrocardiogram during and after inhalation of thirty percent carbon dioxide. J. Appl. Physiol. 6:304-310.
McArdle, L. 1959. Electrocardiographic studies during the inhalation of 30 per cent carbon dioxide in man. Br. J. Anaesth. 31:142-151.
McCarthy, D.S. 1981. Airflow obstruction. P. 16 in Pathophysiology of Respiration, M.H. Kryger, ed. New York: John Wiley & Sons.
Massie, B.M., and M. Sokolow. 1990. Heart and great vessels. Pp. 267-271 in Current Medical Diagnosis and Treatment in 1990, S.A. Schroeder, M.A. Krupp, L.M. Tierney, Jr., and S.J. McPhee, eds. Norwalk, Conn.: Appleton and Lange.
Meessen, H. 1948. Chronic carbon dioxide poisoning experimental studies. Arch. Pathol. 45:36-40.
Menn, S.J., R.D. Sinclair, and B.E. Welch. 1968. Response of Normal Man to Graded Exercise in Progressive Elevations of CO_2. Rep.

No. SAM-TR-68-116. Aerospace Medical Division, USAF School of Aerospace Medicine, Brooks Air Force Base, San Antonio, Tex.

Menn, S.J., R.D. Sinclair, and B.E. Welch. 1970. Effect of inspired pCO_2 up to 30 mm Hg on response of normal man to exercise. J. Appl. Physiol. 28:663-61.

Messier, A.A., and K.E. Schaefer. 1971. The effect of chronic hypercapnia on oxygen affinity and 2,3-diphosphoglycerate. Resp. Physiol. 12:291-296.

Messier, A.A., E. Heyder, and K.E. Schaefer. 1971. Effect of 90-Day Exposure to 1% CO_2 on Acid-Base Status of Blood. Rep. No. 655. U.S. Naval Submarine Medical Center, Submarine Base, Groton, Conn.

Messier, A.A., E. Heyder, W.R. Braithwaite, C. McCluggage, A. Peck, and K.E. Schaefer. 1976. Calcium, magnesium, and phosphorus metabolism, and parathyroid-calcitonin function during prolonged exposure to elevated CO_2 concentrations on submarines. Undersea Biomed. Res. 6(Suppl.):S57-S70.

Mines, A.H. 1981. Pp. 91-99 in Respiratory Physiology. New York: Raven Press.

Mitsuda, H., S. Ueno, H. Mizuno, H. Fujikawa, K. Konaka, and C. Fukada. 1982. Effects of various molecular oxygen levels in mixed gas on acute respiratory insufficiency induced with carbon dioxide inhalation in rats. Kankyo Kagaku Sogo Kenkyusho Nenpo 2:35-46.

Morey, P.R. and D.E. Shattuck. 1989. Role of ventilation in the causation of building-associated illness. Occup. Med. State of the Art Rev. 4:625-642.

Morey-Holton, E.R., H.K. Schnoes, H.F. DeLuca, M.E. Phelps, R.F. Klein, R.H. Nissenson, and C.D. Arnaud. 1988. Vitamin D metabolites and bioactive parathyroid hormone levels during Spacelab 2. Aviat. Space Environ. Med. 59:1038-1041.

Mukherjee, D.P., and S.P. Singh. 1967. Effect of increased carbon dioxide in inspired air on the morphology of spermatozoa and fertility of mice. J. Reprod. Fert. 13:165-167.

Nadel, J.A. and J.G. Widdicombe. 1962. Effect of changes in blood as tensions and carotid sinus pressure on tracheal volume and total lung resistance to airflow. J. Physiol. 163:13-33.

NASA. 1984. $PPCO_2$ history for STS 51-D. P. 4.3.7-7 in Shuttle Operations Data Book. Doc. No. JSC 08934. NASA, Johnson Space Center, Houston, Tex.

NASA. 1988. PPCO$_2$ Constraint. Flight Rule No. 13-8. NASA, Johnson Space Center, Houston, Tex.
Naval Submarine Medical Research Laboratory. 1982. Position Paper: The Toxic Effects of Chronic Exposures to Low Levels of Carbon Dioxide. Rep. No. 973. Naval Submarine Medical Center, Naval Submarine Base, Groton, Conn.
Niemoeller, H., and K.E. Schaefer. 1962. Development of hyaline membranes and atelectasis in experimental chronic respiratory acidosis. Proc. Soc. Exp. Biol. Med. 110:804-808.
Notter, R.H., and J.N. Finkelstein. 1984. Pulmonary surfactant: An interdisciplinary approach. J. Appl. Physiol. 57:1613-1624.
NRC. 1992. Guidelines for Developing Spacecraft Maximum Allowable Concentrations for Space Station Contaminants. Washington, D.C.: National Academy Press.
Okajima, M., and E. Simonson. 1962. Effect of breathing six percent carbon dioxide on ECG changes in young and older healthy men. J. Gerontol. 17:286-288.
Olson, M.S. 1982. Bioenergetics and oxidative metabolism. P. 279 in Textbook of Biochemistry with Clinical Correlations, T.M. Devlin, ed. New York: John Wiley & Sons.
Pak, C.Y.C., C. Skurla, and J.A. Harvey. 1985. Graphic display of urinary risk factors for renal stone formation. J. Urol. 134:867-870.
Pak, C.Y.C., K. Hill, N.M. Cintron, and C. Huntoon. 1989. Assessing applicants to the NASA flight program for their renal stone-forming potential. Aviat. Space Environ. Med. 60:157-161.
Patterson, J.L., H. Heyman, L.L. Battery, and R.W. Ferguson. 1955. Threshold of response of the cerebral vessels of man to increases in blood carbon dioxide. J. Clin. Invest. 34:1857-1864.
Peck, A.S. 1971. The Time Course of Acid-Base Balance While on FBM Patrol. Rep. No. 675. Naval Submarine Medical Center, Naval Submarine Base, Groton, Conn.
Pepelko, W.E. 1970. Effects of hypoxia and hypercapnia, singly and combined, on growing rats. J. Appl. Physiol. 28:646-651.
Phillipson, E.A., G. Bowes, E.R. Townsend, J. Duffin, and J.D. Cooper. 1981. Carotid chemoreceptors in ventilatory responses to changes in venous CO$_2$ load. J. Appl. Physiol. 51:1398-1403.
Pingree, B.J.W. 1977. Acid-base and respiratory changes after prolonged exposure to 1% carbon dioxide. Clin. Sci. Mol. Med. 52:67-74.

Radziszewski, E., R. Guillerm, R. Badre, and C. Abran. 1976. Cinetique de la compensation de l'acidose respiratoire induite par l'hypercapnie chronique experimentale chez l'homme. Bull. Eur. Physiopathol. Resp. 12:-100.

Radziszewski, E., L. Giacomoni, and R. Guillerm. 1988. Effets physiologiques chez l'homme du confinement de longue duree en atmosphere enrichie en dioxyde de carbone. Pp. 19-23 in Proceedings of the Colloquium on Space and Sea. European Space Agency, Brussels, Belgium.

Redding, R.A., T. Arai, W.H.J. Douglas, H. Tsurutani, and J. Oven. 1975. Early changes in lungs of rats exposed to 70% O_2. J. Appl. Physiol. 38:136-142.

Riley, R.L., and B. Barnea-Bromberger. 1976. Acid-Base Changes in Blood of Brewery Workers Exposed to CO_2. An unpublished report cited by NIOSH in Criteria for a Recommended Standard. Occupational Exposure to Carbon Dioxide, Rep. No. NIOSH-76-194. National Institute for Occupational Safety and Health, Cincinnati, Ohio.

Rodkey, F.L., and H.A. Collison. 1979. Effects of oxygen and carbon dioxide on carbon monoxide toxicity. J. Combust. Toxicol. 6: 208-212.

Sax, I. 1984. P. 640 in Dangerous Properties of Industrial Materials. New York: Van Nostrand Reinhold.

Schaefer, K.E. 1949a. [Influence exerted on the psyche and the excitatory processes in the peripheral nervous system under long-term effects of 3% CO_2.] Pfluegers Arch. Gesamte. Physiol. Menschen Tiere 251:716-725.

Schaefer, K.E. 1949b. [Respiratory and acid-base balance during prolonged exposure to a 3% CO_2 atmosphere.] Pfluefers Arch. Gesamte. Physiol. Menschen. Tiere. 251:689-715.

Schaefer, K.E. 1958. Effects of Carbon Dioxide as Related to Submarines and Diving Physiology. Memo. Rep. No. 58-11. Naval Medical Research Laboratory, New London, Conn.

Schaefer, K.E. 1959. Experiences with submarine atmospheres. J. Aviat. Med. 30:350-359.

Schaefer, K.E. 1961a. Blood pH and pCO_2 homeostasis in chronic respiratory acidosis related to the use of amine and other buffers. Ann. N.Y. Acad. Sci. 92:401-413.

Schaefer, K.E. 1961b. A concept of triple tolerance limits based on chronic carbon dioxide toxicity studies. Aerospace Med. 32:197-204.
Schaefer, K.E. 1963a. The effects of CO_2 and electrolyte shifts on the central nervous system. Pp. 101-123 in Selective Vulnerability of the Brain in Hypoxemia, J.P. Schade, ed. Oxford, U.K.: Blackwell Scientific Publications.
Schaefer, K.E. 1963b. Respiratory adaptation to chronic hypercapnia. Ann. N.Y. Acad. Sci. 109:772-782.
Schaefer, K.E. 1963c. Acclimatization to low concentration of carbon dioxide. Ind. Med. Surg. 32:11-13.
Schaefer, K.E. 1979. Physiological stresses related to hypercapnia during patrols on submarines. Undersea Biomed. Res. 6(Suppl.):S15-S47.
Schaefer, K.E., B.J. Hastings, C.R. Carey, and G. Nichols, Jr. 1963a. Respiratory acclimatization to carbon dioxide. J. Appl. Physiol. 18:1071-1078.
Schaefer, K.E., G. Nichols, Jr., and C.R. Carey. 1963b. Calcium phosphorus metabolism in man during acclimatization to carbon dioxide. J. Appl. Physiol. 18:1079-1084.
Schaefer, K.E., M.E. Avery, and K. Bensch. 1964a. Time course of changes in surface tension and morphology of alveolar epithelial cells in CO_2-induced hyaline membrane disease. J. Clin. Invest. 43:2080-2093.
Schaefer, K.E., G. Nichols, Jr., and C.R. Carey, 1964b. Acid-base balance and blood and urine electrolytes of man during acclimatization to CO_2. J. Appl. Physiol. 19:48-58.
Schaefer, K.E., N. McCabe and J. Withers. 1968. Stress response in chronic hypercapnia. Am. J. Physiol. 214:543-548.
Schaefer, K.E., H. Niemoeller, A. Messier, E. Heyder, and J. Spencer. 1971. Chronic CO_2 Toxicity: Species Difference in Physiological and Histopathological Effects. Rep. No. 656. Naval Submarine Medical Research Laboratory, Groton, Conn.
Schaefer, K.E., S.M. Pasquale, A.A. Messier, and H. Niemoeller. 1979a. CO_2-induced kidney calcification. Undersea Biomed. Res. (Submarine Suppl.):S143-S153.
Schaefer, K.E., W.H.J. Douglas, A.A. Messier, M.L. Shea, and P.A.

Gohman. 1979b. Effect of prolonged exposure to 0.5% CO_2 on kidney calcification and ultrastructure of lungs. Undersea Biomed. Res. (Submarine Suppl.):S155-S161.

Schaefer, K.E., S. Pasquale, A.A. Messier, and M. Shea. 1980. Phasic changes in bone CO_2 fractions, calcium, and phosphorus during chronic hypercapnia. J. Appl. Physiol. 48:802-811.

Schwille, P.O., and U. Herrmann. 1992. Environmental factors in the pathophysiology of recurrent idiopathic calcium urolithiasis (RCU), with emphasis on nutrition. Urol. Res. 20:72-83.

Schneider, E.C., and D. Truesdale. 1922. The effects on the circulation and respiration of an increase in the carbon dioxide content of the blood in man. Am. J. Physiol. 63:155-175.

Schulte, J.H. 1964. Sealed environments in relation to health and disease. Arch. Environ. Health 8:438-452.

Sechzer, P.H., L.D. Egbert, H.W. Linde, D.Y. Cooper, R.D. Dripps, and H.L. Price. 1960. Effect of CO_2 inhalation on arterial pressure, ECG and plasma catecholamines and 17-OH corticosteroids in normal man. J. Appl. Physiol. 15:454-458.

Shih, K.C. 1987. P. C-34 in SL-2 Subsystem and Verification Flight Instrumentation Data. Doc. No. TM-SEAD-87004. MacDonald Douglas Huntsville, Huntsville, Ala.

Sinclair, R.D., J.M. Clark, and B.E. Welch. 1969. Carbon dioxide tolerance levels for space cabins. Proceedings of the Fifth Annual Conference on Atmospheric Contamination in Confined Spaces, Sept. 16-18, Wright-Patterson Air Force Base, Dayton, Ohio.

Sinclair, R.D., J.M. Clark, and B.E. Welch. 1971. Comparison of physiological responses of normal man to exercise in air and in acute and chronic hypercapnia. Pp. 409-417 in Underwater Physiology, C.J. Lambertsen, ed. New York: Academic Press.

Skov, P., O. Valbjorn, and the Danish Climate Study Group (DISG) 1987. The "sick" building syndrome in the office environment: The Danish town hall study. Environ. Int. 13:339-349.

Stein, S.N., H.E. Lee, J.H. Annegers, S.A. Kaplan, and D.G. McQuarrie. 1959. The effects of prolonged inhalation of hypernormal amounts of carbon dioxide. I. Physiological effects of 3 percent CO_2 for 93 days upon monkeys. Pp. 527-536 in Research Report NM 24 01 00.01.01, Vol. 17. Naval Medical Research Institute, Bethesda, Md.

Strachova, Z., and F. Plum. 1973. Reproducibility of the rebreathing carbon dioxide response test using an improved method. Am. Rev. Resp. Dis. 107:864-869.

Storm, W.F., and C.L. Giannetta. 1974. Effects of hypercapnia and bedrest on psychomotor performance. Aerosp. Med. 45:431-433.

Sullivan, T.Y., and P.-L. Yu. 1983. Airway anesthesia effects on hypercapnic breathing pattern in humans. J. Appl. Physiol. 55:368-376.

Tansey, W.A., J.M. Wilson, and K.E. Schaefer. 1979. Analysis of health data from 10 years of Polaris submarine patrols. Undersea Biomed. Res. 6(Suppl.):S217-S246.

Tashkin, D.P., and D.H. Simmons. 1972. Effect of carbon dioxide breathing on specific airway conductance in normal and asthmatic subjects. Am. Rev. Resp. Dis. 106:729-739.

Tenney, S.M. 1954. Ventilatory response to carbon dioxide in pulmonary emphysema. J. Appl. Physiol. 6:477-484.

Terrill, J.B., R.R. Montgomery, and C.F. Reinhardt. 1978. Toxic gases from fires. Science 200:1343-1347.

Thomas, J.A. 1991. Toxic responses of the reproductive system. Pp. 484-520 in Cassarett and Doull's Toxicology: The Basic Science of Poisons, M.O. Amdur, J. Doull, and C.D. Klaassen, eds. New York: Pergammon Press.

Thun, M.J., and S. Schober. 1991. Urolithiasis in Tennessee: An occupational window into a regional problem. Am. J. Public Health 81:587-591.

Trinchieri, A., A. Mandressi, P. Luongo, G. Longo, and E. Pisani. 1991. The influence of diet on urinary risk factors for stones in healthy subjects and idiopathic renal calcium stone formers. Br. J. Urol. 67:230-236.

U.S. Navy. 1988. Nuclear Powered Submarine Atmosphere Control Manual. S9510-AB-ATM-010/(U). Department of the Navy, Washington, D.C.

van Aswegen, C.H., P. Hurter, C.A. van der Merwe, and D.J. du Plessis. 1989. The relationship between total urinary testosterone and renal calculi. Urol. Res. 17:181-183.

Vandemark, N.L., B.D. Schanbacher, and W.R. Gomes. 1972. Alterations in testes of rats exposed to elevated atmospheric carbon dioxide. J. Reprod. Fertil. 28:457-459.

Vogel, J.M. 1975. Effect of spaceflight on bone mineral. Acta Astronautica 2:129-140.

Wamsley, J.R., E.W. Youngling, and W.F. Behm. 1969. High fidelity simulations in the evaluation of environmental stress: Acute CO_2 exposure. Aerospace Med. 40:1336-1340.

Wang, T.C. 1975. A study of bioeffluents in a college classroom. ASHRAE Transact. 81:32-33.

Wasserstein, A.G., P.D. Stolley, K.A. Soper, S. Goldfarb, G. Maislin, and Z. Agus. 1987. Case-control study of risk factors for idiopathic calcium nephrolithiasis. Miner. Electrolyte Metab. 13:85-95.

Weitzman, D.O., and J.A.S. Kinney. 1969. Effect on Vision of Repeated Exposure to Carbon Dioxide. Rep. No. 566. Naval Submarine Medical Center, Naval Submarine Base, Groton, Conn.

West, J.B. 1979. Pp. 23, 74 in Respiratory Physiology: The Essentials. Baltimore: Williams & Wilkins.

Whedon, G.D., L. Lutwak, P.C. Rambaut, M.W. Whittle, M.C. Smith, J. Reid, C. Leach, C.R. Stadler, and D.D. Sanford. 1977. Mineral and nitrogen metabolic studies, Experiment M071. Pp. 164-174 in Biomedical Results from Skylab, R.S. Johnston and L.F. Dietlein, eds. National Aeronautics and Space Administration, Washington, D.C.

Whedon, G.D. 1984. Disuse osteoporosis: Physiological aspects. Calcif. Tissue Int. 36:S146-S150.

White, C.S., J.H. Humm, E.D. Armstrong, and N.P.V. Lundgren. 1952. Human tolerance to acute exposure to carbon dioxide. Pp. 439-455 in Rep. No. 1: Six Per Cent Carbon Dioxide in Air and in Oxygen. Aviation Med. (Oct.).

Wilson, A.J., and K.E. Schaefer. 1979. Effect of prolonged exposure to elevated carbon monoxide and carbon dioxide levels on red blood cell parameters during submarine patrols. Undersea Biomed. Res. 6(Suppl.):S49-S56.

Wright, J.R., and J.A. Clements. 1987. Metabolism and turnover of lung surfactant. Am. Rev. Resp. Dis. 135:426-444.

Wooley, W.D., S.A. Ames, and P.J. Fardell. 1979. Chemical aspects of combustion toxicology of fires. Fire Materials 3:110-120.

Yonezawa, A. 1968. [Influence of carbon dioxide inhalation on renal circulation and electrolyte metabolism.] Jpn. Cir. J. 32:1119-1120.

Zharov, S.G., Y.A. Il'in, Y.A. Kovalenko, I.R. Kalinichenko, L.I.

Karpova, N.S. Mikerova, M.M. Osipova, and Y.Y. Simonov. 1963. Effect on man of prolonged exposure to atmosphere with a high CO_2 content. Pp. 155-158 in Proceedings of the International Congress of Aviation and Space Medicine. International Congress of Aviation and Space Medicine, Rome.

Zink, P., and G. Reinhardt. 1975. Carbon dioxide poisoning after prolonged exposure. Beitr. Gerichtl. Med. 33:211-213.

B4 2-Ethoxyethanol

King Lit Wong, Ph.D.
Johnson Space Center Toxicology Group
Biomedical Operations and Research Branch
Houston, Texas

PHYSICAL AND CHEMICAL PROPERTIES

2-Ethoxyethanol is a colorless liquid with a slightly sweet odor (ACGIH, 1986). Without giving concentrations, Waite et al. (1930) reported that the odor of 2-ethoxyethanol is mild and agreeable in low concentrations and disagreeable in high concentrations. It is completely miscible with water and with many organic solvents.

Synonyms: Ethylene glycol monoethyl ether; Cellosolve
Formula: $CH_3CH_2OCH_2CH_2OH$
CAS number: 110-80-5
Molecular weight: 90.1
Boiling point: 135.6°C
Melting point: -70°C
Vapor pressure: 3.7 mm Hg at 20°C
Conversion factors at 25°C, 1 atm: 1 ppm = 3.68 mg/m^3
1 mg/m^3 = 0.27 ppm

OCCURRENCE AND USE

2-Ethoxyethanol is used as a solvent in synthetic resins and nitrocellulose manufacturing and in varnish removers, cleaning solutions, and lacquers (ACGIH, 1986). There is no known use of 2-ethoxyethanol in the spacecraft, but it has been found in the cabin atmosphere during two missions (NASA, 1989-1990). The concentrations of 2-ethoxyethanol detected in these missions varied from 0.8 to 3.0 ppb (NASA, 1989-1990), and off-gassing is probably the source of it. Based on the off-gassing data in the Spacelab, it was estimated that 2-ethoxyethanol will also be generated in the space station (Leban and Wagner, 1989). A possible source is the off-gassing from paint because 2-ethoxyethanol has been detected at a geometric mean air concentration of 2.6 ppm during painting operations in two Belgian shops (Veulemans et al., 1987).

PHARMACOKINETICS AND METABOLISM

Absorption

In general, about 60% of 2-ethoxyethanol vapor inhaled by human subjects are absorbed. Groeseneken et al. (1986a) found that in 10 resting men exposed for 4 h to 2-ethoxyethanol vapor at 10, 20, or 40 mg/m^3 (2.7, 5.4, or 11 ppm), the exhaled concentration was 59-65% lower than the inspired concentration starting at 10 min into the exposure and remained constant throughout the 4-h exposure. Groeseneken et al. also found that the degree of respiratory absorption of 2-ethoxyethanol was not substantially affected by light exercise in the subjects.

Metabolism and Excretion

In human volunteers, 2-ethoxyethanol was oxidized to ethoxyacetic acid (Groeseneken et al., 1986b). Of the 2-ethoxyethanol absorbed by the human respiratory tract during a 4-h exposure at 2.7, 5.4, or 11 ppm, 7-8% or 22-24% were excreted as ethoxyacetic acid within 12 or 48 h, respectively (Groeseneken et al., 1986b). After the termination of the 4-h inhalation exposure in men, the exhaled concentration of 2-ethoxyethanol declined with time bi-exponentially in 4 h (Groeseneken et al., 1986a). The first phase happened very rapidly, since the exhaled 2-ethoxyethanol concentration declined more than 98% in 7.5 min after exposure (Groeseneken et al., 1986a). The half-life of the second phase was calculated to be 102 min. Integration of the exhaled concentration curve with time showed that less than 0.4% of the amount of 2-ethoxyethanol absorbed by the human respiratory tract was exhaled unchanged 4 h after exposure (Groeseneken et al., 1986a). From these data in humans, the exhalation rate of 2-ethoxyethanol after exposure is calculated to be 0.1% of the dose per hour, and the elimination rate of 2-ethoxyethanol as ethoxyacetic acid in urine is 0.6% of the dose per hour. Therefore, it can be concluded that exhalation is less important than metabolism for 2-ethoxyethanol elimination in humans.

The formation of ethoxyacetic acid is important because it was postulated by Foster et al. (1983) to be the active metabolite of 2-ethoxyethanol. Foster et al. found that an equimolar oral dose of ethoxyacetic acid was as toxic as 2-ethoxyethanol to the testis of rats.

There are some species differences in the way ethoxyacetic acid is excreted in the urine. Groeseneken et al. (1986b) discovered that in resting men exposed for 4 h to 2-ethoxyethanol vapor at 10, 20, or 40 mg/m^3 (equivalent to a dose of 0.25, 0.5, or 1.0 mg/kg), ethoxyacetic acid was excreted in the urine entirely in its free form. In rats, 2-ethoxyethanol is also oxidized to ethoxyacetic acid, but, unlike in humans, a part of this metabolite might be first conjugated with glycine before being excreted in the urine (Groeseneken et al., 1986b). Groeseneken et al. (1986b) showed that there were significant amounts of glycine-conjugated ethoxyacetic acid, in addition to the free form, in the urine of rats after an oral dose of 2-ethoxyethanol at 0.5-100 mg/kg. However, Medinsky et al. (1990) failed to find the glycine conjugate of ethoxyacetic acid in the urine of rats exposed to 2-ethoxyethanol in drinking water for 24 h at a dose of 94, 210, or 1216 µg/kg (equivalent to 8.5, 19, or 110 mg/kg). The elimination half-life of ethoxyacetic acid in the urine was 42.0 h in humans and 7.2 h in rats (Groeseneken et al., 1986b).

Ethoxyacetic acid is not the only metabolite of 2-ethoxyethanol in rats. Medinsky et al. (1990) reported that in rats given 2-ethoxyethanol at 8.5, 19, or 110 mg/kg in drinking water for 24 h, 2-ethoxyethanol was metabolized by two competing pathways to either ethoxyacetic acid or ethylene glycol. Parts of the ethoxyacetic acid and ethylene glycol formed were further metabolized to carbon dioxide. Of the absorbed 2-ethoxyethanol, about 18% was excreted as ethylene glycol regardless of the dose. However, the elimination of 2-ethoxyethanol as ethoxyacetic acid in the urine and as exhaled CO_2 depended on the dose. The fractions of 2-ethoxyethanol excreted as CO_2 decreased with the doses: 27%, 22%, and 9% in those given 8.5, 19, or 110 mg/kg, respectively

(Medinsky et al., 1990). In contrast, the fractions of 2-ethoxyethanol excreted as ethoxyacetic acid, the putative active metabolite of 2-ethoxyethanol for testicular toxicity, increased with the doses: 26%, 30%, and 37% in those given 8.5, 19, or 110 mg/kg, respectively (Medinsky et al., 1990). Assuming that the 11-12-week-old male rats used in the study of Medinsky et al. weighed about 240 g, the ventilation of these rats is estimated to be 177 mL/min, according to the data of Leong et al. (1964). It can be calculated that the oral doses of 8.5, 19, or 110 mg/kg used by Medinsky et al. are equivalent to a 6-h inhalation exposure of the 240-g rats to 2-ethoxyethanol at 8.6, 19, or 112 ppm. This calculation allows us to determine if an inhalation study was conducted with high concentrations that overly favored the metabolism of 2-ethoxyethanol to ethoxyacetic acid, the putative active metabolite for testicular toxicity.

A question of practical importance in evaluating the pharmacokinetic and metabolic data is whether 2-ethoxyethanol accumulates in the body during repetitive inhalation exposures. There are no pharmacokinetic and metabolism data on repetitive exposures to 2-ethoxyethanol. However, the data from acute exposures seem to indicate that 2-ethoxyethanol is not likely to accumulate during repetitive exposures. As mentioned above, the exhaled concentration of 2-ethoxyethanol declines in a bi-exponential fashion after an acute inhalation exposure in humans, with the half-lives of 7.5 and 102 min, respectively (Groeseneken et al., 1986a). Since the exhaled concentration correlates with the blood concentration of organic compounds, such as trichloroethylene, benzene, and toluene (Stewart et al., 1962; Sato et al., 1974), the exhaled concentration data on 2-ethoxyethanol suggest that the blood concentration of 2-ethoxyethanol declines with time rather rapidly after an inhalation exposure in humans. In other words, there is little accumulation of 2-ethoxyethanol. The data from the rat study of Medinsky et al. (1990) tend to support this conclusion. Comparing the amount of 2-ethoxyethanol absorbed with the sum of the amounts of it and its metabolites excreted in 48 h after the oral exposure, it appears that the amount absorbed was almost totally eliminated from the body in 48 h (Medinsky et al., 1990).

Finally, it should be noted that, in addition to inhalation, cutaneous absorption is another possible exposure route for 2-ethoxyethanol. There are data showing that 2-ethoxyethanol is absorbed upon dermal application in rats (Sabourin et al., 1990).

TOXICITY SUMMARY

Acute and Short-Term Toxicity

An acute exposure to 2-ethoxyethanol has been known to produce mucosal irritation, lung congestion, lung edema, and death. Inhalation exposure of human subjects to 2-ethoxyethanol at 6000 ppm for a few seconds resulted in moderate eye irritation and a very disagreeable odor, causing a desire to avoid similar exposures (Waite et al., 1930).

A 24-h exposure at 1000 ppm killed one out of six guinea pigs, and it caused acute lung congestion and edema, as well as kidney congestion, but a 16-h exposure at 500 ppm failed to produce any gross pathology or death (Waite et al., 1930). Based on the mortality of guinea pigs exposed for 24 h, the LC_{50} was estimated to be about 1400 ppm. In a study by Werner et al. (1943a), the 7-h LC_{50} of 2-ethoxyethanol in mice was 1830 ppm, which resulted in dyspnea, weakness, moderate to marked follicular phagocytosis in the spleen and marked congestion of the cavernous veins of the spleen. Carpenter et al. (1956) showed that a 4-h exposure at 4000 ppm or an 8-h exposure at 2000 ppm killed one-half of the rats.

Foster et al. (1983) showed that oral administration of 2-ethoxyethanol to rats at concentrations of 250-1000 mg/kg failed to change the weight of the testis, prostate, and seminal vesicles, and it also did not cause any testicular injury in 24 h. However, because 2-ethoxyethanol is known to be toxic to the male reproductive system in subchronic exposure, the finding of Foster et al. cannot rule out the possibility that 2-ethoxyethanol is toxic to the testis after acute exposures. It might take more than 24 h for the testicular toxicity of 2-ethoxyethanol to appear.

Subchronic and Chronic Toxicity

Testicular Toxicity

Subchronic exposures to 2-ethoxyethanol is known to affect the male reproductive system. In a cross-sectional study, Ratcliffe et al. (1987) showed that 80 workers exposed to 2-ethoxyethanol in casting operations had a lower average sperm count per ejaculation than the unexposed controls. The concentration of 2-ethoxyethanol in the breathing zone during a full work shift ranged from not detectable to 23.8 ppm (Ratcliffe et al., 1987).

Welch et al. (1988) conducted a cross-sectional study of 94 shipyard painters exposed to 2-ethoxyethanol and 2-methoxyethanol and 55 unexposed workers. They showed an increased incidence of oligospermia, which was defined as less than 100 million sperm per ejaculate, in the exposed painters who did not smoke compared with the unexposed nonsmokers (36% of 33 exposed nonsmokers had oligospermia versus 16% of 32 unexposed nonsmokers). There was no statistically significant difference in the average sperm count, however, between the exposed and unexposed smokers. There were also no statistically significant differences in sperm morphology, sperm motility, and testicular size. Due to no differences in the levels of luteinizing hormone, follicle-stimulating hormone, and testosterone in the semen of the exposed and the control groups, Welch et al. concluded that the oligospermia was probably not due to any perturbation of the hypothalamic-pituitary axis.

The painters had been employed at that site for an average of 8 y (a standard deviation of 7 y and a range of 0.5 to 33 y) (Welch et al., 1988). The 8-h time-weighted-average (TWA) concentrations of 2-ethoxyethanol and 2-methoxyethanol, measured during the study, were 2.6 ppm (standard deviation (SD), 4.2 ppm) and 0.8 ppm (SD, 1.0 ppm), respectively. Welch et al. admitted that the measured exposure concentrations might underestimate the concentrations of 2-ethoxyethanol in previous exposures. The exposure concentrations in the Welch et al. study were measured with personal samplers, representing concentrations in the breathing zone of the painters (Sparer et al., 1988). The painters wore respirators less than 25% of the time. There were minimal cutaneous exposures of the painters to 2-ethoxyethanol and 2-methoxyethanol, since 94% of the painters had no or only a little paint on them (Sparer et al., 1988). Because these painters had been rotated from job to job involving exposures to either or both of the ethanol ethers, it was impossible to separate the effects of the two in this study.

In the shipyard study of Welch et al. (1988), the painters potentially were exposed to many chemicals: ammonia, carbon black, coal-tar-pitch volatiles, epichlorohydrin, ethyl silicate, formic acid, phosphoric acid, silica, toluene diisocyanate, 38 organic solvents, and 14 metals. Among these chemicals, only lead and epichlorohydrin are known to affect the male reproductive system in humans, but actual sampling and analyses demonstrated that the painters were not exposed to these two reproductive toxicants to any significant extent (Welch et al., 1988). It thus appears that the male reproductive effects detected in the shipyard study might be related to exposure to 2-ethoxyethanol, 2-methoxyethanol, or both (Welch et al., 1988).

Both chemicals have been shown to be reproductive toxicants in male animals. Seminiferous tubule degeneration was detected in rabbits but not in rats exposed to 2-ethoxyethanol at 400 ppm for 6 h/d, 5 d/w, for 13 w (Barbee et al., 1984). Rats exposed to 2-methoxyethanol at 300 ppm for 6 h/d for 3 d showed degenerative changes in spermatocytes of pachytene and meiotic division at spermatogenic stage XIV (Lee and Kinney, 1989). By comparing the results of exposures to 2-ethoxyethanol at 400 ppm and exposures to 2-methoxyethanol at 300 ppm, it appears that, in the rat, 2-ethoxyethanol is less toxic to the testes than 2-methoxyethanol (Barbee et al., 1984; Lee and Kinney, 1989). Such a conclusion was confirmed by Foster et al. (1983) who showed that oral administration of 2-ethoxyethanol to rats at 250 mg/kg/d for 11 d did not affect the testes, and it took an oral dose of 500 mg/kg/d to produce spermatocyte degeneration of similar severity to that seen with administration of 2-methoxyethanol at 100 mg/kg/d.

The adverse effects of 2-ethoxyethanol on the testis do not appear to be permanent. Oudiz et al. (1984) showed that five daily oral intubations of 936, 1872, or 2808 mg/d in rats resulted in sperm-count decreases (starting in the second week after exposure for the highest dose group), with most of the rats exposed to the two highest doses becoming azoospermic by week 7 after exposure. However, partial or complete recovery was seen in the sperm counts and sperm morphology by week 14 and in histological assessment of the testis and epididymis by week 16 (Oudiz et al., 1984). Among the spermatogenesis stages, Oudiz et al. showed that in rats the early-spermatid–late-spermatocyte stages were most sensitive to 2-ethoxyethanol. Foster et al. (1983) showed that in rats, daily oral administration of 2-ethoxyethanol at 1000 mg/kg/d for 7 or 11 d led to degeneration of only the late primary spermatocytes and secondary spermatocytes, with other testicular cell types unaffected. Finally, it should be noted that the reproductive toxicity of 2-ethoxyethanol is confined to males. 2-Ethoxyethanol given in drinking water at 0.5-2% resulted in testicular atrophy and decreased sperm motility in male CD-1 mice, but no reproductive abnormalities were detected in female mice (Lamb et al., 1984).

Hematological Toxicity

Other than testicular toxicity, subchronic or chronic exposures to 2-ethoxyethanol also could affect the blood cells. In the cross-sectional study conducted by Welch et al. (1988), anemia, which was defined as a condition with the blood hemoglobin concentration below 14.0 g/dL, was detected in 9 of 94 exposed shipyard male painters, and anemia was not detected in 55 unexposed subjects in the control group (p concentrations in the shipyard painters with anemia ranged from 12.3 to 13.5 g/dL). The shipyard painters with anemia, however, had normal values for mean corpuscular volume, mean corpuscular hemoglobin, and mean corpuscular hemoglobin concentration (Welch and Cullen, 1988). In addition, the mean blood hemoglobin concentrations in the exposed painters and unexposed controls did not differ statistically (Welch and Cullen, 1988). Therefore, the anemic condition detected in the shipyard painters was relatively mild. Among the chemicals the painters were potentially exposed to, benzene and lead can produce hematological effects, but there were no significant exposures of the shipyard painters to benzene or lead, based on actual sampling and analyses (Welch and Cullen, 1988). The anemia in the painters, therefore, might be related to occupational exposures to 2-ethoxyethanol, 2-methoxyethanol, or both, because both chemicals have been shown to produce hematological abnormalities in laboratory animals. An exposure to 2-ethoxyethanol at 400 ppm for 6 h/d, 5 d/w, for 13 w reduced the leukocyte count in female rats, and it reduced the erythrocyte count in rabbits (Barbee et al., 1984). Exposure of rats to 2-methoxyethanol at 300 ppm for 6 h/d for 9 d resulted in pancytopenia and bone-marrow hypoplasia (Miller et al., 1981).

A study of lithographers by Cullen et al. (1983) also showed that occupational exposures to 2-ethoxyethanol might be related to bone-marrow injury. Myeloid hypoplasia with or without stromal injury was observed in the bone marrow of six of seven workers who had worked for 1-6 y with a five-color press, which used solutions containing, among other chemicals, 2-ethoxyethanol. However, blood counts were normal for these workers. They were exposed, via inhalation and direct skin contacts, to 2-ethoxyethanol, dipropylene glycol monomethyl ether, insoluble pigments, acrylic and epoxy resins, C_4 and C_3 substituted benzene, dichloromethane, 1,2-dichloroethylene, dichloroethane,

1,1,1-trichloroethane, 2-butanone, glycerol triacetate, and *n*-propanol (Cullen et al., 1983). The airborne concentration of dipropylene glycol monomethyl ether ranged from 0.6 to 6.4 ppm, but the exposure concentration of 2-ethoxyethanol was not measured (Cullen et al., 1983).

There is also evidence of the bone-marrow toxicity of 2-ethoxyethanol in laboratory animals. An exposure of rats at a concentration of 370 ppm for 7 h/d, 5 d/w, for 5 w resulted in fat replacement of cells in the bone marrow and a decrease in myeloid cells in the spleen (Werner et al., 1943b). There was also increased hemosiderosis in the spleen, indicating increased red-blood-cell (RBC) destruction, but the 2-ethoxyethanol exposure had no effects on RBC and reticulocyte counts, hemoglobin concentration, total white-blood-cell counts, and the differential counts of granulocytes, lymphocytes, and monocytes (Werner et al., 1943b).

Carcinogenicity

No results of carcinogenicity testing with 2-ethoxyethanol were found in the literature.

Genotoxicity

2-Ethoxyethanol did not cause mutation in four strains of *Salmonella typhimurium* (NIOSH, 1983).

Developmental Toxicity

2-Ethoxyethanol is known to affect the embryo and fetus in laboratory animals. Exposure of pregnant rats at a concentration of 10 ppm for 6 h/d on days 6-15 of gestation led to an increased incidence of limb malrotation in the fetuses, and a similar exposure at 50 ppm reduced the litter size (Doe, 1984). No adverse effects were detected in the mothers in the 10-ppm and 50-ppm groups. An intrauterine exposure of rats at 250 ppm, however, produced late intrauterine death; reduction of the mean fetal weight; increased incidence of minor skeletal defects, including partial or nonossification of the skull, the thoracic centra, the lumbar centra, the lumbar vertebrae, and sternebrae; and increased incidence of 27 presacral vertebrae and sternebrae abnormalities (Doe, 1984). Exposure of the rats at 250 ppm reduced the mean corpuscular volume, hematocrit, and hemoglobin level in the mothers. Exposure of pregnant rabbits to 2-ethoxyethanol at 175 ppm for 6 h/d on days 6-18 of gestation also resulted in an increased incidence of minor skeletal defects in the fetuses and no maternal toxicity (Doe, 1984). Exposure of pregnant rats to 2-ethoxyethanol at 100 ppm for 7 h/d on days 7-13 of gestation resulted in the following conditions in newborns: decreased rotorod performance; increased acetylcholine, dopamine, and norepinephrine in the cerebrum; increased acetylcholine, norepinephrine, and protein in midbrain; increased acetylcholine in the cerebellum; and increased norepinephrine in the brainstem (Nelson et al., 1984). Because SMACs generally are not set on the basis of developmental toxicity, these data on the toxicity of 2-ethoxyethanol in the first trimester are not used here.

Interaction with Other Chemicals

No data on the interaction of 2-ethoxyethanol with other chemicals have been found in the TOXLINE or MEDLINE data bases of the National Library of Medicine (Bethesda, Md.).

TABLE 4-1 Toxicity Summary

Concentration, ppm	Exposure Duration	Species	Effects	Reference
6000	A few s	Human	Moderate eye irritation and very disagreeable odor causing desire to avoid similar exposure	Waite et al., 1930
10	6 h/d on gestation d 6-15 (rat) or d 6-18 (rabbit)	Rabbit, rat	In rabbits, no adverse effects in mother or fetus; in rats, limb malrotation without maternal toxicity	Doe, 1984
25	6 h/d, 5 d/w for 13 w	Rabbit, rat	Lacrimation and mucoid nasal discharge from w 2-10; no histopathology; in rabbits, 5-11% decrease in body-weight gain; in rats, 12% reduction in spleen weight in females	Barbee et al., 1984
50	6 h/d on gestation d 6-15 (rat) or d 6-18 (rabbit)	Rabbit, rat	In rabbits, no adverse effects in mother or fetus; in rats, reduced litter size without maternal toxicity	Doe, 1984
100	7 h/d on gestation d 7-13	Rat	In behavioral tests of pups, decreased rotorod performance at age d 21-29; increased latency of leaving the central area of an open field at age 16-60 d; no effects on activity wheel running, avoidance conditioning, and operant conditioning; in neurochemical tests of pups at age of 21 d, increased acetylcholine, dopamine, and norepinephrine in cerebrum; increased acetylcholine, norepinephrine, and protein in midbrain; increased acetylcholine in cerebellum; increased norepinephrine in brainstem	Nelson et al., 1984
100	7 h/d on gestation d 14-20	Rat	In behavioral tests of pups, decreased activity wheel running at age 32-33 d; in neurochemical tests of pups at age 21 d, increased acetylcholine, serotonin, and dopamine in cerebrum	Nelson et al., 1984

TABLE 4-1 (Continued)

Concentration, ppm	Exposure Duration	Species	Effects	Reference
100	6 h/d, 5 d/w for 13 w	Rabbit, rat	Lacrimation and mucoid nasal discharge from w 2-10; no histopathology; no changes in body, testicular, and pituitary weight; no changes in leukocyte counts, RBC counts, hematocrit, and hemoglobin concentration; in rats, 12% decrease in spleen weight in females, but no change in the spleen/body-weight ratio	Barbee et al., 1984
175	6 h/d on d 6-18 of gestation	Rabbit	Increased incidence of minor skeletal variants (primarily extra ribs) with no maternal toxicity	Doe, 1984
250	6 h/d on d 6-15 of gestation	Rat	Late intrauterine death, reduced fetal weight, increased incidence of minor skeletal variants (27 presacral vertebrae, sternebrae abnormalities, partial or nonossification of the skull, thoracic centra, lumbar centra, lumbar vertebrae, and sternebrae), and dilatation of renal pelvis; reduced mean corpuscular volume, hematocrit, and hemoglobin level in the mother	Doe, 1984
370	7 h/d, 5 d/w for 5 w	Rat	Hemosiderosis in the spleen and increased proportions of immature leukocytes in blood from w 2 of exposure to 3 w after exposure; decrease in myeloid cells in the spleen and fat replacement of cells in the bone marrow 3 w after exposure; no effects on RBC and reticulocyte counts, hemoglobin concentration, total white-blood-cell, lymphocyte, granulocyte, and monocyte counts, body-weight gain; no deaths during exposure and 3 w after exposure	Werner et al., 1943b

400	6 h/d, 5 d/w for 13 w	Rabbit, rat	Lacrimation and mucoid nasal discharge from w 2-10; in rabbits, 8% decrease in body-weight gain, lower testis weight, decreases in hemoglobin, hematocrit, and erythrocyte count; elevated serum globulin levels in males; seminiferous tubule degeneration; in rats, 14% decrease in pituitary weight in males and 15% decrease in spleen weight in females; lower leukocyte count, blood urea nitrogen, and serum cholesterol levels in females; no histopathology in either sex	Barbee et al., 1984
500	16 h	Guinea pig	No deaths, weakness, dyspnea, inactivity, or gross pathology	Waite et al., 1930
500	24 h	Guinea pig	No deaths, weakness, dyspnea, or inactivity; slight gross pathology	Waite et al., 1930
1830	7 h	Mouse	Half died in 1 w; marked dyspnea and weakness; moderate-to-marked follicular phagocytosis and marked congestion of the cavernous veins of the spleen; no evidence of kidney, liver, and heart injury	Werner et al., 1943a
1000	24 h	Guinea pig	1/6 died 2 d after exposure; lung congestion and edema; kidney congestion	Waite et al., 1930
2000	4 h	Rat	0/6 died	Carpenter et al., 1956
2000	8 h	Rat	3/6 died	Carpenter et al., 1956
3000	4 h	Guinea pig	No deaths, weakness, dyspnea, inactivity, or gross pathology	Waite et al., 1930
3000	8 h	Guinea pig	No deaths, weakness, dyspnea, or inactivity; parenchymatous changes in the kidney 4 d after exposure; no gross pathology 8 d after exposure	Waite et al., 1930

TABLE 4-1 (Continued)

Concentration, ppm	Exposure Duration	Species	Effects	Reference
3000	24 h	Guinea pig	5/6 died during exposure; inactivity, weakness, and dyspnea; 1 died 3 d after exposure; lung congestion and edema; hyperemia of the kidneys	Waite et al., 1930
4000	4 h	Rat	3/6 died	Carpenter et al., 1956
4000	8 h	Rat	6/6 died	Carpenter et al., 1956
5000	4 h	Rat	1/12 died	Carpenter et al., 1956
5000	8 h	Rat	10/12 died; hemoglobinuria	Carpenter et al., 1956
6000	1 h	Guinea pig	No deaths, weakness, dyspnea, inactivity, or gross pathology	Waite et al., 1930
6000	4 h	Guinea pig	No deaths, weakness, dyspnea, or inactivity; parenchymatous change in the kidney 4 d after exposure; no gross pathology 8 d after exposure	Waite et al., 1930
6000	24 h	Guinea pig	4/6 died during exposure; 1 died 3 h after exposure; last one was necropsied immediately after exposure; lung congestion and edema; kidney congestion; petechial hemorrhages of stomach mucosa; inactivity, weakness, and dyspnea during exposure	Waite et al., 1930

TABLE 4-2 Exposure Limits Set by Other Organizations

Organization	Concentration, ppm
ACGIH's TLV	5 (TWA)
OSHA's PEL	200 (TWA)
NIOSH's IDLH	6000

TLV = threshold limit value. TWA = time-weighted average. PEL = permissible exposure limit. IDLH = immediately dangerous to life and health.

TABLE 4-3 Spacecraft Maximum Allowable Concentrations

Duration	ppm	mg/m^3	Target Toxicity
1 h	10	40	Hematological toxicity
24 h	10	40	Hematological toxicity
7 d[a]	0.8	3	Hematological toxicity
30 d	0.5	2	Hematological toxicity
180 d	0.07	0.3	Hematological toxicity

[a]There is no official 7-d SMAC. The current official 7-d SMAC for 2-ethoxyethyl acetate is 30 ppm.

RATIONALE FOR ACCEPTABLE CONCENTRATIONS

Although 2-ethoxyethanol is known to cause fetal toxicity and minor teratogenic changes in rats and rabbits (Doe, 1984), SMACs for 2-ethoxyethanol are not established according to its developmental toxicity because pregnant astronauts will not be allowed in space. The major targets of subchronic toxicity of 2-ethoxyethanol are the testes and the hematological system, but very little is known about its acute toxicity.

Mucosal Irritation

Moderate eye irritation was noted by Waite et al. (1930) in human subjects exposed for a few seconds to 2-ethoxyethanol at a concentration of 6000 ppm. However, these data are not relied on in setting the short-term SMACs because the exposure lasted for only several seconds. In a 13-w study conducted by Barbee et al. (1984), increased incidences of lacrimation and mucoid nasal discharge were observed in rabbits and rats after 2-10 w of exposure to 2-ethoxyethanol at 25, 100, or 400 ppm, as compared with the controls. Barbee et al. reported that the lacrimation and nasal discharge failed to show a concentration-response relationship. Nevertheless, the data indicate that 2-ethoxyethanol vapor at 25 ppm could be irritating to the mucous membranes in animals. Because Groeseneken et al. (1986c) did not report any mucosal irritation in 10 human volunteers exposed to 2-ethoxyethanol at up to 10 ppm for 4 h, the nonirritating concentration appears to be 10 ppm in humans. Because the astronauts are expected to be able to tolerate mild mucosal irritation during 1-h or 24-h contingencies, the acceptable concentration (AC), based on irritation, for 1 and 24 h is 25 ppm.

1-h and 24-h ACs based on irritation
= 25 ppm.

7-d, 30-d, and 180-d ACs based on irritation
= 4-h NOAEL × 1/safety factor for small n
= 10 ppm × (square root of n)/10
= 10 ppm × (square root of 10)/10
= 3 ppm.

The 7-d, 30-d, and 180-d ACs are set at the same concentration because mucosal irritation is not expected to increase with time after the first hour of exposure.

Testicular Toxicity

Although Welch and co-workers (1988) showed that occupational exposures to 2-ethoxyethanol at about 2.6 ppm could result in lower sperm count and mild anemia, the SMACs for testicular toxicity should not be set relying solely on the data of Welch and co-workers for three reasons. First, the exact exposure concentrations of 2-ethoxyethanol were not known in that study. The painters were exposed, several years before the study began, to 2-ethoxyethanol at concentrations potentially higher than 2.6 ppm (Sparer et al., 1988; Welch and Cullen, 1988; Welch et al., 1988). Second, in addition to 2-ethoxyethanol, the painters were exposed to 2-methoxyethanol at a concentration of 0.8 ppm, which is also known to produce testicular and hematological toxicity (Miller et al., 1981; Welch and Cullen, 1988; Welch et al., 1988; Lee and Kinney, 1989). Third, the painters were potentially exposed to 59 chemicals other than 2-ethoxyethanol and 2-methoxyethanol (Welch and Cullen, 1988; Welch et al., 1988). Although the investigators did determine the airborne concentrations of the testicular or hematological toxicants among those 59 chemicals to reasonably rule them out, it is possible that some of the chemicals can potentiate the toxicity of 2-ethoxyethanol. For these reasons, the SMACs are not derived entirely from the data on the shipyard painters. Rather, the painter data are used only to check the validity of setting SMACs on the basis of animal data.

The animal data of Barbee et al. (1984) appeared to be suitable for the derivation of SMACs for 2-ethoxyethanol. In their 13-w study, Barbee et al. showed that rabbits were more sensitive than rats to the testicular and hematological toxicity of 2-ethoxyethanol; therefore, the SMACs are established using the rabbit data. A 13-w exposure at 400 ppm for 6 h/d, 5 d/w resulted in testicular damage and decreases in testicular weight in rabbits but not in rats. A similar exposure at 100 ppm failed to cause any changes in the testis in rabbits; thus, the subchronic toxicity study of Barbee et al. (1984) indicates that 100 ppm is the 13-w no-observed-adverse-effect level (NOAEL) in animals.

Applying the traditional interspecies extrapolation factor of 10 to the 13-w NOAEL of 100 ppm, it appears that a 13-w occupational exposure to 2-ethoxyethanol at 10 ppm should be devoid of testicular toxicity in humans. However, compared with the study of Welch et al. (1988) in which an 8-y occupational exposure to a mixture of 2-ethoxyethanol (TWA, 2.6 ppm; SD, 4.2 ppm) and 2-methoxyethanol (TWA, 0.8 ppm; SD, 1.0 ppm) could result in lower sperm counts, the 13-w exposure concentration at 10 ppm, predicted to be safe using the 13-w NOAEL of 100 ppm, does not appear to have any margin of safety. It appears that humans are even more sensitive than rabbits to the testicular toxicity of 2-ethoxyethanol. Although it can be contended that the painter study of Welch et al. had flaws enumerated above, prudence dictates that a 13-w NOAEL lower than 100 ppm should be used to derive the

SMACs. Consequently, the lower 13-w NOAEL of 25 ppm was chosen (Barbee et al., 1984). Even though 25 ppm lowered the body weights of rabbits after a 13-w exposure, 25 ppm is still considered a NOAEL because the higher concentration of 100 ppm did not lower the body weights of rabbits in the same study (Barbee et al., 1984).

7-d AC for testicular toxicity
= 13-w NOAEL × 1/species factor
= 25 ppm × 1/10
= 2.5 ppm.

Haber's rule is used to derive the 30-d AC on the basis of testicular toxicity because it is uncertain whether the testicular injury is reparable. For prudence sake, the testicular injury caused by 2-ethoxyethanol is assumed to be irreparable.

30-d AC = 13-w NOAEL × time adjustment × 1/species factor
= 25 ppm × (6 h/d × 5 d/w × 13 w)/(24 h/d × 30 d) × 1/10
= 25 ppm × 390 h/720 h × 1/10
= 25 ppm × 0.54 × 1/10
= 1.4 ppm.

Because the testicular toxicity in a 180-d exposure is not expected to differ significantly from that in a 30-d exposure, the 180-d AC is set equal to the 30-d AC. Because an oral administration of 2-ethoxyethanol at a dose of 1000 mg/kg failed to cause any male reproductive toxicity in rats (Foster et al., 1983), testicular toxicity is not a consideration in setting the short-term SMACs.

Hematological Toxicity

A subchronic exposure of rabbits to 2-ethoxyethanol at 400 ppm for 6 h/d, 5 d/w, for 13 w reduced the RBC count (Barbee et al., 1984). No change in the count was observed in rats. In the female rats, there were lower leukocyte counts and a 15% decrease in the spleen weight. A similar exposure at 100 or 25 ppm failed to produce any hematological changes in either rabbits or rats. However, a 12% decrease in spleen weight was detected in female rats after the 13-w exposure at 100 or 25 ppm (Barbee et al., 1984). Because there were no hematological and histological changes in female rats exposed to 2-ethoxyethanol at 100 or 25 ppm, the toxicological meaning of the decreases in spleen weight is uncertain. The meaning of the decreases in spleen weight in female rats exposed at 100 ppm is further obscured by the fact that the spleen/body-weight ratio showed no change in that group (Barbee et al., 1984). The spleen/body-weight ratio was lowered only in female rats exposed at 400 or 25 ppm, and body weight showed no statistically significant changes in female rats exposed at 400, 100, or 25 ppm.

All in all, the data of Barbee et al. (1984) indicate that the NOAEL for a 13-w exposure to 2-ethoxyethanol is 100 ppm for hematological toxicity in rabbits and rats. Similar to the analysis of the NOAEL based on testicular toxicity, the comparison of the subchronic NOAEL of 100 ppm from the animal study with the finding of Welch and co-workers (1988) in human workers indicates that a NOAEL lower than 100 ppm should be used. Welch and co-workers found that workers exposed to a mixture of 2-ethoxyethanol (TWA, 2.6 ppm; S.D., 4.2 ppm) and 2-methoxyethanol (TWA, 0.8 ppm; S.D., 1.0 ppm) developed mild anemia. As a result, the lower NOAEL of 25 ppm from the 13-w study is used in setting the SMACs (Barbee et al., 1984).

Microgravity is known to reduce the RBC mass in astronauts (Huntoon et al., 1989). Consequently, a safety factor of 3 is used to account for potential interaction of 2-ethoxyethanol and microgravity on causing anemia.

7-d AC based on hematological toxicity
= 13-w NOAEL × 1/species factor × 1/microgravity factor
= 25 ppm × 1/10 × 1/3
= 10 ppm.

30-d AC based on hematological toxicity
= 13-w NOAEL × 1/species factor × 1/microgravity factor × time adjustment
= 25 ppm × 1/10 × 1/3 × (6 h/d × 5 d/w × 13 w)/(24 h/d × 30 d)
= 0.5 ppm.

180-d AC based on hematological toxicity
= 13-w NOAEL × 1/species factor × 1/microgravity factor × time adjustment
= 25 ppm × 1/10 × 1/3 × (6 h/d × 5 d/w × 13 w)/(24 h/d × 180 d)
= 0.07 ppm.

Werner et al. (1943b) found that repetitive exposures of rats to 370 ppm for 7 h/d, 5 d/w, for 1 w did not cause any changes in the RBC count and hemoglobin concentration in blood. The 1-w NOAEL of 370 ppm is used to derive the 1-h and 24-h ACs.

1-h and 24-h ACs based on hematological toxicity
= 1-w NOAEL × 1/species factor × 1/microgravity factor
= 370 ppm × 1/10 × 1/3
= 10 ppm.

Establishment of SMAC Values

The ACs for all toxic end points are tabulated below. By choosing the lowest AC for each exposure duration, the 1-h, 24-h, 7-d, 30-d, and 180-d SMACs are set at 10, 10, 0.8, 0.5, and 0.07 ppm, respectively.

TABLE 4-4 End Points and Acceptable Concentrations

End Point	Exposure Data	Species and Reference	Time	Species	Small n	1 h	24 h	7 d	30 d	180 d
Mucosal irritation	LOAEL at 25 ppm, 6 h/d, 5 d/w, 2-10 w	Rabbit and rat (Barbee et al., 1984)	—	—	—	25	25	—[a]	—	—
	NOAEL at 10 ppm, 4 h	Human (n = 10) (Groeseneken et al., 1986c)	—	—	10/(sq. rt. of 10)	—	—	3	3	3
Testicular toxicity	NOAEL at 25 ppm for 6 h/d, 5 d/w, 13 w	Rabbit (Barbee et al., 1984)	—	10	—	—	—	2.5	—	—
	NOAEL at 25 ppm for 6 h/d, 5 d/w, 13 w	Rabbit (Barbee et al., 1984)	HR[b]	10	—	—	—	—	1.4	1.4
Hematotoxicity	NOAEL at 370 ppm for 7 h/d, 5 d/w, 1 w	Rat (Werner et al., 1943)	—	10	—	10	10	—	—	—
	NOAEL at 25 ppm for 6 h/d, 5 d/w, 13 w	Rabitt and rat (Barbee et al., 1984)	—	10	—	—	—	0.8	—	—
	NOAEL at 25 ppm for 6 h/d, 5 d/w, 13 w	Rabbit and rat (Barbee et al., 1984)	HR	10	—	—	—	—	0.5	0.07
SMAC						10	10	0.8	0.5	0.07

[a]Extrapolation to these exposure durations produces unacceptable uncertainty in the values.
[b]HR = Haber's rule.

REFERENCES

ACGIH. 1986. P. 237 in Documentation of TLV's and BEI's. Threshold Limit Values Committee, American Conference of Governmental Industrial Hygienists, Cincinnati, Ohio.

Barbee, S.J., J.B. Terrill, D.J. DeSousa, and C.C. Conaway. 1984. Subchronic inhalation toxicology of ethylene glycol monoethyl ether in the rat and rabbit. Environ. Health Perspect. 57:157-163.

Carpenter, C.P., U.C. Pozzani, C.S. Weil, J.H. Nair, G.A. Keck, and H.F. Smyth. 1956. The toxicity of butyl Cellosolve solvent. Arch. Ind. Health 14:114-131.

Cullen, M.R., T. Rado, J.A. Waldron, J. Sparer, and L.S. Welch. 1983. Bone marrow injury in lithographers exposed to glycol ethers and organic solvents used in multicolor offset and ultraviolet curing printing processes. Arch. Environ. Health 38:347-354.

DiVincenzo, G.D., F.J. Yanno, and B.D. Astill. 1972. Human and canine exposures to methylene chloride vapor. Am. Ind. Hyg. Assoc. J. 33:125-135.

Doe, J.E. 1984. Ethylene glycol monoethyl ether and ethylene glycol monoethyl ether acetate teratology studies. Environ. Health Perspect. 57:33-41.

Foster, P.M.D., D.M. Creasy, J.R. Foster, L.V. Thomas, M.W. Cook, and S.D. Gangolli. 1983. Testicular toxicity of ethylene glycol monomethyl and monoethyl ethers in the rat. Toxicol. Appl. Pharmacol. 69:385-399.

Groeseneken, D., H. Veulemans, and R. Masschelein. 1986a. Respiratory uptake and elimination of ethylene glycol monoethyl ether after experimental human exposure. Br. J. Ind. Med. 43:544-549.

Groeseneken, D., H. Veulemans, R. Masschelein, and E. Van Vlem. 1986b. Comparative urinary excretion of ethoxyacetic acid in man and rat after single low doses of ethylene glycol monoethyl ether. Toxicol. Lett. 41:57-68.

Groeseneken, D., H. Veulemans, and R. Masschelein. 1986c. Urinary excretion of ethoxyacetic acid after experimental human exposure to ethylene glycol monoethyl ether. Br. J. Ind. Med. 43:615-619.

Huntoon, C.L., P.C. Johnson, and N.M. Cintron. 1989. Hematology, immunology, endocrinology, and biochemistry. P. 222 in Space Physiology and Medicine, A.E. Nicogossian, C.L. Huntoon, and S.L. Pool, ed. Philadelphia: Lea & Febiger.

Lamb, J.C., IV, D.K. Gulati, V.S. Russell, L. Hommel, and P.S. Sabharwal. 1984. Reproductive toxicity of ethylene glycol monoethyl ether tested by continuous breeding of CD-1 mice. Environ. Health Perspect. 57:85-90.

Leban, M.I., and P.A. Wagner. 1989. Space Station Freedom Gaseous Trace Contaminant Load Model Development. SAE Technical Paper Series 891513. Warrendale, Pa.: Society of Automotive Engineers.

Lee, K.P., and L.A. Kinney. 1989. The ultrastructural and reversibility of testicular atrophy induced by ethylene glycol monomethyl ether (EGME) in the rat. Toxicol. Pathol. 17:759-773.

Leong, K.J., G.F. Dowd, H. MacFarland, and H.M. Noble. 1964. A new technique for tidal volume measurement in unanesthetized small animals. Can. J. Physiol. Pharmacol. 42:189-198.

Medinsky, M.A., G. Singh, W.E. Bechtold, J.A. Bond, P.J. Sabourin, L.S. Birnbaun, and R.F. Henderson. 1990. Disposition of three glycol ethers administered in drinking water to male F344/N rats. Toxicol. Appl. Pharmacol. 102:443-455.

Miller, R.R., J.A. Ayres, L.L. Calhoun, J.T. Young, and M.J. McKenna. 1981. Comparative short-term inhalation toxicity of ethylene glycol monomethyl ether and a propylene glycol monomethyl ether in rats and mice. Toxicol. Appl. Pharmacol. 61:368-377.

NASA. 1989-1990. Postflight reports on the toxicological analyses for STS-27 and STS-32. NASA, Johnson Space Center, Houston, Tex.

Nelson, B.K., W.S. Brightwell, J.V. Setzer, and T.L. O'Donohue. 1984. Reproductive toxicity of the industrial solvent 2-ethoxyethanol in rats and interactive effects of ethanol. Environ. Health Perspect. 57:255-259.

NIOSH. 1983. The glycol ethers, with particular reference to 2-methoxyethanol and 2-ethoxyethanol: Evidence of adverse reproductive effects. P. 7 in Current Intelligence Bulletin No. 39. DHHS (NIOSH) Publ. No. 83-112. National Institute of Occupational Safety and Health, Cincinnati, Ohio.

Oudiz, D.J., H. Zenick, R.J. Niewenhuis, and P.M. McGinnis. 1984. Male reproductive toxicity and recovery associated with acute ethoxyethanol exposure in rats. J. Toxicol. Environ. Health 13:763-775.

Ratcliffe, J.M., D.E. Clapp, S.M. Schrader, T.W. Turner, and J.

Oser. 1987. Health Hazard Evaluation Report HETA 84-415-1688. Precision Castparts Corp., Portland, Oreg. Available from NTIS, Springfield, Va., Doc. No. PB87-108320.

Sabourin, P.J., M.A. Medinsky, L.S. Birbaum, F. Thurmond, and R.F. Henderson. 1990. Dermal absorption and disposition of methoxy-, ethoxy-, and butoxyethanol. Toxicologist 10:236.

Sato, A., T. Nakajima, Y. Fujiwara, and K. Hirosawa. 1974. Pharmacokinetics of benzene and toluene. Int. Arch. Arbeitsmed. 33:169-182.

Sparer, J., L.S. Welch, K. McManus, and M.R. Cullen. 1988. Effects of exposure to ethylene glycol ethers on shipyard painters: I. Evaluation of exposure. Am. J. Ind. Med. 14:497-507.

Stewart, R.D., H.C. Dodd, and H.H. Gay. 1962. Observations on the concentration of trichloroethylene in blood and expired air following exposure of humans. Am. Ind. Hyg. Assoc. J. 23:167-170.

Veulemans, H., D. Groeseneken, R. Masschelein, and E. van Vlem. 1987. Survey of ethylene glycol ether exposures in Belgian industries and workshops. Am. Ind. Hyg. Assoc. J. 48:671-676.

Waite, C.P., F.A. Patty, and W.P. Yant. 1930. Acute response of guinea pigs to vapors of some new commercial organic compounds. Public Health Rep. 45:1459-1466.

Welch, L.S., and M.R. Cullen. 1988. Effect of exposure to ethylene glycol ethers on shipyard painters: III. Hematologic effects. Am. J. Ind. Med. 14:527-536.

Welch, L.S., S.M. Schrader, T.W. Turner, and M.R. Cullen. 1988. Effects of exposure to ethylene glycol ethers on shipyard painters: II. Male reproduction. Am. J. Ind. Med. 14:509-526.

Werner, H.W., J.L. Mitchell, W.F. von Oettinger, and J.W. Miller. 1943a. The acute toxicity of vapors of several monoalkyl ethers of ethylene glycol. J. Ind. Hyg. Toxicol. 25:157-163.

Werner, H.W., J.L. Mitchell, J.W. Miller, and W.F. von Oettinger. 1943b. Effects of repeated exposures of rats to vapors of monoalkyl ethylene glycol ethers. J. Ind. Hyg. Toxicol. 25:347-349.

B5 Hydrazine

Hector D. Garcia, Ph.D., and John T. James, Ph.D.
Johnson Space Center Toxicology Group
Biomedical Operations and Research Branch
Houston, Texas

PHYSICAL AND CHEMICAL PROPERTIES

Hydrazine is a clear, colorless, fuming, oily, hygroscopic, highly polar, flammable liquid with an ammonia-like odor or an extremely irritating gas that is readily adsorbed or condensed onto surfaces and has a high affinity for water (Sevin, 1978).

Synonym:	Diamine
Formula:	N_2H_4
CAS number:	302-01-2
Molecular weight:	32.05
Boiling point:	113.5 °C
Melting point:	1.4-1.5 °C
Liquid density at 25 °C:	1.0045
Vapor pressure:	14.1 mm Hg @ 25 °C (10.4 torr at 20 °C)
Saturated vapor concentration:	18,900 ppm (25 °C) (24,800 mg/m^3)
Solubility:	Miscible with water and methyl, ethyl, propyl, and isobutyl alcohols; insoluble in chloroform and ether
Conversion factors at 25 °C, 1 atm:	1 ppm = 1.31 mg/m^3 1 mg/m^3 = 0.76 ppm

OCCURRENCE AND USE

Hydrazine occurs naturally as a product of nitrogen fixation by *Azotobacter agile*. It has been identified in tobacco grown without the use of maleic hydrazide (Sevin, 1978). Essentially all commercially used hydrazine is chemically synthesized, usually by one of several processes involving chemical oxidation of ammonia.

Hydrazine is used commercially as a polymerization catalyst, a blowing agent, a reducing agent, and an oxygen scavenger in boiler-water treatment; it is also used commerically in the synthesis of maleic hydrazide and in the manufacture of drugs. In combination with water, it is used in the F-16 aircraft emergency power unit as a source of gas to drive a turbine. The hydrazine bases are used in the production of salts and hydrazones that are used in surfactants, detergents, plasticizers, pharmaceuticals, insecticides, and herbicides (Sevin, 1978).

Hydrazine is used as a rocket propellant and for the auxiliary power unit on the space-shuttle orbiters. Hydrazine might be used on the international space station. It is theoretically possible for a small amount (up to a few grams) to be introduced into the spacecraft atmosphere in the unlikely scenario that, during an extravehicular activity, a crew member is in the vicinity of an undetected leak leak of a subtantial amount of hydrazine, which is deposited unnoticed on the surface of the spacesuit and then is vaporized from the suit after the crew member re-enters the spacecraft. During missions in which hydrazine contamination of the airlock is a risk, a real-time chemical monitor using ion-mobility spectrometry has been flown to ensure that unsafe concentrations of propellants, including hydrazine, are not present before the spacecraft is re-entered (Eiceman et al., 1993).

PHARMACOKINETICS AND METABOLISM

Absorption

Hydrazine was detectable in the plasma of anesthetized dogs within 30 s after applying concentrated hydrazine solutions to their intact skin (Smith and Clark, 1972). No data were found in the literature on the absorption of hydrazine by inhalation or ingestion.

Distribution

Hydrazine was distributed rapidly (2 h) to most tissues in mice given single intraperitoneal (i.p.) doses of hydrazine sulfate at 1 mmole/kg of body weight and in rats given 60 mg/kg; the highest concentrations appeared in the kidneys (Dambrauskas and Cornish, 1964).

Excretion

Anesthetized dogs administered hydrazine sulfate intravenously (i.v.) at a concentration of 50 mg/kg excreted 5-11% as hydrazino nitrogen in the urine within 4 h. Unanesthetized dogs administered hydrazine sulfate i.v. at 15 mg/kg excreted 50% as hydrazino nitrogen within 2 d. Rats given subcutaneous (s.c.) injections of hydrazine at 60 mg/kg excreted it unchanged at the rate of 8% after 2 h, 20% after 20 h, and 50% after 40 h.

Metabolism

Hydrazine injected i.p. into rabbits was shown to undergo a two-step acetylation. Using ^{15}N-labeled hydrazine to study the metabolic fate of hydrazine in Sprague-Dawley rats, Springer et al. (1981) found that within 48 h, 25% was exhaled as N_2 gas, 30% appeared in the urine as hydrazine, and 20% appeared as a derivative that was acid hydrolyzable to hydrazine. The disappearance of hydrazine from the blood was biphasic, with half-times of 0.74 and 26.9 h.

TOXICITY SUMMARY

Hydrazine vapor is extremely irritating to the eyes, nose, and throat. Quantitative information on worker exposure, however, can only be estimated from the existing published data (Santodonato et al., 1985). The median concentration of hydrazine detectable by odor is 3-4 ppm (Jacobson et al., 1955). Inhalation can cause dizziness, anorexia (Mac-Ewen et al., 1974), and nausea. Hydrazine can be absorbed dermally

(Smith and Clark, 1972) or orally and can induce contact dermatitis (Evans, 1958; van Ketel, 1964; Høvding, 1967), neurological impairment (Kulagina, 1962; Reid, 1965), and kidney, lung, and liver damage at an i.p. dose of 20 mg/kg for 3 d in rodents (Scales and Timbrell, 1982). Hydrazine is fetotoxic in rats and mice at i.p. doses of 5 mg/kg and is teratogenic in mice at 12 mg/kg.

The summary of the toxic effects of hydrazine given below applies also to the salts of hydrazine, such as sulfate and hydrochloride, because they differ in toxicity from the free base only when the development of toxicity is related to differences in pH, solubility, volatility, or mass (in expression of doses) (Sevin, 1978).

Acute and Short-Term Toxicity

Lethality

The LD_{50} for hydrazine injected i.p. into rats is 80-100 mg/kg of body weight. Rats administered 200 mg/kg went into convulsions within 5 min and died within 3 h of dosing (Scales and Timbrell, 1982). A 2-h exposure to hydrazine vapors at a concentration of 1.0-2.0 mg/L (7600-15,200 ppm) has been reported to induce convulsions, respiratory arrest, and death in mice and rats (Kulagina, 1962). The toxicity of multiple lower doses was cumulative, but surviving animals recovered and lived normal life spans if exposure was discontinued. The LC_{50} for a 4-h exposure to hydrazine is 750 mg/m^3 (570 ppm) for rats and 330 mg/m^3 (250 ppm) for mice (Jacobson et al., 1955). Continuous inhalation exposure to hydrazine at 0.8 ppm was highly lethal to mice within 1-4 w (House, 1964).

Hepatotoxicity

Rats given i.p. injections of hydrazine at 20 mg/kg exhibited vacuolization of liver cells within 24 h of exposure, and those given 10 mg/kg showed no histopathology (Scales and Timbrell, 1982), but they did show increases in triglyceride levels (Timbrell et al., 1982). Those given hydrazine at 60 mg/kg exhibited ultrastructural changes in liver cells 30 min after exposure (Scales and Timbrell, 1982). Methylation of liver DNA has been shown in rats given oral doses of hydrazine at 30-90 mg/kg (Becker et al., 1981).

Subchronic and Chronic Toxicity

Lethality

Continuous inhalation exposure to hydrazine at 0.8 ppm was highly lethal to rats within 7-10 w and moderately lethal (20%) to monkeys within 3-13 w (House, 1964). Dogs, rats, mice, and guinea pigs exposed at 14 ppm for 6 h/d, 5d/w, for up to 6 mo showed varying mortalities, mice being the most sensitive species (Comstock et al., 1954). Rats that survived the exposures were found to have completely recovered with no observable pathology about 6 mo after the last day of exposure (Comstock et al., 1954).

There are a few literature reports of accidental human exposures to high, but unspecified, doses of hydrazine. In one case, a man drank between a swallow and half a cup of liquid hydrazine, which induced vomiting and unconsciousness, followed in a few hours by violent behavior and in a few days by ataxia, nystagmus, and paraesthesia (Roe, 1978). No followup report was found. Another case involving 6 mo of occupational contact exposure once a week to hydrazine by a machinist culminated in conjunctivitis, tremor, cough, fever, vomiting, diarrhea, and death (Sotaniemi et al., 1971). There were no deaths due to acute exposure to hydrazine vapor at a hydrazine factory among all 78 workers exposed for at least 6 mo at an estimated 1-10 ppm (54 of them were exposed for over 2 y) (Wald et al., 1984).

Hepatotoxicity

Fatty changes in the liver have been reported in rats, mice, dogs, and monkeys exposed by inhalation at 0.8 ppm for 90 d (Weatherby and Yard, 1955; House, 1964). Prominent fat vacuoles and pigmentation of liver Kupffer cells were reported in dogs exposed at 14 ppm, for 6 h/d, 5 d/w, for 39 w (Comstock et al., 1954). Exposure at 1 ppm for 6 h/d, 5 d/w, for 1 y produced no hepatotoxicity in mice and male rats but did cause a statistically significant increase (to 60% from an incidence of ≈40%) in focal liver-cell hyperplasia in female rats (MacEwen et al., 1981). "Fatty livers" were induced in rats given hydrazine orally (Weatherby and Yard, 1955).

Carcinogenicity

There are conflicting reports on the carcinogenicity of hydrazine. The National Institute of Environmental Health Sciences finds that there is sufficient evidence for the carcinogenicity of hydrazine in experimental animals but inadequate evidence for its carcinogenicity in humans (NTP, 1989). Inhaled hydrazine has been reported to induce alveolargenic carcinomas in three of eight mice exposed at 0.2 ppm (MacEwen et al., 1974), malignant nasal tumors in 6 of 99 male rats and 5 of 98 female rats exposed at 5 ppm (MacEwen et al., 1981), and benign nasal polyps in 16 of 160 hamsters exposed at 5 ppm (Carter et al., 1981). Unequivocally toxic doses (up to 50 mg/L), however, administered in drinking water for the lifetime of rats were only weakly carcinogenic (Steinhoff and Mohr, 1988). In male rats (the most sensitive species and sex) exposed by inhalation, tumors, which were predominantly benign, occurred only late in life in animals showing many other chronic toxic effects, including a greatly increased inflammatory response of the upper airways (Carter et al., 1981).

The epidemiological data available on hydrazine-related cancers in humans (Karstadt and Bobal, 1982) indicates no excess risk of cancer in workers occupationally exposed to hydrazine vapors, but the number of workers having substantial exposure has been too few to detect anything other than gross carcinogenic hazards (Roe, 1978; Wald et al., 1984). A NASA-sponsored followup study of the Wald cohort of occupationally exposed workers failed to show any increase in cancer-induced mortality 20 y after exposures ended (Morris et al., 1995).

Genotoxicity

Hydrazine has been reported to be mutagenic in several test systems

(Jain and Shukla, 1972; Herbold and Buselmaier, 1976; Kimball, 1977), including the *Escherichia coli* and Ames *Salmonella* (+S9) assays, the *Drosophila-melanogaster*-specific locus test, and L5178Y cultured mouse lymphoma cells. Hydrazine induces sister chromatid exchanges in vitro (MacRae and Stich, 1979). Hydrazine has been shown to saturate the 5,6 double bond and degrade the pyrimidine bases in DNA. DNA from the livers of hydrazine-exposed rats, but not control rats, has been demonstrated to have methylated guanine bases.

Reproductive Toxicity

Hydrazine induces abnormally shaped spermatocytes in treated male mice (Wyrobek and London, 1973). In the mouse dominant lethal test, hydrazine did not induce early fetal deaths or preimplantation losses at single i.p. doses of 42 mg/kg or 52 mg/kg.

Developmental Toxicity

No information was found on the effects of hydrazine exposure on developing embryos.

Interaction with Other Chemicals

No reports of toxicologically relevant interactions with other chemicals were found.

TABLE 5-1 Toxicity Summary

Concentration, ppm	Exposure Duration	Species	Effects	Reference
1-10	≥6 mo, work	Human	No fatalities in 78 workers	Wald et al., 1984
0.2	6 mo, continuous	Mice	Alveolargenic carcinomas in 3/8 exposed and 1/8 controls	MacEwen et al., 1974
0.8	90 d, continuous	Monkeys	Lethality in 2/10, 1 on d 30 and 1 on d 85	House, 1964
0.8	90 d, continuous	Monkeys	Fatty changes of the liver in 7/10 exposed and in 2/10 controls	House, 1964
0.8	90 d, continuous	Rats	48 of 49 died, beginning on d 41	House, 1964
0.8	90 d, continuous	Rats	Fatty changes of the liver in 2/20 exposed	House, 1964
0.8	90 d, continuous	Rats	Lung lesions with leukocyte infiltration in 7/20 exposed	House, 1964
0.8	90 d, continuous	Mice	Lethality in 99/100, beginning on d 6	House, 1964
0.8	90 d, continuous	Mice	Lung lesions with leukocyte infiltration in 29/41 exposed	House, 1964
1	6 h/d, 5 d/w, 1 mo	Mice	Lethality in 16/40 exposed	Haun and Kinkead, 1973
1	6 h/d, 5 d/w, 1 y	Mice	No hepatotoxicity	Vernot et al., 1985
1	6 h/d, 5 d/w, 1 y	Rats, female	Focal liver-cell hyperplasia in 58/97 exposed	Vernot et al., 1985
1	6 h/d, 5 d/w, 1 y	Rats	Nasal adenomatous polyps in 9/97 exposed	Vernot et al., 1985
1	6 h/d, 5 d/w, 1 y	Mice	Pulmonary adenomas in 12/379 exposed and in 4/378 controls	Vernot et al., 1985
1	6 mo, continuous	Mice	Alveolargenic carcinomas in 5/9 exposed and 1/8 controls; lymphosarcomas in 2/9 exposed and 0/8 controls; hepatomas in 1/9 exposed and 0/8 controls	MacEwen et al., 1974

2-6	6 h/d, 5 d/w, 69 d	Guinea pigs	Lung pathology: death of 1/4 exposed	Weatherby and Yard, 1955
5	6 h/d, 5 d/w, 1 y	Rats	Nasal tumors in 58/98 exposed; thyroid carcinoma in 13/98 exposed	Carter et al., 1981; Vernot et al., 1985
5	6 h/d, 5 d/w, 1 y	Syrian hamsters	Nasal tumors in 16/160 exposed	Carter et al., 1981; Vernot et al., 1985
2-5	6 h/d, 5 d/w, 7 d	Dogs	Liver and kidney toxicity and death	Weatherby and Yard, 1955
5	6 h/d, 5 d/w, 6 mo	Rats, male	Sluggishness, emphysema, atelactasis, death (median lethal exposure time = 120 d)	Comstock et al., 1954
14	6 h/d, 5 d/w, 6 mo	Dogs, male	Weight loss, anorexia, tremors, vomiting, fatigue, liver damage, death (median lethal exposure time = 27 d)	Comstock et al., 1954
122-464	6 h/d, 5 d	Rats, male	Initial restlessness, then lethargy, weight loss, pulmonary edema, death (median lethal exposure time = 4.5 d)	Comstock et al., 1954
122-464	6 h/d, 5 d	Mice, female	Initial restlessness, then lethargy, weight loss, pulmonary edema, death (median lethal exposure time = 4.5 d)	Comstock et al., 1954
15,200 (nominal)	0.5-4.0 h	Rats, male	Alternating hyperactivity and lethargy, pulmonary edema, convulsions, death	Comstock et al., 1954

TABLE 5-2 Exposure Limits Set by Other Organizations

Organization	Concentration, ppm
ACGIH's TLV	0.1 (TWA) (0.01 proposed TWA)
OSHA's PEL	1 (TWA, 8 h/d, 40 h/w, lifetime) 0.03 (ceiling)
NIOSH's IDLH	Not established owing to carcinogenicity
NRC's 1-h SPEGL	2
NRC's 24-h SPEGL	0.08

TLV = threshold limit value. TWA = time-weighted average. PEL = permissible exposure limit. IDLH = immediately dangerous to life and health. SPEGL = short-term public emergency guidance level.

TABLE 5-3 Spacecraft Maximum Allowable Concentrations

Duration	ppm	mg/m^3	Target Toxicity
1 h	4	5	Lethality
24 h	0.3	0.4	Hepatotoxicity
7 d[a]	0.04	0.05	Hepatotoxicity
30 d	0.02	0.03	Hepatotoxicity, focal-liver-cell hyperplasia, nasal adenoma
180 d	0.004	0.005	Hepatotoxicity, nasal adenoma

[a]A temporary 7-d SMAC had been set at 0.04 ppm.

RATIONALE FOR ACCEPTABLE CONCENTRATIONS

Hydrazine induces a variety of toxic effects including irritation of the eyes, nose, and throat, contact dermatitis, and, at higher concentrations, dizziness, anorexia, nausea, vomiting, cough, fever, diarrhea, temporary blindness, hepatotoxicity, tremors, hyperexcitability, convulsions, and death. Prolonged exposures to relatively low concentrations have been reported to induce hepatotoxicity and carcinogenicity (Comstock et al., 1954; House, 1964; Haun and Kinkead, 1973; MacEwen et al., 1974; Carter et al., 1980; MacEwen et al., 1981; Vernot et al., 1985). Of these end points, quantitative data are available only for hepatotoxicity, carcinogenicity, and lethality. SMACs were determined following the guidelines of the National Research Council for exposure durations of 1 h, 24 h, 7 d, 30 d, and 180 d by establishing acceptable concentrations (ACs) for each adverse effect at each exposure duration and selecting the lowest AC at each exposure duration to be the SMAC (NRC, 1992).

Hepatotoxicity

Liver effects induced by inhalation of airborne hydrazine include fatty changes in the liver in several species after 90 d of continuous exposure at 0.8 ppm (House, 1964). In calculating an AC, these data were used rather than data on hyperplasia because the fatty changes were seen at a much earlier time for similar concentrations of hydrazine.

AC values for liver toxicity were calculated for 180-d and 30-d exposures by using Haber's rule on data for fatty changes induced in 7 of 10 monkeys exposed continuously at 0.8 ppm for 90 d (also seen in 2 of 10 control monkeys). An uncertainty factor of 10 was applied for extrapolation from animals to humans, and another factor of 10 was applied for extrapolation from an effect level to a no-effect level. The use of Haber's rule to extrapolate from a 90-d exposure to exposure durations of 7 d or less (i.e., $\geq$ 12-fold extrapolation) was deemed inappropriate.

180-d AC = 0.8 ppm × (90 d/180 d) ÷ 10 ÷ 10 = 0.004 ppm.
30-d AC = 0.8 ppm × (90 d/30 d) ÷ 10 ÷ 10 = 0.02 ppm.

An AC for a 24-h exposure can be derived from the data of Timbrell et al. (1982), which showed that an i.p. dose of 10 mg/kg in rats induced the lowest-observed-adverse-effect level (LOAEL) for increased liver triglycerides when examined 24 h after dosing. For a 70-kg man inhaling 20 m^3/d, assuming 100% absorption, one can convert from an i.p. dose to an equivalent inhalation dose and derive a 24-h AC as follows:

10 mg/kg × 70 kg ÷ 20 m^3 ÷ 10 (NOAEL) ÷ 10 (species)
× 0.76 (to convert mg/m^3 to ppm) = 0.3 ppm.

An AC for a 7-d exposure can be calculated from the 24-h AC by using Haber's rule. Thus,

7-d AC = 24-h AC × (1 d /7 d) = 0.3 ppm × 1/7 = 0.04 ppm.

Focal-Liver-Cell Hyperplasia

Liver effects induced by inhalation of hydrazine include a significant increase in the incidence of focal-liver-cell hyperplasia in female rats (but not in male rats, mice, hamsters, or dogs) exposed at 1 ppm or 5 ppm for 6 h/d, 5 d/w, for 1 y (Vernot et al., 1985). (Hyperplasia occurred in 58 of 97 exposed female rats compared with 57 of 147 controls.) Female rats exposed at 0.25 ppm under the same conditions had no significant increase in focal-liver-cell hyperplasia (36 of 100). Thus, 0.25 ppm is a NOAEL for focal-liver-cell hyperplasia for an exposure of 1560 h or 65 d. ACs can be calculated for 30 d and 180 d by using a factor of 10 for species differences and Haber's rule for extrapolating from 65 d to 180 d. The concentration is not increased for extrapolating from 65 d to 30 d:

30-d AC = 0.25 ppm ÷ 10 = 0.03 ppm.
180-d AC = 0.25 ppm × (65 d/180 d) ÷ 10 = 0.009 ppm.

Carcinogenicity

Hydrazine has been found to be carcinogenic in animal model systems (MacEwen et al., 1981; Carter et al., 1981; Vernot et al., 1985). The oncogenic changes were mostly benign and observable only at the microscopic level, producing little or no impairment of respiratory function and no effect on life expectancy. The nononcogenic toxicities of hydrazine exposure in animals were more severe in producing debilitation and lethal effects. There are, moreover, no known reports of hydrazine-induced human tumors. Most human exposures to hydrazine have been accidental or job-related, and dose-response data are not available. The only clearly demonstrated effect induced by inhalation of hydrazine is nasal polyps in rodents and, at higher exposures, nasal adenocarcinomas. The most sensitive species for this effect is rats;

hence, they are used to make the risk estimate at a risk level of 10^{-4}. ACs were calculated by using the linearized multistage model described below (NRC, 1992).

Based on the data of MacEwen et al. (1981) and using linearized multistage extrapolation, the NRC Committee on Toxicology (COT) calculated that the lower 95% confidence limit on the inhalation dose of hydrazine that would produce a 1% lifetime tumor incidence in rats is 0.055 ppm for a 6 h/d, 5 d/w, 52 w/y for a 1-y exposure (NRC, 1985). This dose extrapolates to 0.005 ppm for a continuous 2-y exposure:

$$0.055 \text{ ppm} \times (6 \text{ h}/24 \text{ h}) \times (5 \text{ d}/7 \text{ d}) \times (1 \text{ y}/2 \text{ y}) = 0.005 \text{ ppm}.$$

Extrapolating from the 1% tumor incidence for a continuous 2-y exposure at 0.005 ppm to calculate the expected incidence for a 24-h exposure at the same concentration (0.005 ppm), the COT estimated that the tumor risk for rats should be less than 10^{-6} (NRC, 1985, 1992).

The linearized multistage model is considered sufficiently conservative so that an additional species extrapolation factor is unnecessary (J. Doull, Committee on Toxicology, personal commun., 1989). Therefore, the following equation, based on Crump and Howe's (1984) multistage model with only the first stage dose-related, was used to calculate the exposure concentrations, D, which would yield a tumor risk of 10^{-4} for exposure durations of 24 h, 7 d, 30 d, and 180 d:

$$D = \frac{d \times (25{,}600)^k \times (10^{-4}/\text{risk})}{(25{,}600 - (365 \times \text{age}))^k - (25{,}600 - (365 \times \text{age}) - t)^k},$$

where d is the concentration during a lifetime exposure (0.005 ppm in this case); 25,600 is the number of days in a 70-y human lifetime; k is the number of stages in the model (three in this case); 10^{-4} is the acceptable risk level; age is the minimum age of an astronaut in years (30 y in this case); t is the exposure duration in days (1, 7, 30, or 180); risk is the risk of tumor for lifetime exposure to d (10^{-2} in this case).

This equation yields values of

$D_{24h} = 1.0$ ppm.
$D_{7d} = 0.2$ ppm.
$D_{30d} = 0.04$ ppm.
$D_{180d} = 0.007$ ppm.

Lethality

Analysis of the lethality data is a difficult and frustrating process. The data for repeated exposures to hydrazine are highly scattered. A major factor in the poor reproducibility of results between laboratories might be the propensity of hydrazine to adhere to surfaces. An early report (Comstock et al., 1954) showed that, at a nominal concentration of 20,000 ppm, the recovery of hydrazine vapor from the chamber atmosphere decreases from 26% to 4% with dead rats in the chamber; if the rats are alive, the recovery is decreased to 2%. This clearly indicates that a large fraction of the airborne hydrazine adheres to the rat fur, probably about 10 times the amount retained in the respiratory system. Examination of the methods used for many of the experiments reveals serious short-comings in some of the study designs (number of animals tested; not sham-exposing control animals, etc.). Despite these potential problems in experimental designs and highly scattered rodent data, the conservative approach would be to use the data for the most sensitive species (mice). Epidemiological data on workers occupationally exposed to hydrazine vapors over periods of months to years lead to the conclusion that humans are much less susceptible to hydrazine toxicity than are mice (Wald et al., 1984). Personal communication with Dr. Nick Wald (April 1992) confirmed that there were no deaths seen in his epidemiological study among the 78 factory workers exposed to hydrazine at 1-10 ppm over a period of at least 6 mo. ACs can be calculated from these data as follows:

For a 1-h exposure: The assumption is made that if the hydrazine concentrations were between 1 and 10 ppm for ≥6 mo, it would be highly likely that there would be at least one period during those ≥6 mo when the hydrazine concentration was 5 ppm for at least 1 h. Thus, multiplying 5 ppm by the square root of 78 divided by 10 to adjust for the use of less than 100 subjects,

$$1\text{-h AC} = 5 \text{ ppm} \times \frac{\sqrt{78}}{10} = 4 \text{ ppm}$$

is a NOAEL for lethality.

For exposures of 24 h, 7 d, and 30 d: AC values were calculated using the lower end of the concentration range (1 ppm) and assuming a work schedule of 6 h/d, 5 d/w, for 6 mo (equivalent to 32.5 d continuous exposure) and not increasing the 30-d AC for exposure durations shorter than 30 d. Thus,

$$\text{24-h, 7-d, 30-d ACs} = 1 \text{ ppm} \times \frac{\sqrt{78}}{10} = 0.9 \text{ ppm}.$$

For exposures of 180 d: Using Haber's rule, the AC is

$$\text{180-d AC} = 1 \text{ ppm} \times \frac{30}{80} \times \frac{\sqrt{78}}{10} = 0.15 \text{ ppm}.$$

Spaceflight Effects

The susceptibility of astronauts to the toxic effects of hydrazine would not be expected to be affected by the physiological changes induced by spaceflight.

RECOMMENDATIONS

Studies are needed to definitively determine the fate of the large fraction of the hydrazine vapor that disappears during controlled laboratory exposures. This determination would help eliminate the uncertainty in the total dose received by exposed animals. Currently, it is not known whether hydrazine vapor might be undergoing some reaction in the air or on surfaces that converts it to a form that, although not measured as hydrazine in analytical measurements of airborne concentrations, might be the ultimate toxin or carcinogen or that might be metabolized by the body to the ultimate toxin or carcinogen.

A study on the relative absorption rates for dermal versus inhalation absorption of hydrazine vapor would aid in estimating the total absorbed dose during exposures.

A carcinogenicity study using a continuous exposure protocol including concentrations that do not produce nasal inflammation or necrosis would be helpful.

A carcinogenicity study in which animals are continuously exposed to high concentrations of hydrazine vapor for 1 h, 24 h, or 7 d and then observed for their lifetime would be useful.

TABLE 5-4 End Points and Acceptable Concentrations

End Point	Exposure Data	Species and Reference	Uncertainty Factors				Acceptable Concentrations, ppm				
			NOAEL	Time	Species	Space-flight	1 h	24 h	7 d	30 d	180 d
Lethality	NOAEL at 1ppm, 6 h/d, 5 d/w, 6 mo	Human (Wald, 1984)	1	HR[a]	1	1	4	0.9	0.9	0.9	0.15
Nasal adenoma	At 1.0 ppm, 6 h/d, 5 d/w, 1 y	Rat (MacEwen et al., 1981; Carter et al., 1981; Vernot et al., 1985)	1	[b]	1	1	—[c]	1	0.2	0.04	0.007
Hepatotoxicity	At 20 mg/kg, i.p.	Rat (Timbrell et al., 1982)	10	HR	10	1	—	0.3	0.04	—	—
	At 0.78 ppm, 90 d continuous	Monkey, rat, and mouse (House, 1964)	10	HR	10	1	—	—	—	0.02	0.004
Focal liver-cell hyperplasia	At 1 ppm, 6 h/d, 5 d/w, 1 y	Female rat (Vernot et al., 1985)	10	HR	10	1	—	—	—	0.03	0.009
SMAC							4	0.3	0.04	0.02	0.004

[a]HR = Haber's rule.
[b]Calculated on the basis of COT's equation (NRC, 1985) derived from Crump and Howe's multstage carcinogenicity model and using a lifetime cancer risk of 10^{-4}. This model was not used to calculate acceptable concentrations for exposures shorter than 24 h.
[c]Extrapolation to these exposure durations produces unacceptable uncertainty in the values.

REFERENCES

Becker, R.A., L.R. Barrows, and R.C. Shank. 1981. Methylation of liver DNA guanine in hydrazine hepatotoxicity: Dose-response and kinetic characteristics of 7-methylguanine and O^6-methylguanine formation and persistence in rats. Carcinogenesis 2:1181-1188.

Carter, V.L.J., K.C. Back, and J.D. MacEwen. 1981. The Oncogenic Hazard From Chronic Inhalation of Hydrazine. AGARD-CP-309, B5/1-B5-9, Advisory Group for Aerospace Research and Development, AGARD Conference Proceedings, Sept. 15-19, 1980, Toronto.

Comstock, C.C., L.H. Lawson, E.A. Greene, and F.W. Oberst. 1954. Inhalation toxicity of hydrazine vapor. AMA Arch. Ind. Hyg. Occup. Med. 10:476-490.

Crump, K.S., and R.B. Howe. 1984. The multistate model with a time-dependent dose pattern: Applications to carcinogenic risk assessment. Risk Anal. 4:163-176.

Dambrauskas, T., and H.H. Cornish. 1964. The distribution, metabolism, and excretion of hydrazine in rat and mouse. Toxicol. Appl. Pharmacol. 6:653-663.

Doull, J. 1989. Letter to Col. Thayer J. Lewis, M.C., Headquarters, U.S. Air Force, Bolling Air Force Base, Washington, D.C.

Eiceman, G.A., M.R. Salazar, M.R. Rodriguez, T.F. Limero, S.W. Beck, J.R. Cross, R. Young, and J.T. James. 1993. Ion mobility spectrometry of hydrazine, monomethylhydrazine, and ammonia in air with 5-nonanone reagent gas. Anal. Chem. 65:1696-1702.

Evans, D. 1958. Two cases of hydrazine hydrate dermatitis without systemic intoxication. Br. J. Ind. Med. 16:126-127.

Haun, C.C., and E.R. Kinkead. 1973. Pp. 351-363 in Chronic Inhalation Toxicity of Hydrazine. AMRL-TR-73-125, Paper No. 25. Aerospace Medical Research Laboratory, Wright-Patterson Air Force Base, Dayton, Ohio.

Herbold, B., and W. Buselmaier. 1976. Induction of point mutations by different chemical mechanisms in the liver microsomal assay. Mutat. Res. 40:73-84.

House, W.B. 1964. Tolerance Criteria for Continuous Inhalation Exposure to Toxic Materials, III. Effects on Animals of 90-Day Exposure to Hydrazine, Unsymmetrical Dimethylhydrazine, Decaborane,

and Nitrogen Dioxide. ASD-TR-61-519 (III). Wright-Patterson Air Force Base, Dayton, Ohio.
Høvding, G. 1967. Occupational dermatitis from hydrazine hydrate used in boiler protection. Acta Derm.-Venereol. 47:293-297.
Jacobson, K.H., J.H. Clem, H.J. Wheelwright, Jr., W.E. Rinehart, and N. Mayes. 1955. The acute toxicity of the vapors of some methylated hydrazine vapors. AMA Arch. Ind. Health 12:609-616.
Jain, H.K. and P.T. Shukla. 1972. Locus specificity of mutagens in *Drosophila*. Mutat. Res. 14:440-442.
Karstadt, M., and R. Bobal. 1982. Availability of epidemiologic data on humans exposed to animal carcinogens: 2. Chemical uses and production volume. Teratogen. Carcinogen. Mutagen. 2:151-168.
Kimball, R.F. 1977. The mutagenicity of hydrazine and some of its derivatives. Mutat. Res. 39:111-126.
Kulagina, N.K. 1962. The toxicological characteristics of hydrazine. Toxicology of new industrial chemical substances. Acad. Med. Sci. USSR 4:65-81.
MacEwen, J.D., E.E. McConnell, and K.C. Back. 1974. Pp. 225-235 in The Effects of 6-Month Chronic Low Level Inhalation Exposures to Hydrazine on Animals. AMRL-TR-74-125. Paper No. 16. Aerospace Medical Research Laboratory, Wright-Patterson Air Force Base, Ohio.
MacEwen, J.D., E.R. Kinkead, E.H. Vernot, C.C. Haun, and A.I. Hall. 1981. Chronic Inhalation Toxicity of Hydrazine: Oncogenic Effects. Rep. ISS AFAMRL-TR-81-56, Order AD-A101847. Aerospace Medical Research Laboratory, Wright-Patterson Air Force Base, Dayton, Ohio.
MacRae, W.D., and H.F. Stich. 1979. Induction of sister-chromatid exchanges in Chinese hamster ovary cells by thiol and hydrazine compounds. Mutat. Res. 68:351-365.
Morris, J., J.W. Densem, N.J. Wald, and R. Doll. 1995. Occupational exposure to hydrazine and subsequent risk of cancer. Occup. Environ. Med. 52:43-45.
NRC. 1985. P. 5-21 in Emergency and Continuous Exposure Guidance Levels for Selected Airborne Contaminants, Vol. 5. Washington, D.C.: National Academy Press.
NRC. 1992. Guidelines for Developing Spacecraft Maximum Allow-

able Concentrations for Space Station Contaminants. Washington, D.C.: National Academy Press.
NTP. 1989. Fifth Annual Report on Carcinogens: Summary. NTP 89-239: 166-168. National Institute of Environmental Health Sciences, Research Triangle Park, N.C.
Reid, F.J. 1965. Hydrazine poisoning. Br. Med. J. 2:1246.
Roe, F.J.C. 1978. Hydrazine. Ann. Occup. Hyg. 21:323-326.
Santodonato, J., S. Bosch, W. Meylan, J. Becker, and M. Neal. 1985. P. ii in Monograph on Human Exposure to Chemicals in the Workplace: Hydrazine. AMRL-TR-84-533. Aerospace Medical Research Laboratory, Wright-Patterson Air Force Base, Dayton, Ohio.
Scales, M.D.C., and J.A. Timbrell. 1982. Studies on hydrazine hepatotoxicity. 1. Pathological Findings. J. Toxicol. Environ. Health 10:941-953.
Sevin, I.F. 1978. Criteria for a Recommended Standard Occupational Exposure to Hydrazines. Publ. No. DHEW-78-172. National Institute for Occupational Safety and Health, Rockville, MD.
Smith, E.B., and D.A. Clark. 1972. Absorption of hydrazine through canine skin. Toxicol. Appl. Pharmacol. 21:186-193.
Sotaniemi, E.J., J. Hirvonen, H. Isomaki, J. Takkunen, and J. Kaila. 1971. Hydrazine toxicity in the human: Report of a fatal case. Ann. Clin. Res. 3:30-33.
Springer, D.L., B.M. Krivak, D.J. Broderick, D.J. Reed, and F.N. Dost. 1981. Metabolic fate of hydrazine. J. Toxicol. Environ. Health 8:21-29.
Steinhoff, D. and U. Mohr. 1988. The question of carcinogenic effects of hydrazine. Exp. Pathol. 33:133-143.
Timbrell, J.A., M.D. Scales, and A.J. Streeter. 1982. Studies on hydrazine hepatotoxicity. 2. Biochemical findings. J. Toxicol. Environ. Health 10:955-968.
van Ketel, W.G. 1964. Contact dermatitis from a hydrazine derivative in a stain remover: Cross sensitization to apresoline and isoniazide. Acta Derm.-Venereol. 44:49-53.
Vernot, E.H., J.D. MacEwen, R.H. Bruner, C.C. Haun, E.R. Kinkead, D.E. Prentice, A. Hall III, R.E. Schmidt, R.L. Eason, G.B. Hubbard, and J.T. Young. 1985. Long-term inhalation toxicity of hydrazine. Fundam. Appl. Toxicol. 5:1050-1064.
Wald, N., J. Boreham, R. Doll, and J. Bonsall. 1984. Occupational

exposure to hydrazine and subsequent risk of cancer. Br. J. Ind. Med. 41:31-34.

Weatherby, J.H., and A.S. Yard. 1955. Observations on the subacute toxicity of hydrazine. AMA Arch. Ind. Health 11:413-419.

Wyrobek, A.J., and S.A. London. 1973. Pp. 417-432 in Effect of Hydrazines on Mouse Sperm Cells. AMRL-TR-73-125. Paper No. 30. Aerospace Medical Research Laboratory, Wright-Patterson Air Force Base, Dayton, Ohio.

B6 Indole

Chiu-Wing Lam, Ph.D., and John T. James, Ph.D.
Johnson Space Center Toxicology Group
Biomedical Operations and Research Branch
Houston, Texas

PHYSICAL AND CHEMICAL PROPERTIES

Indole is a colorless crystalline solid. It has an intense fecal odor at moderate concentrations; however, the odor at very low concentrations is pleasant (*Merck Index*, 1989).

Synonym:	2,3-Benzopyrrole
Formula:	C_8H_7N
CAS number:	120-72-9
Molecular weight:	117.1
Boiling point:	130°C
Melting point:	52°C
Vapor pressure:	Not found
Conversion factors at 25°C, 1 atm:	1 ppm = 4.8 mg/m^3 1 mg/m^3 = 0.21 ppm

OCCURRENCE AND USE

Indole is a naturally occurring compound, constituting about 2.5% of jasmine oil and 1% of orange-blossom oil, and in both cases, it contributes to their fragrances (*Kirk-Othmer Concise Encyclopedia of Chemical Technology*, 1985). Indole is a component of perfumes (*Kirk-Othmer Concise Encyclopedia of Chemical Technology*, 1985; *Merck Index*,

1989). Paradoxically, indole has an intense fecal odor (*Merck Index*, 1989), presumably at higher concentrations. It is also a bacterial decomposition product of tryptophan in the gut (Hammond et al., 1984; Eisele, 1986). Indole is one of the odorous components found in sewage and animal wastes, including human feces (Veber, 1967, as cited in Sgibnew and Orlova, 1971; Karlin et al., 1985). The compound is also found in animal tissues where putrefactive processes have occurred, presumably by the decomposition of tryptophan (Eisele, 1986). Trace levels of indole are expected to be present in manned spacecraft.

PHARMACOKINETICS AND METABOLISM

Indole, an anaerobic metabolite of tryptophan, can be condensed with serine by microorganisms to recover tryptophan (Meister, 1965). However, this biosynthetic pathway for tryptophan has not been observed in humans and rats. Absorbed indole from the gut is hydroxylated to form indoxyl, which conjugates with sulfate to produce indican (indoxylsulfuric acid) in the liver (Meister, 1965). Indoxyl and indican are found in human plasma and urine. A mean plasma indican concentration of 3 mg/L was reported in 56 males (range 1.2-4.8 mg/L) and 44 females (range 0.6-5.4 mg/L) (Geigy Pharmaceuticals, 1974). The daily urinary excretion of indoxylsulfate in normal adults was reported to average 200 mg (range 140-250 mg) (Haddox and Saslaw, 1963).

Sgibnew and Orlova (1971) reported that indole was not detected in the blood of rabbits exposed at 10 mg/m^3 for 3 h. When 10 mg of indole was injected intravenously into each of the five rabbits, an average plasma indole concentration of 0.3 mg% was detected only in the blood samples taken 15 min after the injection. Because of the inability to detect indole in the blood thereafter, Sgibnew and Orlova concluded that indole was quickly removed and rendered harmless. However, it should be pointed out that 0.3 mg% was close to the detection limit of the calormetric method employed by these authors. Indole concentrations in the blood were investigated by Hammond et al. (1980) in cows dosed orally with the compound at 50, 100, or 200 mg/kg. Plasma indole concentrations reached peak concentrations of 4.5, 8, and 20 mg/mL, respectively, 3 h after dosing. The plasma concentrations decreased to 4.4%, 7.4%, and 38% of the corresponding peak concentrations after 12 h and to 2%, 0.6%, and 1.4% after 24 h. Indole was not detected (<0.02 mg/mL) in plasma of these cows 72 h after injection. These results indicate that indole was rapidly eliminated from the blood of the cows.

TOXICITY SUMMARY

Indole at a few parts per million has an unpleasant odor and can elicit toxic symptoms, such as nausea. The consistent toxicological property of indole, an aromatic amine, observed in animal studies is its ability to cause the formation of Heinz bodies, which are known to be produced by other aromatic amine compounds, such as aniline (Smith, 1986). Chronic studies by the subcutaneous route have shown that indole might have a weak leukemogenic activity in mice, but not in hamsters. The toxicity of indole is summarized in the Table 6-1.

Acute and Short-Term Toxicity

The human odor threshold of indole was reported to be 0.45 mg/m^3 ($\approx$0.1 ppm) (Sgibnew and Orlova, 1971). Very unfavorable odor was perceived in concentrations approaching 9.0 mg/m^3; 2 of the 12 test subjects complained of nausea. According to the authors, brief exposures at this concentration did not produce electrocardiographic and electroencephalographic changes (exposure length not specified). Exposing 20 mice and 15 rats to indole at a concentration of 9-10 mg/m^3 for 2 h produced no toxic signs except some unrest during the first 15 min; no deaths occurred during the 14-day post-exposure observation period (Sgibnew and Orlova, 1971).

The hematological effects and subsequent renal lesions induced by indole were observed in four cows given the compound at a single oral dose of 100 mg/kg and then 200 mg/kg 2 w later (Hammond et al., 1980). Hemolysis was observed 24 h after each chemical exposure. At the high dose, all four cows had blood-colored urine. Necropsy 1 w later revealed renal tubular epithelial degeneration attributable to hemoglobinuric nephrosis. There was a grayish-brown discoloration of the endothelial surfaces of the aorta and other elastic arteries, and a mild cloudy swelling in hepatocytes. No lesions were found in the lung. The dose of indole that produced no observable clinical signs of toxicity, including hemolysis, in these cows was 50 mg/kg given in a single dose 1 mo before the 200-mg/kg dose.

Subchronic and Chronic Toxicity

Inhalation studies were conducted by Sandage (1961) in 10 rhesus monkeys, 50 rats, and 100 mice exposed continuously to indole at a concentration of 10.5 ppm (50 mg/m^3) for 90 d. Hematological examination of the exposed rodents revealed that numerous Heinz bodies were present in the blood. Heinz-body occurrence was most prominent in mice and was observed after 3-4 d of exposure. It was slightly less prominent in the rats and was observed after 7-10 d of indole exposure. No Heinz bodies were observed in monkeys until 70 d of exposure. In the mice, the appearance of Heinz bodies was accompanied by anisocytosis (presence in blood of erythrocytes showing excessive variation in size) and polychromatophilia, with many erythrocytes exhibiting diffuse basophilia. There was a general leukocytosis and marked reticulocytosis with about 80% of the reticulated cells appearing morphologically atypical. Instead of the fine thread-like reticulum of normal reticulocytes, there was a profuse, densely staining nuclear material that filled the cell, which appeared to consist of coarse granules. Pathological studies on 25% of the exposed mice revealed 95% of the animals had pigment in the renal tubular cells. However, renal abnormalities were not found in any exposed monkeys or rats examined (about 25% of those exposed).

Significant elevation of serum sodium, cholinesterase, and amylase levels in monkey blood serum was also noted by Sandage (1961), who suggested that such elevated levels provided a clue to neurological effects. However, pathological examination showed no brain damage. Histopathological studies of the heart, lung, liver, and kidney from the exposed monkeys revealed no statistical difference from that of control monkeys. Two monkeys from the exposed group and none from the controls died. Sandage concluded that the death rate (2 of 10) of the monkeys exposed to indole could not be considered "significant." The cause of death was not given.

Carcinogenicity

There are three studies in mice that provide some evidence that indole might be leukemogenic. Eckhart and Stich (1957) reported that three of the 50 RFH-bred white mice given indole at 0.5 mg (in 50% propylene glycol) subcutaneously every 3-4 d for 140 d (weekly dose approximately 40 mg/kg) developed leukemia and three developed aleukemic myelosis; 10 of the initial 50 mice died in the first 4 w of indole administration (4 mg per mouse). Leukemia was not observed among the 100 control mice; however, no indication was given that the control mice were given sham injections of the vehicle. According to these authors, the spontaneous rate of leukemia observed in 1000 RFH mice was 1:500.

A similar study was conducted by Dzioev (1974), who injected indole subcutaneously in mice at a weekly dose of 1 mg for 9 mo. The author reported that of 60 surviving mice, 13 showed leukosis and 21 had pulmonary adenomas. Exposure to tryptophan (presumably serving as controls) resulted in tumors in 8 of the 28 surviving animals. Subcutaneous injection of indole in mice (C-57) was also conducted by Rauschenbach et al. (1963), who administered a dose of 2.5 mg per mouse once a week for 5-10 mo. No leukemia was observed in 17 mice that survived longer than 1 y; however, one mouse developed adenocarcinoma of the breast and seven were leukemoid. Cancer or hematopoietic changes were not observed in the 30 mice given tryptophan (presumably serving as controls) and surviving longer than 1 y (10 died before 1 y). These carcinogenicity results of indole reported in the latter two studies (Rauschenbach et al., 1963; Dzioev, 1974) were classified by NIOSH as equivocal (NIOSH, 1985-86). In addition, the leukemogenic effect of indole was not observed in hamsters chronically exposed to indole. In a study to investigate the role of indole in the carcinogenicity of 2-acetylaminofluorene, 23 male and 30 female Syrian golden hamsters were given 1.6% indole alone in the diet as controls for 10 mo (Oyasu et al., 1972). These animals showed no tumors of the bladder or liver.

TABLE 6-1 Toxicity Summary

Concentration	Exposure Duration	Species	Effects	Reference
0.45 mg/m^3 ($\approx$ 0.1 ppm)	Brief	Human	Odor detected by some subjects	Sgibnew et al., 1971
9 mg/m^3 or less	Brief	Human	Unfavorable odor, nausea, no changes in EKG and EEG	Sgibnew et al., 1971
9 mg/m^3	3 h	Rabbit	Restlessness	Sgibnew et al., 1971
10 mg/m^3	2 h	Mouse, rat	No clinical signs or deaths in 14 d	Sgibnew et al., 1971
50 mg/m^3 (10.5 ppm)	90 d continuous	Mouse	Hematological abnormality appeared 3-4 d after exposure begun, Heinz bodies, anisocytosis, leukocytosis, marked reticulocytosis, atypical reticulocytes	Sandage, 1961
50 mg/m^3 (10.5 ppm)	90 d continuous	Rat	Hematological abnormality appeared 7-10 d after exposure begun, Heinz bodies	Sandage, 1961
50 mg/m^3	90 d continuous	Rhesus monkey	Hematological abnormality appeared 70 d after exposure begun, Heinz bodies, some altered clinical test parameters, no exposure-induced pathology in heart, liver, lung, and kidney	Sandage, 1961
50 mg/kg, i.p.	Single dose	Cow	None	Hammond et al., 1980
100 mg/kg, i.p.	Single dose	Cow	Heinz bodies	Hammond et al., 1980
200 mg/kg, i.p.	Single dose	Cow	Heinz bodies, blood-colored urine, kidney lesions, heart lesions, microscopic degeneration of liver	Hammond et al., 1980
1 mg/w per mouse ($\approx$ 40 mg/kg/w), s.c.	140 d	Mouse	Myeloic leukemia (leukemic and aleukemic melosis) incidence of 6/40, leukemia incidence in controls of 0/100, 20% of mice died in the first 4 w after a mean total dose of 40 mg per mouse	Eckhart and Stich, 1957

1 mg/w per mouse	9 mo	Mouse	Cancer equivocal	Dzioev, 1974
2.5 mg/w per mouse	5-10 mo	Mouse	Cancer equivocal	Rauschenbach et al., 1963
1.6% in diet	10 mo	Hamster	No bladder or liver cancer (other organs not examined)	Oyasu et al., 1972

TABLE 6-2 Exposure Limits Set by Other Organizations

Organization	Concentration, ppm
ACGIH's TLV	None set
OSHA's PEL	None set
NRC	1.0 ppm (10 and 60 min)[a]
NRC	0.1 ppm (90 and 180 d)[a]

TLV = threshold limit value. PEL = permissible exposure limit.
[a]Values recommended to NASA by the NRC in 1972.

TABLE 6-3 Spacecraft Maximum Allowable Concentrations

Duration	ppm	mg/m^3	Target Toxicity
1 h	1	5	Nausea
24 h	0.3	1.5	Nausea, hematological changes
7 d	0.05	0.25	Nausea, hematological changes
30 d	0.05	0.25	Mortality
180 d	0.05	0.25	Mortality

RATIONALE FOR ACCEPTABLE CONCENTRATIONS

The SMACs for indole depend not only on the toxicological effects induced, but also on the interactions with spaceflight-induced changes and the normal turnover of indole in man. The toxicological effects induced by indole include nausea, hematological changes, mortality, and possibly leukemogenic effects. Because spaceflight was known to induce approximately a 10% reduction in red-cell mass (Huntoon et al., 1989), this finding was considered when the hematological effects of indole were used to set an acceptable concentration (AC) for effects on red cells. Inhaled indole is not known to induce effects on the respiratory system; therefore, systemically mediated effects, produced by an absorbed dose from inhalation or other route of administration, might be considered equivalent. Moderate amounts of indole are normally absorbed from the gastrointestinal tract, so there is a significant systemic body burden. If the amount of indole entering the body via respiration is small compared with the normal load, then there will be no toxic effects from the inhaled indole.

Nausea

Brief exposures at 2 ppm indole induced nausea in 2 of 12 test subjects (Sgibnew and Orlova, 1971). Slight nausea is tolerable for brief periods as long as crew-member performance is not affected; however, interpretation of the duration of exposure is difficult from the original study. It was estimated that reducing the exposure concentration by a factor of 2 would have caused no more than slight nausea in a few people even if the exposure were for 1 h. Hence, the 1-h AC (nausea) was set at 1 ppm. Since nausea induced by odorous compounds is a threshold effect, and nausea would be less tolerable for a 24-h period, the 24-h AC (nausea) was set at 0.5 ppm to minimize the chance of nausea. For longer exposures, nausea is not acceptable, so the 7-d, 30-d, and 180-d ACs were calculated from the 2-ppm lowest-observed-adverse-effect level (LOAEL) as follows:

$$7\text{-d, }30\text{-d, }180\text{-d ACs} = 2 \text{ ppm} \times \frac{1}{10} \times \frac{\sqrt{12}}{10} = 0.07 \text{ ppm}.$$

The factor of 10 was applied to estimate a no-observed-adverse-effect level (NOAEL) from the LOAEL, and the factor of the square root of 12 divided by 10 was applied to account for a small number of test subjects (12 subjects). This value is about one-third the concentration that gave a sweet odor when inhaled by test subjects for the first time, but not during the second exposure (Sgibnew and Orlova, 1971).

Hematological Effects

The key study for this toxic end point was the continuous 90-d inhalation study reported by Sandage (1961). Many toxic end points were assessed in the exposures of mice, rats, and monkeys to indole at a concentration of 10.5 ppm. For human risk assessment, the study has several shortcomings, but represents the best available data. The control animals were not sham exposed; they were kept in a separate room that was not environmentally the same as the exposure chambers; and the findings were reported statistically as increases or decreases compared with control groups rather than numerically (except mortality). Favorable points of the study were that several species were exposed, the concentrations were measured analytically (the actual measurements were not reported), and the data were apparently subjected to statistical analysis.

In this 90-d continuous inhalation study (Sandage, 1961), exposures at 10.5 ppm, which was the only concentration tested, produced Heinz bodies in mice, rats, and monkeys after 3, 7, and 70 d of exposure, respectively. Because only a single concentration was used, the conventional derivation of a LOAEL from a dose-response curve could not be applied. However, employing an unconventional species-based approach, it was noted that 3 × 10.5, 7 × 10.5, and 70 × 10.5 d-ppm induced Heinz bodies in one, two, and three animal species, respectively. Hence, 3 × 10.5 d-ppm was estimated to be the LOAEL. Using the NRC guideline, the LOAEL was divided by 10 to get a NOAEL of 1 ppm. The 7-d NOAEL of 0.4 ppm, which is equal to 1 ppm × 3/7, was then divided by 10 (species factor) and 3 (spaceflight factor, see below) to obtain the 7-d AC of 0.015 ppm. The hematological effect was not observed 24 h after the exposure; therefore, 10.5 ppm was the NOAEL for the 24-h exposure. By applying the same safety factors (species and spaceflight), the 24-h AC would be 0.3 ppm. Because only one exposure concentration was tested in the study, the LOAEL for 30-d and 180-d exposures could not be obtained or extrapolated, and 30-d and 180-d ACs could not be established based on these data.

The factor of 3 for spaceflight-induced effects was appropriate because the magnitude of the red-cell mass changes observed in astronauts has been about 10% (Huntoon et al., 1989). This change has caused no clinically detectable effects and could be considered adaptive in nature. The mechanism of this change might be due to microgravity effects on the kidney resulting from fluid shifts. Kidneys, which produce erythropoietin, play an important role in red-cell production. However, the mechanism of the red-cell-mass changes has not been clearly elucidated at this time. The hematology factor should be smaller than the factor of 5 used for cardiac effects because cardiac effects have actually caused a cosmonaut to be returned early from a mission (Gazenko et al., 1990).

Mortality

The same study used to set the ACs for hematology was used to set the ACs for mortality because it is the only long-term inhalation study available (Sandage, 1961). The author reported that there was no statistically significant increase in mortality in the exposed groups compared with the control groups; however, there were more deaths in the exposed groups by the end of the 90-d study. A comparison is given below for mortality incidences.

TABLE 6-4 Mortality in Control and Exposed Groups

Species	Control Deaths	Deaths at 10.5-ppm Exposure
Mice	16 of 100	22 of 100
Rats	2 of 50	5 of 50
Monkeys	0 of 9	2 of 10

Because the study did not provide information on the cause of death, and it appears that more deaths occurred in the exposed groups, mortality was considered in setting the SMACs. The 30- and 180-d ACs for mortality were set by dividing the exposure at a concentration of 10.5 ppm by factors of 10 to get to a NOAEL for mortality and 10 for interspecies extrapolation. Haber's rule (90 d/180 d) was used to decrease the 180-d exposure concentration. The ACs for 30 and 180 d were 0.1 and 0.05 ppm, respectively.

Leukemogenic Effects

As described in the Toxicity Summary, there are several reports that indole is leukemogenic when given by injection. NIOSH (1985-1986) has classified two of the studies (Rauschenbach et al., 1963; Dzioev, 1974) as equivocal, and the remaining study (Eckhart and Stich, 1957) appears positive for leukemogenic effects. However, the study control groups might not have been given injections of the vehicle used when indole was administered, and the results of the study are difficult to extrapolate to inhalation exposures. The injection dosages were highly toxic bolus quantities, whereas inhalation dosages of an equivalent amount would be through long-term, low-level administrations. Furthermore, it is difficult to estimate the risk of carcinogenesis from a study in which only one exposure group was used. It also appears that mortality is a more important end point because in the study in which 6 of 40 mice showed leukemogenic changes, 10 of 50 died before any animals showed these changes. Because the inhalation data have been used to estimate ACs for mortality, it was not necessary to establish ACs for cancer based on injection data, which were obtained for purposes other than estimating an inhalation risk.

Normal Indole Uptake

Daily urinary excretion of indican averages 200 mg in normal adults, and this is equivalent to 110 mg of indole absorbed from the gastrointestinal tract. It is reasonable to assume that a 5% increase in this indole input from an inhalation source would be toxicologically insignificant. Hence, 5 mg/d could enter through the respiratory system without causing an effect. For a 70-kg man breathing 20 m^3/d, an indole concentration of 0.25 mg/m^3 (0.05 ppm) would cause an additional uptake of 5 mg/d, assuming 100% of the inhaled dose is absorbed. Thus, 0.05 ppm should be a lower bound on any SMAC value selected. The 7-, 30-, 180-d SMACs were set at the metabolic load threshold of 0.05 ppm based on these considerations.

TABLE 6-5 End Points and Acceptable Concentrations

End Point	Exposure Data	Species and Reference	Uncertainty Factors				Acceptable Concentrations, ppm				
			NOAEL	Time	Species	Micro-gravity	1 h	24 h	7 d	30 d	180 d
Nausea	NOAEL at 2 ppm, 2/12 exposed briefly	Humans (Sgibnew et al., 1971)	2	1-10	1	1	1	0.5	0.07	0.07	0.07
Hematological changes	NOAEL at 10 ppm, 3 d continuous	Mice, rats, monkeys[a] (Sandage, 1961)	1	1	10	3	—[b]	0.3	—	—	—
	LOAEL at 10 ppm, 3.5 d continuous	Mice, rats, monkeys[a] (Sandage, 1961)	10	2	10	3	—	—	0.015[c]	—	—
Mortality	At 10 ppm, 90 d continuous	Mice, rats, monkeys (Sandage, 1961)	10	1-2	10	1	—	—	—	0.1	0.05
Normal Uptake	5%, with 100% absorption[d]	Humans (Haddox and Saslaw, 1963)	Lower bound[c]				0.05	0.05	0.05	0.05	0.05
SMAC							**1**	**0.3**	**0.05**	**0.05**	**0.05**

[a]Monkeys and rats did not show effects before 7 d; slight effects were seen in mice after 3-4 days of exposure.
[b]Not applicable.
[c]Lower bound of 0.05 indicates that 0.015 is too low.
[d]Inhaling indole at 0.05 ppm (assuming 100% absorption) would amount to 5% gastrointestinal uptake.

REFERENCES

Dzioev, F.K. 1974. [Study of carcinogenic action of tryptophan and some of its metabolites]. Voprosy Onkologh 20:75-81.

Eckhart, H., and W. Stich. 1957. Untersuchungen uber experimentelle leukamien. II. Mitteilung die indole-leukamie bei der weiben maus. Klinische Wochenschrift 35:504-511.

Eisele, G.R. 1986. Distribution of indole in tissues of dairy cattle, swine and laying pullets. Bull. Environ. Contam. Toxicol. 37:246-262.

Gazenko, O.G., A.I. Grigor'yev, S.A. Bugrov, V.V. Yegorov, V.V. Bogomolov, I.B. Kozlovskaya, and I.K. Tarasov. 1990. [Review of the major results of medical research during the flight of the second prime crew of the Mir Space Station.] Kosmicheskaya Biologiya i Aviakosmicheskaya Meditsina 23(4):3-11; translated and abstracted in USSR Space Life Sciences Digest, L.R. Stone and R. Tetter, eds., NASA Contractor Report 3922(34):119-120.

Geigy Pharmaceuticals. 1974. P. 575 in Documenta Geigy Scientific Tables. Geigy Pharmaceuticals, Ardsley, N.Y.

Haddox, C.H., and M.S. Saslaw. 1963. Urinary 5-methoxytryptamine in patients with rheumatic fever. J. Clin. Invest. 4:435-441.

Hammond, A.C., J.R. Carlson, and R.G. Breeze. 1980. Indole toxicity in cattle. Vet. Record 107:344-346.

Hammond, A.C., B.P. Glenn, G.B. Huntington, and R.G. Breeze. 1984. Site of 3-methylindole and indole absorption in steers after ruminal administration of L-tryptophan. Am. J. Vet. Res. 45:171-174.

Huntoon, C.L., P.C. Johnson, and N.M. Cintron. 1989. Hematology, immunology, endocrinology, and biochemistry. P. 222 in Space Physiology and Medicine, 2nd Ed., A. Nicogossian, C. Huntoon, and S. Pool, eds. Philadelphia: Lea & Febiger.

Karlin, D.A., A.J. Mastromarino, R.D. Jones, J.R. Stroehlein, and O. Lorentz. 1985. Fecal skatole and indole and breath methane and hydrogen in patients with large bowel polyps or cancer. J. Cancer Res. Clin. Oncol. 109:135-141.

Kirk-Othmer Concise Encyclopedia of Chemical Technology. 1985. P. 639 in Kirk-Othmer Concise Encyclopedia of Chemical Technology. New York: Wiley-Interscience.

Meister, A. 1965. Pp. 841-883 in Biochemistry of the Amino Acids, 2nd Ed., Vol 2. New York: Academic Press.

Merck Index. 1989. P. 786 in Merck Index, 11th Ed. Rahway, N.J.: Merck & Co.

NIOSH. 1985-86. Indole. In Registry of Toxic Effects of Chemical Substances. National Institute of Occupational Safety and Health, Cincinnati, Ohio.

Oyasu, R., T. Kitajima, M.L. Hopp, and H. Sumie. 1972. Enhancement of urinary bladder tumorigenesis in hamsters by coadministration of 2-acetylaminofluorene and indole. Cancer Res. 32:2027-2033.

Rauschenbach, M.O., E.I. Jarova, and T.O. Protasova. 1963. Blastomogenic properties of certain metabolites of tryptophane. Acta Unio Int. Cancrum 19:660-662.

Sandage, C. 1961. Tolerance Criteria for Continuous Inhalation Exposure to Toxic Material. II. Effects on Animals of 90-Day Exposure to H_2S, Methyl Mercaptan, Indole, and a Mixture of H_2S, Methyl Mercaptan, Indole and Skatole. ASD Tech. Rep. 61-519 (II). Biomedical Laboratory, Aerospace Medical Laboratory, Wright-Patterson Air Force Base, Dayton, Ohio.

Sgibnew, A.K., and T.A. Orlova. 1971. K voprosu izucheniia toksichnostic. Pp. 190-195 in Problemy Kosmicheskoi Biologii, Vol. 16 [Translation: Problem of studying the toxicity of indole. Pp. 233-239 in Problems of Space Biology, Vol. 16], V.N. Chernigovskiy, ed. Academy of Sciences of the USSR, Department of Physiology. Moscow: Nauka Press.

Smith, R. 1986. Toxic responses of the blood. P. 239 in Casarett and Doull's Toxicology: The Basic Science of Poisons, 3rd Ed., C.D. Klaassen, M.O. Amdur, and J. Doull, eds. New York: Macmillan.

Veber, T.V. 1967. P. 276 in Chelovek Pod Vodoy i v Kosmose. Moscow: Voyenizdat Press.

B7 Mercury

John T. James, Ph.D., and Harold L. Kaplan, Ph.D.
Johnson Space Center Toxicology Group
Biomedical Operations and Research Branch
Houston, Texas

PHYSICAL AND CHEMICAL PROPERTIES

Elemental mercury is a heavy, silvery-white, slightly volatile liquid at room temperature (Stokinger, 1981; ACGIH, 1986).

Synonym:	Quicksilver
Formula:	Hg
CAS number:	7439-97-6
Atomic weight:	200.6
Boiling point:	356.6°C
Melting point:	-38.9°C
Vapor pressure:	0.0018 torr (25°C)
Solubility:	Insoluble in H_2O, soluble in non-polar solvents, vapor soluble in blood than in H_2O
Conversion factors at 25°C, 1 atm:	1 ppm = 8.2 mg/m^3 1 mg/m^3 = 0.12 ppm

OCCURRENCE AND USE

Mercury occurs in three chemical forms in the environment: (1) elemental (metallic) mercury (Hg°); (2) inorganic mercurous (Hg$^+$) and mercuric (Hg^{++}) compounds or ions; and (3) organic mercury compounds (Stokinger, 1981). Although this document is concerned with mercury vapor, data on inorganic or organic forms also are included when relevant to the toxicity of mercury vapor. The major sources of mercury vapor and compounds of mercury are both natural (degassing of earth's crust, emissions from volcanoes, evaporation of natural bodies of water) and man-made (mining, smelting, refuse incineration, combustion of fossil fuel) (WHO, 1991).

Mercury has many diverse uses because of its properties. Liquid elemental mercury is a common component of thermometers, barometers, manometers, and other laboratory and medical measuring instruments (Stokinger, 1981). It is also widely used in electrical devices, including lamps, switches, rectifiers and batteries. The breakage of any of these instruments or devices inside the spacecraft cabin could result in the release and volatilization of liquid mercury and the subsequent exposure of crew members to potentially toxic levels of the vapor. Containment of mercury in such devices is strictly controlled by NASA based on mercury's SMACs and the ability of the air revitalization system to remove mercury vapor from spacecraft air. No method has been developed to monitor mercury vapor in spacecraft air. Recently, there has been much debate over possible health hazards from the inhalation, and ingestion, of elemental mercury released from dental amalgams in the mouth (WHO, 1991). Reported intraoral vapor concentrations range from 0.003 to 0.029 mg/m^3 (Vimy and Lorscheider, 1985).

PHARMACOKINETICS AND METABOLISM

Inhalation of mercury vapor is the most important route of uptake for elemental mercury (WHO, 1991). Human subjects retained approximately 70-80% of inhaled mercury vapor, retention occurring almost entirely in the alveoli (Nielsen-Kudsk, 1965; Hursch et al., 1976). Oxidation of elemental mercury to the mercuric ion is the primary metabolic pathway (Hursch et al., 1976). Mercury accumulates in many tissues, but the most important are the brain and kidneys. Clearance half-times of mercury inhaled by human test subjects vary from 1.7 d for the lungs to 64 d for the kidney region (Hursch et al., 1976).

Absorption

Experimental results indicated that absorption of mercury vapor by the skin poses a very minor hazard compared with that by inhalation. In human volunteers exposed via the forearm skin, the rate of uptake by the total body skin was estimated to be 2.2% of that by the lung (Hursch et al., 1989). Approximately one-half of the mercury retained by the skin was shed by desquamation of epidermal cells, and the remainder was slowly released into the body (Hursch et al., 1989).

Metabolism and Distribution

Mercury vapor rapidly passes from the inspired air in the alveoli into the bloodstream because of its high lipophilicity (Aschner and Aschner, 1990). The dissolved elemental mercury (Hg°) is soon oxidized to mercuric ions (Hg^{++}), partly in red blood cells and partly after diffusion into other tissues of the body (Hursch et al., 1988). This oxidation occurs under the influence of the catalase-hydrogen peroxide complex (Complex I) in mammalian tissues (Nielsen-Kudsk, 1969). Catalase inhibitors, such as ethanol and aminotriazole, inhibit the oxidation reaction, which can change the distribution, retention, and excretion of inhaled mercury vapor (Magos et al., 1974; Hursch et al., 1980; Khayat and Dencker, 1984).

Following the inhalation of mercury vapor, mercury quickly accumulates within the brain, but to a much lesser extent than in the kidneys (Magos, 1967). Despite its rapid oxidation by red blood cells, some solubilized vapor persists in the bloodstream sufficiently long to reach and diffuse across the blood-brain barrier into the brain, where it is oxidized by the catalase-hydrogen peroxide system to the divalent mercuric form (Dunn et al., 1978). The mercuric ions, which traverse the blood-brain barrier less freely than elemental mercury, bind to sulfhydryl-containing ligands and are retained within the brain. Because of this greater diffusibility of the vapor, the mercury content in the brains of animals exposed to the vapor was ten times greater than that of animals injected with an equivalent dose of mercuric salts (Berlin et al., 1969).

Excretion

The principal routes of elimination of inhaled mercury vapor are the urine and feces; a small portion (7%) of the retained mercury is excreted in the expired air as elemental mercury vapor (Dunn et al., 1981). A small amount of the exhaled vapor is formed by the reduction of divalent mercury produced by the oxidation of the elemental mercury. This reduction occurs both in animals (mice and rats) and humans (Berlin et al., 1969; Sugata and Clarkson, 1979; Dunn et al., 1981). The exhalation of mercury vapor is increased in catalase-deficient mice and by ethanol in mice and humans (Dunn et al., 1981; Ogata et al., 1987). The clearance of inhaled mercury vapor from tissues of the body follows a complicated pattern; biological half-times differ according to the tissue and the time after exposure (WHO, 1991). Tracer studies on human volunteers and animals indicate that, after a short exposure to mercury vapor, the first phase of elimination from blood has a half-time of approximately 2-4 d and accounts for about 90% of the retained mercury (WHO, 1991). This is followed by a second phase with a half-time of 15-30 d. After inhalation by human volunteers of a mixture of stable and radioactive mercury vapor for 14-24 min, elimination from the body followed a single exponential process, with an average half-time of 58 d (Hursch et al., 1976). Average half-times for mercury clearance from different parts of the body were the following: lungs, 1.7 d; brain, 21 d; kidneys, 64 d; and chest, 43 d. It is probable that a fraction of the mercury in the brain and the kidneys has a longer biological half-life, particularly when exposures are prolonged (WHO, 1991).

TOXICITY SUMMARY

Acute Toxicity

In humans, acute inhalation of mercury vapor might cause irritation and inflammation of the respiratory tract, resulting in tracheo-bronchitis, bronchiolitis, pneumonitis, and various neuropsychiatric reactions or symptoms (Milne et al., 1970; McFarland and Reigel, 1978; WHO, 1991). Accidental exposure of four workers for 2-5 h to mercury vapor at 1.1-2.9 mg/m^3 (determined by simulating the exposure conditions) caused minimal discomfort during exposure, but fever, cough, dyspnea, and mild chest tightness developed 4 h later (Milne et al., 1970). In another accident, exposure of workers for less than 8 h to the vapor at a calculated maximum concentration of 44 mg/m^3 caused fever, chills, chest pain, and weakness, and, in some, impaired pulmonary function and evidence of interstitial pneumonitis (McFarland and Reigel, 1978). Various symptoms characteristic of chronic mercurialism developed later, including tremor, nervousness, irritability, lack of ambition, and loss of sexual drive.

Animal studies of the acute toxicity of mercury vapor are limited in number and scope and were mostly conducted with high concentrations. In a pharmacokinetic study in which monkeys, rabbits, and rats were exposed to mercury vapor at 1 mg/m^3 for 4 h, signs of toxicity (increased irritability, aggressiveness, decreased food intake for 24 h) were observed only in rats (Berlin et al., 1969). An 8-h exposure of a dog at 20 mg/m^3 caused dyspnea and weakness; death occurred on the day of exposure (Fraser et al., 1934). In rabbits, a 1-h exposure at 28.8 mg/m^3 caused moderate histopathological changes to the kidneys and brain and mild changes to the lungs and heart (Ashe et al., 1953). With 2-, 4-, or 6-h exposures, damage to the kidneys and brain was severe (extensive cellular degeneration and necrosis) and mild to moderate in the lungs, liver, colon, and heart. In rats exposed to mercury vapor at 0.55 mg/m^3 for 12, 14, or 24 h (four rats per group), it was reported that one animal from each group died with neurological signs. Only small peripheral hemorrhages were found in lung tissues (Møller-Madsen, 1992). These deaths might not have been from mercury since this result is at variance with many other reports (Ernst et al., 1993).

Short-Term and Subchronic Toxicity

In squirrel monkeys exposed to mercury vapor at 1 and 2 mg/m^3, 5 d/w (hours per day not specified), for 6-69 d, no pathological changes were evident in the brain, although some brain structures contained up to 8 ppm of mercury (Berlin, 1976). Exposure of dogs at 15-20 mg/m^3, 8 h/d, for 2 or 3 d caused dyspnea, often with cyanosis, extreme weakness or lassitude, occasional vomiting and diarrhea, and death (Fraser et al., 1934). After exposures at 12.5 and 6 mg/m^3, these effects were less severe and deaths were delayed. Daily 8-h exposures at 3 and 4.4 mg/m^3 for 20-32 d caused gingivitis, loss of appetite, and diarrhea, without death. In two dogs exposed at 1.9 mg/m^3, 8 h/d, for 40 d, no toxic effects were evident except for a transient redness of the gums of one animal (Fraser et al., 1934). In rabbits, repeated exposures at 28.8 mg/m^3 for 1, 2, or 4 h for 5 d or for 6 h for 2 or 3 d caused severe damage to the kidneys and brain and less damage to the lungs, liver, colon, and heart (Ashe et al., 1953). After exposures at 6 mg/m^3, 7 h/d, 5 d/w, for 1-5 w, tissue damage to the kidneys, brain, heart and lungs was mild to moderate and more severe when exposures were 6-11 w. At 0.86 mg/m^3, 7 h/d, 5 d/w, for 2-5 w, histopathological changes were generally mild and evident only in the kidneys and brain; for exposures of 6-12 w, changes were moderate in the kidneys and mild in the brain, liver, heart, and lungs. Localized concentrations of mercury in the brain have been studied in rats exposed at up to 0.4 mg/m^3, 6 h/d, for 2 w (Ernst et al., 1993). In Wistar rats exposed to mercury vapor at 0.1 mg/m^3, 6 h/d, 5 d/w, for 7 w, the mercury content in the kidney was found to be about 50-fold higher than in any other organ (Eide and Wesenberg, 1993). Autoimmune glomerulonephritis and proteinuria were found in Brown-Norway rats exposed to mercury vapor at 1 mg/m^3, 24 h/d for 5 w (Hua et al., 1993). A less-exposed group from the same study (5 h/d for 5 w) showed autoimmune glomerulonephritis without proteinuria.

Repeated exposure to mercury vapor also causes decrements in operant behavior of the pigeon and the rat. In eight pigeons exposed to mercury vapor at 17 mg/m^3, 2 h/d, 5 d/w, response rates (key pecking for food reward) were reduced 50% in one of the eight in the first week, in six of the eight in the fifth week, and in all eight by the fourteenth week (Armstrong et al., 1963). Tremors in the head, neck, and wings appeared during the fifth week of exposure and were the only toxic signs.

In rats, exposure to mercury vapor at 17 mg/m^3, 2 h/d, 5 d/w, over 30 d decreased shock-avoidance responses and increased escape-response latency time in a conditioned operant test (Beliles et al., 1968). Escape-response latency increased after 15 d of exposure and was 7 times that of controls at 30 d. Tremor and weight loss were evident during the last 5 d. Recovery of operant behavior began at 14 d post-exposure and was almost complete at 45 d. Histopathological changes were not evident in the kidneys, liver, or lungs; but in the brain, two of three rats had lymphocytic cuffing around capillaries of the medulla oblongata. The same exposure regimen exacerbated shock-induced reflexive fighting behavior of rats and increased spontaneous fighting, requiring termination of the experiment after 18 d (Beliles et al., 1968).

Chronic Toxicity

In humans, the principal target organs of chronic exposure to mercury vapor are the central nervous system (CNS) and the kidneys. The classic symptoms of chronic inhalation of the vapor include (1) intention tremor; (2) erethism, a neuropsychiatric syndrome that includes irritability, excitability, loss of memory, loss of self-confidence, insomnia, and depression; and (3) gingivitis (Cragle et al., 1984). Less-common effects on the kidneys, including the nephrotic syndrome, proteinuria, and other signs of renal dysfunction, have been attributed to chronic exposure (Kazantzis et al., 1962; Foa et al., 1976; Buchet et al., 1980; Roels et al., 1982; Rosenman et al., 1986).

One major study reported no significant signs or symptoms of chronic mercury exposure in 479 workers whose time-weighted average exposures were at or below 0.1 mg/m^3, some symptoms (tremor, insomnia, loss of appetite and weight) among 88 workers exposed at 0.11-0.27 mg/m^3, and no evidence of kidney damage or other organ injury (Smith et al., 1970). In a followup study of these and other workers, no correlation was found between adverse health effects in workers and exposure to mercury vapor at levels between 0.05 and 0.1 mg/m^3 (McGill et al., 1964; Danzinger and Possick, 1973). Other investigators also reported no cases of classic mercury toxicity or evidence of significant exposure-related abnormalities when average concentrations did not exceed 0.1 mg/m^3 (McGill et al., 1964; Danzinger and Possick, 1973). In contrast, other studies attributed various toxic signs, including erethism, tremor, decreased nerve conduction velocity, decreased red-blood-cell cholinesterase, and renal dysfunction, in workers to exposure to mercury vapor below 0.1 mg/m^3, or even below 0.05 mg/m^3 (Bidstrup et al., 1951; Fawer et al., 1983; Verberk et al., 1986; WHO, 1991).

Peak or time-integrated average urinary mercury levels in workers were reported to be associated with neurological dysfunction, increased tremor, impaired psychomotor performance, decreased coordination, verbal intelligence and memory, renal dysfunction, and other effects (Langolf et al., 1978, 1981; Buchet et al., 1980; Roels et al., 1982; Smith et al., 1983; Rosenman et al., 1986).

Based on its review of the literature, the World Health Organization (WHO, 1991) concluded that, at mercury vapor concentrations above 0.08 mg/m^3 (corresponding to urinary mercury of 0.1 mg/g creatinine), the probability of developing neurological signs (tremor, erethism) is high. At 0.025-0.08 mg/m^3 (corresponding to urinary mercury of 0.03-0.10 mg/g creatinine), the incidence of certain less severe toxic effects is increased. These subtle effects include impaired psychomotor performance, measurable tremor, impaired nerve conduction velocity, fatigue, irritability, loss of appetite, and possibly proteinuria.

Continuous low-level exposure to mercury vapor also occurs as a result of the release of vapor from amalgam fillings in the mouth (WHO, 1991). Several studies have shown a correlation between the number of amalgam fillings or surfaces with mercury content in brain and kidney tissue from human autopsy (WHO, 1991). However, in an epidemiological study of 1024 women, there were no positive correlations between the number of fillings and the symptoms reported (Ahlqwist et al., 1988).

In 2 dogs, 18 rabbits, and 25 rats exposed to mercury vapor at 0.1 mg/m^3, 7 h/d, 5 d/w, for up to 83 w, there was no histopathology in organs of rats at 72 w, in organs of rabbits and a dog at 83 w, or in kidney biopsy samples from 2 dogs taken at 38 and 48 w (Ashe et al., 1953). Kidney function tests in the 2 dogs at 41 or 43 w and 60 or 83 w also were normal.

Mercury vapor at 3 mg/m^3, 5 d/w, caused a 50% reduction in shock-avoidance response rate in two of seven rats after 15 w and in avoidance after 41 w and escape rates after 35 w in all seven rats (Kishi et al., 1978). Response rates were normal within 12 w after termination of exposure. No histological changes were evident in the brain, lungs, or liver, but there was a slight degenerative change to the tubular epithelium of the kidneys.

Exposure to mercury vapor at 4 mg/m^3, 6 h/d, 4 d/w, for 13 w caused occasional tremor and clonus of the fore- and hind-legs of two of six rabbits (Fukuda, 1971). Highest mercury concentrations in brain structures were in the cerebellum, tegmentum, and thalamus.

Genotoxicity and Carcinogenicity

An increased incidence of chromosome aberrations was reported in four asymptomatic workers who had been exposed to mercury vapor at 0.15-0.44 mg/m^3 and had elevated urinary mercury levels during the preceding year (Popescu et al., 1979). Another study did not find an increase of structural aberrations in peripheral blood lymphocytes in workers exposed to the mercury vapor (Mabille et al., 1984).

There are no reports, to our knowledge, that mercury vapor is carcinogenic. An epidemiological study of 2400 workers exposed to mercury vapor for several years did not find excess deaths from diseases or cancers of the brain and CNS, kidneys, liver, or lungs due to mercury vapor exposure (Cragle et al., 1984).

Reproductive and Developmental Toxicity

The reproductive and developmental toxicities of mercury vapor will be considered together since studies often address both end points. The database consists primarily of epidemiological studies of workers exposed as a result of employment in a factory or as dental professionals and rodent inhalation studies. The potential for developmental toxicity is high because elemental mercury readily crosses the placenta and accumulates in many fetal organs, but at concentrations below those typically found in the mother (Clarkson et al., 1972; Khayat and Dencker, 1982).

The reproductive toxicity of mercury has been evaluated in both male and female industrial workers exposed to concentrations that were often incompletely characterized. Female workers (n = 153) in a mercury vapor lamp factory exposed to concentrations mostly below a time-weighted average of 0.05 mg/m^3 reported higher rates of menstrual disorders and adverse pregnancy outcomes than unexposed workers did; however, the authors conclude that their findings neither proved nor excluded the possibility that mercury causes adverse effects on reproduction (De Rosis et al., 1985). Male workers (n = 103) from various industrial plants where mercury vapor exposures increased urinary concentrations to 50 µg/g of creatinine (1 µg/g of creatinine in controls) showed no statistically significant difference in the observed number of offspring compared with a matched control population (Lauwerys et al, 1985). In another study of male workers (n = 247) with exposures to mercury vapor and average urinary concentrations over a 20-y period ranging from 27 to 107 µg/L, there was no association between the father's mercury exposure and decreased fertility, increased malformations in offspring, or childhood illness. Furthermore, the father's mercury exposure was not a significant risk factor for miscarriage after controlling for previous history (Alcser et al., 1989). In a result that seems to contradict this finding, a doubling of the rate of spontaneous abortions was found in the wives of 152 workers with average urinary mercury concentrations above 50 µg/L before pregnancy (Cordier et al,. 1991).

A number of studies have focused on potential adverse effects in dental professionals who receive exposure to mercury vapor during amalgam restorations. In a questionnaire study of nearly 60,000 dental workers divided into two groups according to the frequency of amalgam restorations performed, there was no significant increase in the rate of spontaneous abortion or congenital malformations with the presumed increase in exposure to mercury vapor (Brodsky et al, 1985). In contrast, Sikorski et al. (1987) found that the rate of reproductive failures and menstrual cycle disorders in 81 female dental workers was associated with the mercury content of their hair. In a study of over 8000 infants born to dental workers in Sweden, the frequency of perinatal death, low birth weight, and malformations was comparable to the incidence in the general population (Ericson and Kallen, 1989). On the whole, data from worker studies must be considered inconclusive about the potential for mercury vapor to cause reproductive or developmental toxicity at concentrations experienced in occupational settings.

Studies in rodents exposed to concentrations well above those experienced by workers have demonstrated mercury vapor's potential for reproductive and developmental toxicity. The estrus cycle was prolonged in rats exposed at 2.5 mg/m^3, 6 h/d, 5 d/w for 21 d; however, during the latter weeks of the exposure, CNS signs were observed (Baranski and Szymczyk, 1973). In the same study, offspring of females exposed 3 w before mating and during gestational days 7-20 were reduced in number and all died by the sixth day after birth. In a study reported by abstract only, rats were exposed on gestational days 15-20 or 1-20 to concentrations of 0.1, 0.5, and 1.0 mg/m³ (Steffek et al., 1987). No effects were seen at 0.1 mg/m³; however, increased resorptions and cranial defects (2 of 84) were found in offspring from the 0.5-mg/m³ group. Relative to the control group, maternal toxicity was evident in the high-exposure group as decreased weight gain (Steffek et al., 1987). Several behavioral effects were found in offspring of dams exposed to mercury vapor for 1 or 3 h/d at 1.8 mg/m³ for 7 d (days 11-14 and 17-20 of gestation) (Danielsson et al., 1993).

Interactions with Other Chemicals

Catalase inhibitors, such as ethanol and aminotriazole, inhibit the oxidation of elemental mercury to mercuric ion in blood and tissues (Nielsen-Kudsk, 1969; Magos et al., 1974). Pretreatment with ethyl alcohol (rat and marmoset monkey) or aminotriazole (rat) caused decreased mercury retention in most organs and in the whole body, increased blood concentrations of elemental mercury, and increased retention of mercury in most liver and adrenal cells (Khayat and Dencker, 1984). Ingestion of moderate amounts of ethanol by three human volunteers decreased mercury uptake by red blood cells and retention of mercury by the body and increased the rapid phase of vapor expiration and mercury storage in the liver (Hursch et al., 1980).

TABLE 7-1 Toxicity Summary[a]

Concentration, mg/m³ [b]	Exposure Duration	Species	Effects	Reference
N.S.	N.S.	Human (workers)	Decreased short-term memory with increasing exposure (based on urinary Hg)	Smith et al., 1983
N.S.	y (N.S.)	Human (workers)	No neurotoxic effects detected by routine medical exam; changes in tremor spectra and decreased memory with increased urinary Hg	Langolf et al., 1981
N.S.	0.5-34 y (mean 7 y)	Human (workers)	Glomerular dysfunction, with prevalence of increased urinary excretion of high molecular-weight proteins and increased β-galactosidase activity in plasma and urine increasing with urinary Hg	Buchet et al., 1980
N.S.	<2 to >10 y	Human (workers)	Elevated urine and blood Hg; routine clinical tests (physical exam, blood chemistries, urinalysis) normal; neuropsychiatric symptoms, elevated urinary glucosaminidase, decreased motor nerve conduction, lenticular opacities	Rosenman et al., 1986
N.S.	0.5-20 y	Human (workers)	Neurological and neuromuscular abnormalities, decreased psychomotor performance at elevated urine and blood Hg levels	Miller et al., 1975
N.S.	13-16 y	Human (workers)	No excess death from diseases or cancers of brain and CNS, kidneys, liver, or lungs due to mercury exposure	Cragle et al., 1984
0.01-0.27	1-20 y	Human (workers)	Loss of appetite, weight loss, tremors, insomnia at >0.1 mg/m³; no significant toxic signs/symptoms at <0.1 mg/m³	Smith et al., 1970
0.02 (estimate)	0.5-19 y	Human (workers)	Increased finger tremor; 0.017 mg/m³ of vapor estimated from urinary Hg	Verberk et al., 1986

0.026 (average)	15 y average (1-41 y range)	Human (workers)	Changes in hand tremor indicative of neurological effects	Fawer et al., 1983
0.028 (average)	1-15 y	Human (workers)	Increased plasma galactosidase and catalase, decreased rbc cholinesterase, increased blood and urine Hg	Lauwerys and Buchet, 1973
0.043 (average)	N.S.	Human (workers)	Minor, subclinical, functionally insignificant effects in tests for tremor, EMG, and psychomotor performance when urine Hg >0.5 mg/lL	Langolf et al., 1978
0.05-0.10	3.5, 20 y	Human (workers)	No evidence of weight loss or other adverse effects	Bunn et al., 1986
0.08-0.10	Up to 6 y	Human (workers)	No evidence of Hg toxicity by physical examinations	McGill et al., 1964
0.08 (average)	0.5-40 y	Human (workers)	No cases of classical mercury poisoning by physical exam/medical history, but symptoms (insomnia, gingivitis, salivation, or irritability) in a few workers	Danzinger and Possick, 1973
N.S. (possibly 0.15-0.44)	9 y (average)	Human (workers)	Increased incidence of chromosomal aberrations in peripheral blood	Popescu et al., 1979
1.1-2.9	2.5-5 h	Human (workers)	Fever, cough, dyspnea and tightness in chest developed 4 h after exposure	Milne et al., 1970
44 (maximum)	<8 h	Human (workers)	Fever, chills, chest pain, weakness, impaired pulmonary function; later, chronic neuropsychiatric symptoms and tremor	McFarland and Reigel, 1978
0.1	7 h/d, 5 d/w, up to 83 w	Dog, rabbit, rat	No histopathology; kidney function tests normal in 2/2 dogs at up to 83 w	Ashe et al., 1953
0.55	12-24 h	Rat	3/12 animals died, may not be exposure-related	Moller-Madsen, 1992
0.86	7 h/w, 5 d/w, 2-5 w	Rabbit	Mild histopathological changes only in kidneys and brain	Ashe et al., 1953

TABLE 7-1 (Continued)

Concentration, mg/m^3	Exposure Duration	Species	Effects	Reference
0.86	7 h/d, 5 d/w, 6-12 w	Rabbit	Moderate histopathological changes in kidneys; mild changes in brain, liver, heart, and lungs	Ashe et al., 1953
0.86	7 h/w, 5 d/w, 2-5 w	Rabbit	Histopathological changes mild and only in kidneys and brain	Ashe et al., 1953
1	4 h	Monkey, rabbit, rat	Toxic signs (increased irritability and aggressiveness, decreased food intake 24 h) only in the rat	Berlin et al., 1969
1	24 h/d, 35 d or 6 h/d, 35 d	Rat	Autoimmune glomerulonephritis proteinuria in long-term exposure group	Hua et al., 1993
1, 2	h/d N.S., 5 d/w, 6-69 d	Monkey (squirrel)	No pathological changes in brain	Berlin, 1976
1.9	8 h/d, 40 d	Dog	Only toxic effect was transient gum redness in 1/2	Fraser et al., 1934
3	3 h/d, 5 d/w, up to 41 w	Rat	50% reduction in shock-avoidance response rates in 2/7 and 7/7 in 15 and 41 w, respectively; and in escape rates in 7/7 in 35 w; no histopathology, except mild changes in kidneys	Kishi et al., 1978
3, 4	8 h/d, 20-32 d	Dog	Gingivitis, loss of appetite, diarrhea, no deaths	Fraser et al., 1934
4	6 h/d, 4 d/w, 13 w	Rabbit	Occasional tremor and clonus of fore- and hindlegs in 2/6	Fukuda, 1971
6	7 h/d, 5 d/w, 1-11 w	Rabbit	Mild-to-moderate histopathological changes in kidneys, brain, heart, and lungs for 1-5 w exposure, more severe for 6-11 w	Ashe et al., 1953
15-20	8 h/d, 2-3 d	Dog	Dyspnea, cyanosis, weakness, or lassitude; occasional vomiting and diarrhea and death	Ashe et al., 1953

17	2 h/d, 5 d/w, 30 d	Rat	Shock-avoidance responses decreased, escape response latency time increased after 15 d; tremor and weight loss at 25-30 d; no pathology in kidneys, liver, and lungs; mild pathology in medulla oblongata	Beliles et al., 1968
17	2 h/d, 5 d/h, 30 d	Pigeon	Response rate (key pecking) reduced 50% in 1/8 1st w, 6/8 5th w, and 8/8 14th w; tremor appeared at 5th w	Armstrong et al., 1963
20	8 h	Dog	Dyspnea and weakness, death on exposure day	Fraser et al., 1934
28.8	1 h	Rabbit	Moderate histopathological changes in kidneys and brain; mild in lungs and heart	Ashe et al., 1953
28.8	2, 4, or 6 h	Rabbit	Severe histopathological damage to kidneys and brain; mild-to-moderate in lungs, liver, colon, and heart	Ashe et al., 1953
28.8	1, 2, or 4 h/d, 5 d; 6 h/d 2 or 3 d	Rabbit	Severe tissue damage to kidneys and brain; lesser damage to lungs, liver, heart, and colon	Ashe et al., 1953

[a] Only results of inhalation studies are included.
[b] N.S. = not specified.

TABLE 7-2 Exposure Limits Set by Other Organizations

Organization	Concentration, mg/m^3
ACGIH's TLV	0.05, TWA (skin)
OSHA's PEL	0.1, TWA (ceiling)
NIOSH's REL	0.05, TWA (skin)
NIOSH's IDLH	28
NRC's EEGL	0.2, 24 h[a]
NRC's CEGL	0.01, 90 d[a]

[a]NRC, 1984.
TLV = threshold limit value. TWA = time-weighted average. PEL = permissible exposure limit. REL = recommended exposure limit. IDLH = immediately dangerous to life and health. EEGL = emergency exposure guidance level. CEGL = continuous exposure guidance level.

TABLE 7-3 Spacecraft Maximum Allowable Concentrations

Duration	ppm	mg/m^3	Target Toxicity
1 h	0.01	0.08	Respiratory tract
24 h	0.002	0.02	Respiratory tract
7 d	0.001	0.01	CNS, kidney
30 d	0.001	0.01	CNS, kidney
180 d	0.001	0.01	CNS, kidney

RATIONALE FOR ACCEPTABLE CONCENTRATIONS

In setting SMAC values for mercury vapor, the toxic effects on the respiratory tract, the brain and CNS, and the kidney must be considered. Few well-controlled animal studies have been conducted with observations and measurements of toxic end points in adequate numbers of animals and at more than one concentration of vapor. In most reports of human exposures, analytical data are lacking. Guidelines from the Committee on Toxicology have been used to structure the rationale below (NRC, 1992).

Respiratory System Toxicity

For single, brief exposures to mercury vapor, it appears that the respiratory system is the most sensitive target organ in human beings (Milne et al., 1970; McFarland and Reigel, 1978). A few hours after exposure, cough, shortness of breath, and tightness of the chest develop, and when clinical evaluations have been conducted, a diffuse pulmonary infiltrate has been found. There are no well-characterized human exposures; however, data from two accidental industrial exposures were used to estimate an acceptable concentration (AC) for the lung. In one accident, nine workers were exposed for about 5 h to a mercury concentration that could have been as much as 40 mg/m^3 (McFarland and Reigel, 1978). Even though three of the workers complained of no illness, six had moderate respiratory symptoms like those indicated above. It is possible that human genetic variations in catalase activity and enzymes related to endogenous peroxide supply might affect human responses to inhaled mercury (Clarkson et al., 1980). All six injured workers recovered from respiratory injury; however, in some there were lingering subjective symptoms including fatigue, irritability, and sexual disinterest. Two of the exposed workers had lingering tremors, which never fully disappeared in one subject. In another report of accidental human exposures at 1-3 mg/m^3 for 2.5-5 h, similar respiratory symptoms were reported in three-fourths of the exposed workers and one-fourth of the workers reported minimal respiratory symptoms (Milne et al., 1970). Authors of the first study assert that all nine workers had approximately equal exposures to mercury vapor even though there was a wide range in apparent lung injury. An estimate of short-term human exposure limits was made as follows: A LOAEL was estimated from the 13 exposed workers as an exposure of 2 mg/m^3 for 5 h. For the 1-h AC to protect the lung, the estimate was 2 mg/m^3 × 0.4 (the square root of 13 divided by 10 for the small number of subjects) × 1/10 (LOAEL to NOAEL), or 0.08 mg/m^3. For the 24-h AC, the estimate was 2 mg/m^3 × 0.4 × 1/10 × 5/24 (Haber's rule), or 0.02 mg/m^3.

The above estimates of safe mercury concentrations were based on incomplete human data and should not be adopted without comparison with available animal data on lung injury. From short-term exposures of rabbits at concentrations of 30 mg/m^3 for 1-4 h, the degree of histologically detected injury was reported as mild (Ashe et al., 1953). From these data, the 1-h AC to protect the lung was 30 mg/m^3 × 1/10 (species extrapolation) × 1/10 (LOAEL to NOAEL), or 0.3 mg/m^3. This was about fourfold above the estimate from human data, so the lower 1-h AC of 0.08 mg/m^3 was adopted. For the 24-h AC to protect the lung, as estimated from the rabbit data, the value was 30 mg/m^3 × 1/10 × 1/10 × 4/24 (Haber's rule), or 0.05 mg/m^3. Again, this was above the estimate from human data, so the 24-h AC for the lung of 0.02 mg/m^3 was adopted.

Nephrotoxicity

Data were available for subchronic and chronic exposures of animals to mercury vapor. No more than mild histopathological changes were seen in the kidneys of rabbits exposed at 0.86 mg/m^3, 7 h/d, 5 d/w, for up to 4 w, and no histopathological changes were seen in the kidneys of rabbits, rats, and two dogs exposed at 1 mg/m^3 for 83 w (Ashe et al., 1953). Applying a species factor of 10 to the long-term NOAEL of 0.1 mg/m^3 gave an AC for nephrotoxicity of 0.01 mg/m^3 for exposures of 7, 30, and 180 d. Haber's rule was not applied because mercury concentrations in kidneys of exposed rabbits did not increase after the fourth week of exposures (Ashe et al., 1953).

Neurotoxicity

Animal data for estimating potential neurotoxicity in humans were available from the same study that provided data on nephrotoxicity. Mild histopathological changes in the brains of rabbits resulted from exposure at 0.86 mg/m^3, 7 h/d, 5 d/w, for 2-4 w, but no histopathological changes were detected when rabbits, rats, and two dogs were exposed at 0.1 mg/m^3 for 83 w (Ashe et al., 1953). In the 83-w study, the tissue concentrations of mercury in the brain were roughly an order of magnitude below the concentrations found in the kidney, but do not conclusively show that mercury is not accumulating in brain tissue. However, the exposures at 0.86 mg/m^3 show no increased accumulation between the fifth and twelfth weeks of exposure, nor does the extent of tissue damage increase (Ashe et al., 1953). Hence, the concentration of 0.1 mg/m^3 was considered a NOAEL, and a species factor of 10 was applied, without using Haber's rule, to derive an AC for neurotoxicity of 0.01 mg/m^3 for 7-, 30-, and 180-d exposures.

A large epidemiological study reported no significant toxic effects below 0.1 mg/m^3 in workers chronically exposed to mercury vapor, but there were some complaints of symptoms (Smith et al., 1970). A few much smaller studies suggest occasional complaints, symptoms, or subclinical effects at exposures below 0.1 mg/m^3 and even possibly below 0.05 mg/m^3 (Bidstrup et al., 1951; Fawer et al., 1983; Verberk et al., 1986). Consequently, 0.1 mg/m^3 is considered a LOAEL and a factor of 10 was applied to estimate a human NOAEL.

No adjustments to the AC values were necessary for any microgravity-induced physiological changes.

RECOMMENDATIONS

The most important SMACs for mercury in spacecraft air are the short-term values because this chemical would be removed from the air after release and because a continuous slow-release source is unlikely to be accidentally created. The long-term effects of mercury have been studied thoroughly in occupationally exposed populations; however, the short-term effects in human beings can only be approximated from the few accidental exposures that have occurred. Only one animal study was found on short-term effects of mercury vapor inhalation and it left unanswered several questions important to the setting of short-term SMACs. The time-vs.-concentration relationships need better definition for brief continuous exposures lasting from 1 h to a few days. The relationships need to be defined for each apparent target site in animal models: brain, kidney, liver, colon, heart, and lung. In addition, the mechanism of the damage needs exploration to improve extrapolations from animal models to humans. Because the lung appears to be the target site in humans after acute exposure, future research should be focused on understanding biochemical and microscopic changes in that organ. Appropriate animal models should be selected carefully, with the initial experiments involving exposure of several species.

TABLE 7-4 End Points and Acceptable Concentrations

End Point	Exposure Data	Species and Reference	Uncertainty Factors				Acceptable Concentrations, mg/m³				
			NOAEL	Time	Species	N	1 h	24 h	7 d	30 d	180 d
Respiratory tract toxicity	At 2 mg/m³, ≈5 h; 9 of 13 moderately severe responses	Human, n = 13 (Milne et al., 1970; McFarland and Reigel, 1978)	10	1 or 5 HR[a]	1	2.5	0.08	0.02	—	—	—
	At 30 mg/m³, 1-4 h; 4 of 4 mild responses	Rabbit (Stokinger, 1981; WHO, 1991)	10	1 or 6 HR	10	—	0.3	0.05	—	—	—
Nephrotoxicity	NOAEL at 0.1 mg/m³, 7 h/d, 5 d/w, 83 w	Dog, rabbit, rat (Ashe et al., 1953)	1	1[b]	10	—	—	—	0.01	0.01	0.01
Neurotoxicity	NOAEL at 0.1 mg/m³, 7 h/d, 5 d/w, 1-20 y	Dog, rabbit, rat (Smith et al., 1970)	1	1[c]	10	—	—	—	0.01	0.01	0.01
	LOAEL at 0.1 mg/m³, 8 h/d, 5 d/w, 1-20 y	Human (Smith et al., 1970)	10	1[c]	1	—	—	—	—	—	0.01
SMAC							0.08	0.02	0.01	0.01	0.01

[a]HR = Haber's rule.
[b]A time factor was not used because mercury concentrations did not increase in kidneys after 4 w of exposure.
[c]A time factor of 1 for prolonged exposures is supported by mercury concentrations in brains of exposed animals.

REFERENCES

ACGIH. 1986. Mercury. In Documentation of the Threshold Limit Values and Biological Exposure Indices. American Conference of Governmental Industrial Hygienists, Cincinnati, Ohio.

Ahlqwist, M., C. Bengtsson, B. Furunes, L. Hollender, and L. Lapidus. 1988. Number of amalgam tooth fillings in relation to subjectively experienced symptoms in a study of Swedish women. Community Dent. Oral Epidemiol. 16:227-231.

Alcser, K.H., K.A. Brix, L.J. Fine, L.R. Kallenbach, and R.A. Wolfe. 1989. Occupational mercury exposure and male reproductive health. Am. J. Ind. Med. 15:517-529.

Armstrong, R.D., L.G. Leach, P.R. Bellusio, E.A. Maynard, H.G. Hodge, and J.K. Scott. 1963. Behavioral changes in the pigeon following inhalation of mercury vapor. Am. Ind. Hyg. Assoc. J. 24:366-375.

Aschner, M., and J.L. Aschner. 1990. Mercury neurotoxicity: mechanisms of blood-brain barrier transport. Neurosci. Behav. Rev. 14:169-176.

Ashe, W.F., E. Largent, F. Dutra, D. Hubbard, and M. Blackstone. 1953. Behavior of mercury in the animal organism following inhalation. A.M.A. Arch. Ind. Hyg. Occup. Med. 7:19-43.

Baranski, B., and I. Szymczyk. 1973. Effects of mercury vapor upon reproductive functions of female white rats [in Polish]. Med. Pr. 24:248.

Beliles, R.P., R.S. Clark, and C.L. Yulie. 1968. The effects of exposure to mercury vapor on behavior of rats. Toxicol. Appl. Pharmacol. 12:15-21.

Berlin, M. 1976. Dose-response relations and diagnostic indices of mercury concentrations in critical organs upon exposure to mercury and mercurials. Pp. 235-245 in Effects and Dose-Response Relationships of Toxic Metals, G.F. Nordberg, ed. Amsterdam: Elsevier.

Berlin, M., J. Fazackerley, and G. Nordberg. 1969. The uptake of mercury in the brains of mammals exposed to mercury vapor and to mercuric salts. Arch. Environ. Health 18:719-729.

Bidstrup, P.L., J.A. Bonnel, D.G. Harvey, and S. Locket. 1951. Chronic mercury poisoning in men repairing direct-current meters. Lancet 2:856-861.

Brodsky, J.B, E.N. Cohen, C. Whitcher, B.W. Brown, Jr., and M.L. Wu. 1985. Occupational exposure to mercury in dentistry and pregnancy outcome. J. Am. Dent. Assoc. 111:779-780.

Buchet, J.P., H. Roels, A. Bernard, Jr., and R. Lauwerys. 1980. Assessment of renal function of workers exposed to inorganic lead, cadmium or mercury vapor. J. Occup. Med. 22:741-750.

Bunn, W.B., C.M. McGill, T.E. Barber, J.W. Cromer, Jr., and L.J. Goldwater. 1986. Mercury exposure in chloralkali plants. Am. Ind. Hyg. Assoc. J. 47:249-254.

Clarkson, T.W., L. Magos, and M.R. Greenwood. 1972. The transport of elemental mercury into fetal tissues. Biol. Neonate 21:239-244.

Clarkson, T.W., S. Halbach, L. Magos, and Y. Sugata. 1980. On the mechanism of oxidation of inhaled mercury vapor. Pp. 419-427 in the Molecular Basis of Environmental Toxicity. R.S. Bhatnagar, ed. Symposium at the 176th National Meeting of the American Chemical Society. Ann Arbor, Mich.: Ann Arbor Science Publishers.

Cordier, S., F. Deplan, L. Mandereau, and D. Hemon. 1991. Paternal exposure to mercury and spontaneous abortions. Br. J. Ind. Med. 48:375-81.

Cragle, D.L., D.R. Hollis, J.R. Qualters, W.G. Tankersley, and S.A. Fry. 1984. A mortality study of men exposed to elemental mercury. J. Occup. Med. 26:817-821.

Danielsson, B.R.G., A. Fredriksson, L. Dahlgren, A. T. Gardlund, L. Olsson, L. Dencker, and T. Archer. 1993. Behavioural effects of prenatal metallic mercury inhalation exposure in rats. Neurotoxicol. Teratol. 15:391-396.

Danzinger, S.J., and P.A. Possick 1973. Metallic mercury exposure in scientific glassware manufacturing plants. J. Occup. Med. 15:15-20.

De Rosis, F., S.P. Anastasio, L. Selvaggi, A. Beltrame, and G. Moriani. 1985. Female reproductive health in two lamp factories: effects of exposure to inorganic mercury vapour and stress factors. Br. J. Ind. Med. 42:488-494.

Dunn, J.D., T.W. Clarkson, and L. Magos. 1978. Ethanol-increased exhalation of mercury in mice. Br. J. Ind. Med. 35:241-244.

Dunn, J.D., J.W. Clarkson, and L. Magos. 1981. Interaction of ethanol and inorganic mercury: generation of mercury vapor in vivo. J. Pharmacol. Exp. Ther. 216:19-23.

Eide, R., and G.R. Wesenberg. 1993. Mercury contents of indicators and target organs in rats after long-term, low-level, mercury vapor exposure. Environ. Res. 61:212-222.

Ericson, A., and B. Kallen. 1989. Pregnancy outcome in women working as dentists, dental assistants or dental technicians. Int. Arch. Occup. Environ. Health 61:329-333.

Ernst, E., M-K Christensen, and E.H. Poulsen. 1993. Mercury in the rat hypothalmic arcuate nucleus and median eminence after mercury vapor exposure. Exp. Mol. Pathol. 58:205-214.

Fawer, R.F., Y. DeRibaupierre, M.P. Guillemin, M. Berode, and M. Lob. 1983. Measurement of hand tremor induced by industrial exposure to metallic mercury. Br. J. Ind. Med. 40:204-208.

Foa, V., L. Caimi, L. Amante, C. Antonini, A. Gattinoni, G. Tettamanti, A. Lombardo, and A. Giuliani. 1976. Patterns of some lysosomal enzymes in the plasma and of proteins in urine of workers exposed to inorganic mercury. Int. Arch. Occup. Environ. Health 37:115-124.

Fraser, A.M., K.I. Melville, and R.L. Stehle. 1934. Mercury-laden air: The toxic concentration, the proportion absorbed, and the urinary excretion. J. Ind. Hyg. 16:77-91.

Fukuda, K. 1971. Metallic mercury induced tremor in rabbits and mercury content of the central nervous system. Br. J. Ind. Med. 28:308-311.

Hua, J., L. Pelletier, M. Berlin, and P. Druet. 1993. Autoimmune glomerulonephritis induced by mercury vapor exposure in the Brown Norway rat. Toxicology 79:119-129.

Hursch, J.B., T.W. Clarkson, M.G. Cherian, J.V. Vostal, and R. V. Mallie. 1976. Clearance of mercury (Hg-197, Hg-203) vapor inhaled by human subjects. Arch. Environ. Health 31:302-309.

Hursch, J.B., M.R. Greenwood, T.W. Clarkson, J. Allen, and S. Demuth. 1980. The effect of ethanol on the fate of mercury vapor inhaled by man. J. Pharmacol. Exp. Therap. 214:520-527.

Hursch, J.B., S.P. Sichak, and T.W. Clarkson. 1988. In vitro oxidation of mercury by the blood. Pharmacol. Toxicol. 63:266-273.

Hursch, J.B., T.W. Clarkson, E. Miles, and L.A. Goldsmith. 1989. Percutaneous absorption of mercury vapor by man. Arch. Environ. Health 44:120-127.

Kazantzis, G., K.F.R. Schiller, A.W. Asscher, and R.G. Drew. 1962. Albuminuria as the nephrotic syndrome following exposure to mercury and its compounds. Q. J. Med. 31:403-418.

Khayat, A., and L. Dencker. 1982. Fetal uptake and distribution of metallic mercury vapor in the mouse: Influence of ethanol and aminotriazole. Biol. Res. Pregn. 3:1:38-46.

Khayat, A., and L. Dencker. 1984. Organ and cellular distribution of inhaled metallic mercury in the rat and Marmoset monkey (Callithrix jacchus): Influence of ethyl alcohol pretreatment. Acta Pharmacol. Toxicol. 55:145-152.

Kishi, R., K. Hashimoto, S. Shimizu, and M. Kobayashi. 1978. Behavioral changes and mercury concentrations in tissues of rats exposed to mercury vapor. Toxicol. Appl. Pharmacol. 46:555-566.

Langolf, G.D., D.B. Chaffin, and R. Henderson. 1978. Evaluation of workers exposed to elemental mercury using quantitative tests of tremor and neuromuscular functions. Am. Ind. Hyg. Assoc. J. 39:976-984.

Langolf, G.D., P.J. Smith, R. Henderson, and H.P. Whittle. 1981. Measurements of neurological functions in the evaluations of exposure to neurotoxic agents. Ann. Occup. Hyg. 24:293-296.

Lauwerys, R.R., and J.P. Buchet. 1973. Occupational exposure to mercury vapors and biological action. Arch. Environ. Health 27:65-68.

Lauwerys, R., H. Roels, P. Genet, G. Toussaint, A. Bouckaert, and S. De Cooman. 1985. Fertility of male workers exposed to mercury vapor or to manganese dust: a questionnaire study. Am. J. Ind. Med. 7:171-176.

Mabille, V., H. Roels, P. Jacquet, A. Leonard, and R. Lauwerys. 1984. Cytogenic examination of leucocytes of workers exposed to mercury vapor. Int. Arch. Occup. Environ. Health 53:257-260.

Magos, L. 1967. Mercury blood interaction and mercury uptake by brain. Environ. Res. 1:323-327.

Magos, L., Y. Sugata, and T.W. Clarkson. 1974. Effects of 3-amino-1,2,4-triazole on mercury uptake by in vitro human blood samples and by whole rats. Toxicol. Appl. Pharmacol. 28:367-373.

McFarland, R.B., and H. Reigel. 1978. Chronic mercury poisoning from a single brief exposure. J. Occup. Med. 20:532-534.

McGill, C.M., A.C. Ladd, M.B. Jacobs, and L.J. Goldwater. 1964. Mercury exposure in a chlorine plant. J. Occup. Med. 6:335-337.

Miller, J.M., D.B. Chaffin, and R.G. Smith. 1975. Subclinical psychomotor and neuromuscular changes in workers exposed to inorganic mercury. Am. Ind. Hyg. Assoc. J. 36:725-733.
Milne, J., A. Christophers, and P. DeSilva. 1970. Acute mercurial pneumonitis. Br. J. Ind. Med. 27:334-338.
Møller-Madsen, B. 1992. Localization of mercury in CNS of the rat. V. Inhalation exposure to metallic mercury. Arch. Toxicol. 66:79-89.
NRC. 1984. Emergency and Continuous Exposure Limits for Selected Airborne Contaminants. Vol. 1. Washington, D.C.: National Academy Press.
NRC. 1992. Guidelines for Developing Spacecraft Maximum Allowable Concentrations for Space Station Contaminants. Washington, D.C.: National Academy Press.
Nielsen-Kudsk, F. 1965. Absorption of mercury vapor from the respiratory tract in man. Acta Pharmacol. Toxicol. 23:250-262.
Nielsen-Kudsk, F. 1969. Factors influencing the in vitro uptake of mercury in blood. Acta Pharmacol. Toxicol. 27:161-172.
Ogata, M., K. Kenmotsu, N. Hirota, T. Meguro, and H. Aikoh. 1987. Reduction of mercuric ion and exhalation of mercury in acatalasemic and normal mice. Arch. Environ. Health 42:26-30.
Popescu, H.I., L. Negru, and I. Lancranjan. 1979. Chromosome aberrations induced by occupational exposure to mercury. Arch. Environ. Health 34:461-463.
Roels, H., R. Lauwerys, J.P. Buchet, A. Bernard, A. Barthels, M. Oversteyns, and J. Gaussin. 1982. Comparison of renal function and psychomotor performance in workers exposed to elemental mercury. Int. Arch. Occup. Environ. Health 50:77-93.
Rosenman, K.D., J.A. Valciukas, L. Glickman, B.R. Meyers, and A. Cinotti. 1986. Sensitive indicators of inorganic mercury toxicity. Arch. Environ. Health 41:208-215.
Sikorski, R., T. Juszkiewicz, T. Paszkowski, and T. Szprengier-Juszkiewicz. 1987. Women in dental surgeries: Reproductive hazards in occupational exposure to metallic mercury. Int. Arch. Occup. Environ. Health 59:551-557.
Smith, R.G., A.J. Vorwald, L.S. Patil, and T.F. Mooney, Jr. 1970. Effects of exposure to mercury in the manufacture of chlorine. Am. Ind. Hyg. Assoc. J. 31:687-700.
Smith, P.J., G.D. Langolf, and J. Goldberg. 1983. Effects of occupa-

tional exposure to elemental mercury on short term memory. Br. J. Ind. Med. 40:413-419.

Steffek, A.J., R. Clayton, C. Siew, and A.C. Verrusio. 1987. Effects of elemental mercury vapor exposure on pregnant Sprague-Dawley rats [abstract]. J. Dent. Res. 66:239.

Stokinger, H.E. 1981. The metals. Pp. 1493-2060 in Patty's Industrial Hygiene and Toxicology, Vol. 2A, 3rd Rev. Ed. New York: John Wiley & Sons.

Sugata, Y., and T.W. Clarkson. 1979. Exhalation of mercury-further evidence for an oxidation-reduction cycle in mammalian tissues. Biochem. Pharmacol. 28:3474-3476.

Verberk, M.M., H.J.A. Salle, and C.H. Kemper. 1986. Tremor in workers with low exposure to metallic mercury. Am. Ind. Hyg. Assoc. J. 47:559-562.

Vimy, M.J., and F.L. Lorscheider. 1985. Serial measurements of intra-oral air mercury: estimation of daily dose from dental amalgam. J. Dent. Res. 64:1072-1085.

WHO. 1991. Environmental Health Criteria 118: Inorganic Mercury. Geneva: World Health Organization.

B8 Methylene Chloride

King Lit Wong, Ph.D.
Johnson Space Center Toxicology Group
Biomedical Operations and Research Branch
Houston, Texas

PHYSICAL AND CHEMICAL PROPERTIES

Methylene chloride is a volatile and colorless liquid (ACGIH, 1986). Its vapor is not flammable or explosive (Merck, 1989).

Synonyms:	Dichloromethane
Formula:	CH_2Cl_2
CAS number:	75-09-2
Molecular weight:	84.9
Boiling point:	39.8°C
Melting point:	-96.7°C
Vapor pressure:	440 torr 25°C
Conversion factors at 25°C, 1 atm:	1 ppm = 3.47 mg/m^3
	1 mg/m^3 = 0.29 ppm

OCCURRENCE AND USE

Methylene chloride is a widely used solvent (NTP, 1986). Examples of its use are as a paint remover and a degreasing agent. There is no known use of methylene chloride in spacecraft, but methylene chloride has been shown to off-gas in space shuttles reaching typically 0.1 ppm in a few days (NASA, 1989).

PHARMACOKINETICS AND METABOLISM

Absorption

The blood equilibrates with inhaled methylene chloride sooner at rest than at exercise. DiVincenzo et al. (1972) showed that the methylene chloride concentration in blood in 11 human volunteers, who exercised during one-third of the exposure duration, did not plateau 2 h into an exposure to methylene chloride at a concentration of 100 or 200 ppm. Similarly, in a study conducted by Astrand et al. (1975) with five human subjects who were exposed to methylene chloride at 500 ppm for 2 h, with the first 30 min at rest, followed by 30 min of exercise at a 50-watt workload, 30 min of exercise at a 100-watt workload, and 30 min of exercise at a 150-watt workload, both the arterial and venous concentrations of methylene chloride did not reach a plateau in 2 h (Astrand et. al., 1975). In contrast, DiVincenzo and Kaplan (1981) showed that the methylene chloride concentration in venous blood reached a plateau in 2 h during a 7.5-h exposure of four to six sedentary human volunteers to methylene chloride at 50-150 ppm. However, when exposed to methylene chloride at 200 ppm, the blood concentration failed to plateau in 7.5 h (DiVincenzo and Kaplan, 1981).

Experiments demonstrated that methylene chloride is quite well absorbed during inhalation exposures. DiVincenzo et al. (1972) reported that methylene chloride vapor was rapidly absorbed by the lung during the first few minutes of exposure of 11 human volunteers to methylene chloride at 100 or 200 ppm. Astrand et al. (1975) showed that human subjects at rest absorbed 55% of the amount of methylene chloride inhaled in a 30-min exposure at 250 or 500 ppm. The absorption decreased to 40% when the subjects were working at a load of 50 watts, which is equivalent to light exercise (Astrand et al., 1975). In a study conducted by DiVincenzo and Kaplan (1981), up to 70% of the methylene chloride inhaled in a 7.5-h exposure at 50-200 ppm was absorbed by resting human subjects.

Distribution

In rats, methylene chloride is distributed, after a 1-h exposure at 560 ppm, mainly to white adipose tissue (Carlsson and Hultengren, 1975). The tissues, ranked according to methylene chloride concentrations in decreasing order, are white adipose tissue, liver, kidneys, and brain.

Metabolism

Methylene chloride is metabolized by two enzyme pathways in rodents (Kubic et al., 1974; Ahmed and Anders, 1978). The glutathione transferase pathway metabolizes methylene chloride into hydrogen chloride, formaldehyde, and carbon dioxide. Methylene chloride is also metabolized by the cytochrome P-450 system into hydrogen chloride, carbon monoxide, and carbon dioxide. McKenna et al. (1982) showed that metabolism of methylene chloride was saturable in rats; the percentage that was metabolized in 48 h after a 6-h methylene chloride exposure decreased from 95% to 69% to 45% as the exposure concentration increased from 50 ppm to 500 ppm to 1500 ppm, respectively. McKenna et al. reported that the major metabolites of methylene chloride in rats were carbon monoxide and carbon dioxide, which were exhaled.

DiVincenzo and Kaplan (1981) showed that, in a 7.5-h exposure of four to six sedentary human subjects to methylene chloride at 50-200 ppm, about 30% of the absorbed methylene chloride was converted into carbon monoxide, leading to a carboxyhemoglobin (COHb) concentration of 1.9-6.8% in blood. Even though the methylene chloride concentration in blood was approaching a plateau 2 h into the exposure at 50-150 ppm, the increase in COHb concentration did not slow down in the same period. According to Stewart et al. (1972), formation of 2.6-8% COHb in blood occurred in 11 men after a 1-2-h inhalation exposure to methylene chloride at 515-986 ppm.

Excretion

DiVincenzo and Kaplan (1981) reported that, after a 7.5-h methylene chloride exposure at 50-200 ppm in four to six human subjects, less than 5% of the absorbed methylene chloride was excreted unchanged in the expired air, and 25-34% was excreted as carbon monoxide during and after the exposure. DiVincenzo et al. (1972) showed that methylene chloride's blood concentration follows a bi-exponential decay in humans. The first phase of the decay is very rapid, followed by a slower phase with a half-life of about 40 min. A physiologically based pharmacokinetic model has been developed by Andersen et al. (1987). The model's predictions of the blood concentration of methylene chloride in mice, rats, hamsters, and humans agreed quite well with experimental data. Peterson (1978) also modeled the uptake, metabolism, excretion of methylene chloride in man. The model was used to predict the exhaled concentration of methylene chloride and the blood COHb concentration after an acute methylene chloride exposure.

After a methylene chloride exposure ends, the COHb concentration in blood might continue to rise, depending on the length of the exposure. In two studies in which humans were exposed to methylene chloride at 250-986 ppm for 1 or 2 h, the COHb concentration rose an average of 33% within 1 or 2 h after the exposure ended and then decreased with time (Stewart et al., 1972; Astrand et al., 1975) This suggests that, after the 1-2 h exposure, methylene chloride is released from some of the tissues and metabolized into carbon monoxide, leading to a temporary accumulation of COHb in blood. It is interesting that such a phenomenon does not occur in longer methylene chloride exposures. Serial samplings failed to demonstrate any further increase in COHb concentrations after a 7.5-8-h exposure of methylene chloride in two human studies (DiVincenzo et al., 1981; Andersen et al., 1987).

TOXICITY SUMMARY

Acute and Short-Term Toxicity

Acute exposures to methylene chloride are known to adversely affect the central nervous system (CNS) and the liver. These adverse effects are summarized below.

CNS Effects

Because carbon monoxide is one of methylene chloride's metabolites, the acute toxicity of methylene chloride resembles that of carbon monoxide. Putz et al. (1979) showed that a 4-h exposure to methylene chloride at 200 ppm, resulting in 5% COHb in the fourth hour, impaired the hand-eye coordination and increased the reaction time in 12 human volunteers. In the same study, these CNS effects were reproduced by a carbon monoxide exposure that yielded 5% COHb.

Acute methylene chloride exposures could impair vigilance performance in humans. In the 4-h exposure of 12 human volunteers to methylene chloride at 200 ppm conducted by Putz et al. (1979), impaired auditory vigilance was found. Winneke and Fodor (1976) also studied visual vigilance in eight women by measuring their abilities to correctly detect random drops in the intensity of a train of pulses of white noise. The vigilance performance started to deteriorate 1 h into the exposure to methylene chloride at 500 ppm. The eight women also subjectively felt a more rapid decline in their soberness, and they felt tired more rapidly during the 2-h and 20-min exposure to methylene chloride at 500 ppm than during the sham exposure to air. Winneke (1981) reported that visual vigilance was impaired by acute methylene chloride exposures as low as 300 ppm, so he concluded that "prolonged monotonous observation-tasks are easily disturbed by" methylene chloride.

Methylene chloride also could impair visual or CNS alertness in human subjects. In the study of Winneke and Fodor (1976), there was decreased visual or CNS alertness as early as 50 min into an exposure of 12 women to methylene chloride at 500 ppm for 2 h and 20 min, as measured by a drop in the monocular critical flicker frequency. Similar drops in the critical flicker frequency were detected by Winneke (1981) in a 95-min exposure to methylene chloride at 300 ppm. Stewart et al. (1972) also reported that a 2-h exposure to methylene chloride at 986 ppm, resulting in 10.1% COHb in the blood, changed the amplitude of visual-evoked potentials triggered by 100 strobe flashes in three out of three human volunteers.

Unlike other aspects of CNS function, Winneke's group showed that cognitive performances of human subjects were quite resistant to methylene chloride's depressive effect on the CNS (Winneke and Fodor, 1976; Winneke, 1981). DiVincenzo et al. (1972) exposed 11 men to methylene chloride at 100 and 200 ppm for 2 to 4 h, with the men exercising approximately one-third of the exposure duration. The exposure did not change the time for the men to complete tests of adding single-digit numerals. Winneke and Fodor (1976) showed that, in an exposure of 12 women to methylene chloride at 500 ppm for 2 h and 20 min, there were no differences in their performances in an addition test and a letter-canceling test. Even a 2-h exposure at 1000 ppm failed to reduce cognitive performances in human subjects, as determined by an addition test, the learning and retention of nonsense syllables, and the reproduction of visual patterns (a test of short-term memory) (Winneke, 1981).

As the exposure concentration increases, methylene chloride produces more overt CNS depression. Winneke (1981) reported that a 4-h methylene chloride exposure at 800 ppm results in depressive mood and motor impairment. As the concentration approached 1000 ppm, Stewart et al. (1972) reported that two of three human subjects complained of mild light-headedness after 1 h of exposure; one of the two developed a sensation of "thick tongue." Moskowitz and Shapiro (1952) reported four cases of accidental exposures to unknown but presumably very high concentrations of methylene chloride for 1-3 h, which produced unconsciousness in all the victims; three men finally recovered after 3-6 h and one man never regained consciousness and died.

Hepatic Effect

Other than acting on the CNS, methylene chloride might also affect the liver. A 6-h exposure at 5000 ppm or higher increases the hepatic triglyceride concentration in guinea pigs (Balmer et al., 1976).

Subchronic and Chronic Toxicity

Subchronic exposures to methylene chloride have been reported to produce COHb and toxic effects in the liver, kidney, and the respiratory system.

Carboxyhemoglobin Formation

Kim and Carlson (1986) compared COHb formation in rats exposed to the same concentrations of methylene chloride at 200, 550, or 960 ppm for either 8 h/d for 5 d or 12 h/d for 4 d. They found no significant difference in the COHb concentrations in rats exposed to the same concentration for 8 h/d or 12 h/d after the first and last exposures of the week. Similar results were found in mice. The half-lives of the disappearance of COHb in the 8-h or 12-h groups of rats also did not differ. However, the half-lives depended on the exposure concentration: the half-lives were 50 and 130 min in rats exposed to 550 ppm for 8 h or 960 ppm for 8 h, respectively. They concluded that unusual workshift would probably not change methylene chloride's toxicity mediated via COHb formation.

Non-neoplastic Effects on the Liver and Kidney

Subchronic exposure to methylene chloride could produce liver and kidney toxicity. MacEwen et al. (1972b) reported cellular vacuolization, nuclear enlargement, and iron pigmentation in portal areas of the liver and cortical tubular-cell degeneration in the kidney of rats exposed to methylene chloride at 1000 ppm, 24 h/d, for 100 d. In similarly exposed mice, ductal proliferation and large masses of brown pigment were found in or around the portal areas. In addition, a mild ballooning degeneration of cytoplasm and chromatin clumping were noted in the livers of these mice. In the kidneys, a very faint granular staining with hemosiderin was observed in some tubules of half of the mice examined. MacEwen et al. (1972b) also reported marked fatty liver in four dogs and mild fat accumulation in the liver of four monkeys exposed to methylene chloride at 1000 ppm for 100 d, but the kidney was not affected in the dogs and monkeys.

In 20 rats continuously exposed to methylene chloride at 25 ppm for 100 d, Haun et al. (1972) detected fatty changes and cytoplasmic vacuolization in the liver, as well as nonspecific tubular degeneration and regeneration in the kidney. In 20 mice exposed at 100 ppm, the only pathology discovered was fatty liver. No histopathology was found in any tissues of four dogs and four monkeys exposed at 100 ppm. At a lower concentration of 25 ppm, the only species affected was rats, which had fatty liver and nonspecific tubular degeneration in their kidneys.

Evaluation of these data indicates that the liver is more sensitive than the kidney to methylene chloride. These data also revealed species differences in the sensitivity toward methylene chloride's hepatic effects. The sensitivity of four test species ranked in decreasing order as rats, mice, dogs, and monkeys is shown in Table 8-1.

TABLE 8-1 Species Differences in Sensitivity for Hepatic Effects[a]

100-d Exposure Concentration, ppm	Hepatic Changes			
	Rat	Mouse	Dog	Monkey
1000	Marked	Marked	Marked	Mild
100	Mild	LOAEL	NOAEL	NOAEL
25	LOAEL	NOAEL	None	None

[a]Data from Haun et al. (1972) and MacEwen et al. (1972b).

The National Toxicology Program (NTP, 1986) sponsored subchronic and chronic toxicity studies conducted at exposure concentrations much higher than those in studies performed by MacEwen et al. (1972b) and Haun et al. (1972). The NTP's studies failed to show that rats were clearly more sensitive toward the non-neoplastic effects of methylene chloride than mice. In the NTP's subchronic toxicity study, rats and mice were exposed to methylene chloride at 1000, 2100, or 4200 ppm, 6 h/d, 5 d/w, for 90 d (NTP, 1986). The exposure at 1000 or 2100 ppm did not cause any histopathology, and the exposure at 4200 ppm produced mild centrilobular hydropic degeneration in mice but not in rats.

In the NTP's chronic toxicity study, rats were exposed at 1000, 2000, or 4000 ppm, and mice were exposed at 2000 or 4000 ppm, 6 h/d, 5 d/w, for 2 y (NTP, 1986). Rats and mice suffered different types of histopathology in the liver; rats were afflicted with more types of histopathology than mice. Mice in both the 2000- and 4000-ppm groups developed only cytological degeneration in the liver. In comparison, several types of hepatic pathology were found in the 1000-, 2000-, and 4000-ppm groups: focal granulomatous inflammation, focal necrosis, hemosiderosis, and cytoplasmic vacuolization.

In a chronic toxicity study sponsored by several chemical companies, Burek et al. (1984) exposed rats and hamsters to methylene chloride at 500, 1500, or 3500 ppm, 6 h/d, 5 d/w, for 2 y, and showed differences in the sensitivities between the two species. The methylene chloride exposures failed to increase the incidence of liver or kidney histopathology in hamsters, and some of the exposures affected the liver and kidney in rats. Similar to the findings of the NTP (1986), Burek et al. found that chronic methylene chloride exposures were more damaging to the liver than the kidney in rats. Methylene chloride did not cause any concentration-dependent increases in the incidence of glomerulonephropathy in female rats. In male rats, however, chronic methylene chloride exposures led to glomerulonephropathy at 1500 and 3500 ppm. In terms of liver injuries, chronic methylene chloride exposures at 500, 1500, or 3500 ppm produced vacuolization consistent with fatty liver in both male and female rats and they also caused multinucleared hepatocytes in female rats. Exposures at 1500 or 3500 ppm resulted in necrosis of individual hepatocytes in male rats, and the exposure at 3500 ppm produced coagulation necrosis and foci of altered hepatocytes in female rats.

Non-neoplastic Effects on the Respiratory System

Repetitive exposures of mice to methylene chloride at 4000 ppm, 6 h/d, 5 d/w, for up to 13 w, have been shown by Foster et al. (1992) to produce cytoplasmic vacuoles in bronchiolar Clara cells. The lesion appeared only on the second day of each week of exposure and resolved after the second day. The disappearance of the lesion correlated with a decrease in cytochrome P-450 monooxygenase activity in Clara cells, suggesting that Clara cells developed tolerance to methylene chloride with time by the inactivation of one of the pathways of methylene chloride metabolism. In contrast to mice, rats are not susceptible to this toxicity of methylene chloride (Foster et al., 1986). Since the Clara cell lesion did not appear to be too serious and disappeared with time, SMACs are not set according to the Clara cell lesion.

In the chronic toxicity study by the NTP (1986), exposures at 4000 ppm, 6 h/d, 5 d/w, for 2 y have been shown to cause squamous metaplasia in the nasal cavities of female rats but not those of male rats or female and male mice. Similar exposures at 2000 ppm failed to produce such a change. Squamous metaplasia in the nose is not relied on in setting methylene chloride's SMACs because it is a toxic effect seen only at very high exposure concentrations.

Neoplastic Effects

Two 2-y bioassays showed that methylene chloride was carcinogenic in rats and mice, but not in hamsters (Burek et al., 1984; NTP, 1986). In a chronic exposure of rats and hamsters to methylene chloride at 3500 ppm conducted by Burek et al. (1984), salivary gland sarcomas were the only kind of tumor found, and these sarcomas were observed only in the male rats. In the NTP's chronic toxicity study (1986), methylene chloride produced leukemia and benign mammary tumors in female rats and alveolar and bronchiolar adenomas and carcinomas in mice, as well as hepatocellular adenomas and carcinomas in mice. The incidences of lung tumors in female mice were 3 of 50, 16 of 48, and 40 of 48 in the 0-, 2000-, and 4000-ppm groups, respectively. The corresponding incidences of liver tumors were 3 of 50, 30 of 48, and 41 of 48. According to the NTP, methylene chloride shows clear evidence of carcinogenicity in female F344/N rats and male and female B6C3F$_1$ mice. An epidemiological study did not find any significant increase in cancer-related mortality in workers exposed to methylene chloride at 30-1200 ppm for up to 30 y (Friedlander et al., 1978). The ACGIH (1986) has classified it as a suspected human carcinogen.

The methylene chloride metabolites via the glutathione transferase pathway have been postulated to be the active metabolites in causing its carcinogenicity (Andersen et al., 1987). One of the metabolites formed is formaldehyde. Casanova et al. (1992) studied DNA-protein cross-links in rodents exposed to methylene chloride. They exposed mice and hamsters to methylene chloride at 4000 ppm, 6 h/d, for 2 d and then to ^{14}C-methylene chloride on the third day for 6 h at a concentration decaying from 4500 to 2500 ppm. They found DNA-protein cross-links in mouse liver, but not in mouse lung, while the cross-links failed to show up in either organs of hamsters. Casanova et al. stated that the failure to detect DNA-protein cross-links in mouse lung did not rule out the possibility that the cross-links existed in subpopulations of lung cells. They attributed the DNA-protein cross-links to formaldehyde formed from methylene chloride's metabolism via the glutathione transferase pathway.

Genotoxicity

Methylene chloride is mutagenic in *Salmonella typhimurium* (Jongen et al., 1978). It has been shown to cause chromosomal aberrations, but not sister chromatid exchange, in Chinese hamster ovary cells in vitro (Thilager and Kumaroo, 1983). It also failed to produce micronuclei in mice (Gocke et al., 1981).

Developmental Toxicity

It should be noted that methylene chloride has not been found to be teratogenic. Schwetz et al. (1975) exposed rats and mice to methylene chloride at 1225 ppm, 6 h/d, on gestation days 6-15 and failed to find any malformations in the fetuses. Because the exposure duration used by Schwetz et al. might not be long enough for a chemical that acts via its metabolites, Hardin and Manson (1980) exposed five female rats to methylene chloride at 4500 ppm, 6 h/d, 7 d/w, for 12-14 d before breeding and on days 1-17 of gestation. Hardin and Manson did not detect any increases in the incidence of skeletal or soft-tissue malformations or external anomalies.

Interaction with Other Chemicals

No evidence of interaction involving methylene chloride and other chemicals has been found in the literature.

TABLE 8-2 Toxicity Summary

Concentration, ppm	Exposure Duration	Species	Effects	Reference
30-1200	40 h/w, up to 30 y	Human (workers)	No increase in cancer-related mortality	Friedlander et al., 1978
200	4 h	Human	No effect on the speed of completing a pegboard test	DiVincenzo et al., 1972
200	4 h	Human	Impairment in hand-eye coordination (55% increase in tracking error); a 19% increase in response time; impaired auditory vigilance	Putz et al., 1979
250 for 30 min followed by 500 for 30 min, 750 for 30 min, and 1000 for 30 min	120 min	Human	No effects on the ability to add, short-term memory, and reaction time.	Gamberale et al., 175
300	95 min	Human	Visual flicker fusion affected indicating mental fatigue	Winneke, 1981
500	2.5 h	Human	Critical flicker fusion threshold affected and impaired vigilance	Winneke and Fodor, 196
500	2.5 h	Human	Complaints about general uneasiness	Winneke, 1981
80	4 h	Human	Depressed mood: feeling weak, slow, disorganized, and passive. Slower reaction times for simple and choice reactions, reduced tapping speed, impaired steadiness, precision of aiming movements, and coordination	Winneke, 1981
986	2 h	Human	Changes in the visual evoked potential amplitudes, light-headedness, and speaking difficulty	Stewart et al., 1972
1000	2.5 h	Human	No effect on cognitive performance	Winneke, 1981
25	24 h/d, 100 d	Mouse	No effects on spontaneous activity, hexobarbital sleep time, and weight gain; no histopathology	Haun et al., 1972; Thomas et al., 1972

25	24 h/d, 100 d	Rat	Normal weight gain; fatty liver and renal degeneration	Haun et al., 1972
100	24 h/d, 70 d	Mouse	Fatty liver starting at d 7 microscopically; increased hepatic triglyceride level starting after 2 w	Weinstein and Diamond, 1972
100	24 h/d, 100 d	Monkey, dog	No effects on body weight, gross and microscopic pathology, hematology, and clinical chemistry	Haun et al., 1972
100	24 h/d, 100 d	Mouse	No effects on spontaneous activity, hexobarbital sleep time, and weight gain; fatty liver	Haun et al., 1972; Thomas et al., 1972
500	6 h/d, 5 d/w, 2 w	Rat	Decreased succinate dehydrogenase activity in cerebellum	Savolainen et al., 1981
500	6 h/d, 5 d/w	Rat, hamster	Cytoplasmic vacuolization in liver cells in rats; multinucleated hepatocytes in female rats; carboxyhemoglobin increased by 13% in rats and 25% in hamsters	Burek et al., 1984
560	6 h	Guinea pig	14.3% carboxyhemoglobin.	Balmer et al., 1976
1000	6 h/d, 5 d/w, 2 w	Rat	Decreased succinate dehydrogenase activity in cerebellum and increased acid proteinase activity in cerebrum; both effects reversed 7 d after exposure	Savolainen et al., 1981
1000	6 h/d, 5 d/w, 2 y	Rat	Hemosiderosis, hepatocytomegaly, cytoplasmic vacuolization, focal necrosis, focal granulomatous inflammation in the liver; no tumors	NTP, 1986
1000	24 h/d, 100 d	Monkey, dog, rat, mouse	Dogs died with fatty liver and splenic atrophy; fatty liver in all species; kidney injury in rats	MacEwen et al., 1972a
1500	6 h/d, 5 d/w, 2 y	Hamster, rat	No injury in hamsters; cytoplasmic vacuolization in hepatocytes and necrosis of individual hepatocytes in male rats; multinucleated hepatocytes in female rats; chronic glomerulonephropathy in male rats; carboxyhemoglobin increased by 25% in hamsters	Burek et al., 1984

TABLE 8-2 (Continued)

Concentration, ppm	Exposure Duration	Species	Effects	Reference
2000	6 h/d, 5 d/w, 2 y	Rat	Mononuclear cell leukemia in females, but no tumors in males; hemosiderosis and cytoplasmic vacuolization in liver; focal necrosis and focal granulomatous inflammation in the liver of female rats; hepatocytomegaly and focal granulomatous inflammation of bile ducts in male rats	NTP, 1986
2000	6 h/d, 5 d/w, 2 y	Mouse	Alveolar/bronchiolar adenomas and carcinomas. Hepatocellular adenomas or carcinomas combined in females; cytoplasmic vacuolization in liver.	NTP, 1986
2100	6 h/d, 5 d/w, 90 d	Rat, mouse	No effect on body weight; no histopathology	NTP, 1986
3500	6 h/d, 5 d/w, 2 y	Hamster, rat	No injury in hamsters darcomas of salivary gland in male rats; reduced survival in female rats; cytoplasmic vacuolization in hepatocytes and multinucleated hepatocytes in female rats; carboxyhemoglobin increased by 11% in rats, 27% in hamsters	Burek et al., 1984
3700	5 h/d, 5 d/w, 4 w	Rat	Increases in protein, sialic acid, lactate dehydrogenase, acid and alkaline phosphatase, and hexose in lung lavage	Sahu et al., 1980
4000	6 h/d, 5 d/w, 2 y	Rat	Benign mammary tumors; mononuclear cell leukemia and focal granulomatous inflammation in liver in females; mesothelioma from tunica vaginalis and focal granulomatous inflammation of bile ducts in males; focal necrosis, cytoplasmic vacuolization, hemosiderosis in liver; reduced survival; squamous metaplasia of nasal cavity	NTP, 1986

4000	6 h/d, 5 d/w, 2 y	Mouse	Alveolar/bronchiolar adenomas and carcinomas; hepatocellular adenomas or carcinomas combined; testicular, ovarian, and uterine atrophy; hepatic cytologic degeneration	NTP, 1986
4200	6 h/d, 5 d/w, 90 d	Rat, mouse	No effect on body weight. Hydropic degeneration in the liver of mice	NTP, 1986
4500	6 h/d, d 1-17 of gestation	Rat	No teratogenicity	Hardin and Manson, 1980
5000	6 h	Guinea pig	Increased hepatic triglyceride level; 16.3% carboxyhemoglobin.	Balmer et al., 1976
5000	7 h/d, 5 d/w, 6 mo	Dog, rabbit, guinea pig, rat	No histopathology; retardation of growth in guinea pigs	Heppel et al., 1944
5000	24 h/d, 1 w	Mouse	Liver degeneration and fatty change	Weinstein et al., 1972
8400	6 h/d, 5 d/w, 90 d	Rat, mouse	Reduced body weight gain; hepatic hydropic degeneration in mice, foreign body pneumonia in rats, and deaths	NTP, 1986
10,000	4 h/d, 5 d/w, 8 w	Dog, rabbit, guinea pig, rat	Fatty liver in dogs and guinea pigs; no histopathology in rabbits and rats	Heppel et al., 1944
11,100	6 h	Guinea pig	Increased hepatic triglyceride level; 17.6% carboxyhemoglobin; lung congestion and hemorrhage	Balmer et al., 1976
11,600	6 h	Guinea pig	Half of the animals died within 18 h	Balmer et al., 1976
18,000	6 h	Rat	Half of the rats died	Laham et al., 1978

[a]Only results of inhalation studies are included.
[b]N.S. = not specified.

TABLE 8-3 Exposure Limits Set by Other Organizations

Organization	Concentration, ppm
ACGIH's TLV	50 (TWA)
OSHA's PEL	500 (TWA)
	1000 (ceiling)
NIOSH's REL	75 (TWA)
	500 (ceiling)
NIOSH's IDLH	5000

TLV = threshold limit value. TWA = time-weighted average. PEL = permissible exposure limit. REL = recommended exposure limit. IDLH = immediately dangerous to life and health.

TABLE 8-4 Spacecraft Maximum Allowable Concentrations

Duration	ppm	mg/m^3	Target Toxicity
1 h	100	350	CNS depression
24 h	35	120	CNS depression
7 d[a]	15	50	CNS depression
30 d	5	20	Liver
180 d	3	10	Liver

[a]Former 7-d SMAC = 25 ppm.

RATIONALE FOR ACCEPTABLE CONCENTRATIONS

The acceptable concentrations (ACs) for the three major toxic end points of CNS depression, liver toxicity, and carcinogenicity are estimated for continuous exposures lasting 1 h, 24 h, 7 d, 30 d, or 180 d. The lowest AC among the three end points will be chosen as the SMAC for each exposure duration.

CNS Depression

As discussed in the Toxicity Summary, one of the major acute effects of methylene chloride is CNS depression, which appears to be due to carbon monoxide formed from methylene chloride's metabolism. A 4-h exposure to methylene chloride at 200 ppm, which yields 5% COHb in blood, impairs the hand-eye coordination and auditory vigilance (Peterson et al., 1978), but there are no data on the no-observed-adverse-effect level (NOAEL) of methylene chloride. It makes sense to adopt the NOAEL of COHb used in setting the 1-h and 24-h SMACs of carbon monoxide as a potential basis for setting the 1-h and 24-h SMACs of methylene chloride. Three percent COHb is the target COHb concentration used to set both the 1-h and 24-h SMACs for carbon monoxide (Wong, 1990). The task here is to determine the methylene chloride concentrations that produce about 3% COHb in 1 and 24 h. Assuming a baseline COHb concentration of 0.6% due to endogenous CO production, the task is to determine the methylene chloride concentrations that would increase the COHb concentration by 2.4%. The increases in COHb concentrations produced by various methylene chloride exposure scenarios are shown in Table 8-5.

To derive the 1-h AC based on CO formation, a linear regression line was fitted through the data of percent COHb increase versus C × T by forcing the fitted line through the origin. All the data in the above table were used except the data points at 3750 and 1972 ppm-h because their corresponding responses of 10% and 9.3% increases in COHb were too far away from the region of interest, 2.4%. The linear regression yielded a line with a slope of 0.0038, r^2 of 0.74, and a 95% confidence limit of 100 ppm-h at a 2.4% increase in COHb. Accordingly, 100 ppm is selected as the 1-h AC based on CO formation.

To derive the ACs based on CO formation for a 24-h, 7-d, 30-d, or 180-d methylene chloride exposure, the physiologically based pharmacokinetic (PB-PK) model of Andersen et al. (1991) was used. For a 70-kg man with a starting COHb of 0.6%, this model predicted that an exposure to methylene chloride at 35 ppm would produce a final COHb of 3% in 24 h, so 35 ppm is chosen to be the 24-h AC based on CO formation.

To calculate the acceptable 7-d, 30-d, and 180-d methylene chloride concentrations based on the carbon monoxide metabolite, the target COHb concentrations of 1.6% were adopted from the 7-d, 30-d, and 180-d carbon monoxide SMACs. According to the PB-PK model of Andersen et al. (1991), a continuous exposure to methylene chloride at 14 ppm would raise the COHb concentration from 0.6% to 1.6% in a 70-kg man in 7, 30, or 180 d. The 7-d, 30-d, and 180-d ACs based on CO formation are all set at 14 ppm.

TABLE 8-5 Increases in COHb Produced by Methylene Chloride Exposures

MeCl Concentration, ppm	Exposure Time, h	C × T[a] (ppm × h)	No.	Increase in % COHb	Reference
50	7.5	375	11	1.6	Peterson, 1978
50	7.5	375	4-6	0.8	DiVincenzo and Kaplan, 1981
100	5	500	6	3.9	Andersen et al., 1981
100	7.5	750	11	3.2	Peterson, 1978
100	7.5	750	4-6	2.2	DiVincenzo and Kaplan, 1981
150	7.5	1125	4-6	4.0	DiVincenzo and Kaplan, 1981
180	8	1440	4	4.5	Ratney et al., 1974
200	1	200	4-6	2.2	DiVincenzo and Kaplan, 1981
200	2	400	4-6	3.3	DiVincenzo and Kaplan, 1981
200	4	800	12	4.2	Putz et al., 1979
200	7.5	1500	4-6	5.8	DiVincenzo and Kaplan, 1981
250	2	500	4	3.2	Astrand et al., 1975
250	7.5	1875	35	7	Peterson, 1978
350	5	1750	6	6.0	Andersen et al., 1981
500	2	1000	5	4.0	Astrand et al., 1975
500	7.5	3750	5	10	Peterson, 1978
515	1	515	8	1.8	Stewart et al., 1972
691	2	1382	3	4.8	Stewart et al., 1972
986	2	1972	3	9.3	Stewart et al., 1972

[a]Concentration × Time.

It should be noted that the National Research Council's Subcommittee on SMACs recognized the potential that methylene chloride's toxicity due to COHb formation could be aggravated by a reduction in the mass of red blood cells (RBCs) in the bodies of astronauts in microgravity. However, the ACs based on CO formation derived from data gathered on earth should be valid because microgravity reduces astronauts' RBC mass by only about 10%. Thus, any aggravation on the

COHb concentration formed in astronauts during methylene chloride exposures will not be significant.

Liver Toxicity

Subchronic or chronic exposures of rats to methylene chloride have been shown to produce these hepatic changes: fatty liver in the studies of MacEwen et al. (1972a,b) and Haun et al. (1972); cytoplasmic vacuolization (which might reflect fatty changes) and necrosis in the studies of Burek et al. (1984) and NTP (1986); and hemosiderosis and focal granulomatous inflammation in the NTP (1986) study.

Methylene chloride's hepatic effects depend on exposure duration to a certain extent. MacEwen et al. (1972b) reported that livers of rats exposed to methylene chloride at 5000 ppm, 24 h/d, 7 d/w, for 4 w exhibited the same degree of cellular vacuolization, nuclear enlargement, and iron pigmentation as that in rats exposed at 5000 ppm, 24 h/d, 7 d/w, for 14 w. However, the NTP (1986) study showed that a subchronic exposure of rats at 2100 ppm, 6 h/d, 5 d/w, for 13 w was not hepatotoxic, but a chronic exposure at 2000 ppm, 6 h/d, 5 d/w, for 2 y produced liver injuries. Similarly, in a comparison of methylene chloride's hepatic effects in mice exposed at 100 ppm, 24 h/d, for 3 d, for 1, 2, 3, 4, or 10 w, Weinstein and Diamond (1972) showed that liver histopathology became somewhat more severe as the exposure was lengthened, but the hepatic triglyceride concentration did not increase linearly with the exposure duration. Their results are summarized in Table 8-6.

Therefore, the bulk of the data indicate that methylene chloride's liver toxicity is somewhat time-dependent in rats. As a result, the prudent approach in deriving an AC based on hepatic toxicity would be to lower the AC as the exposure time is lengthened.

Instead of using the data gathered by MacEwen et al. (1972b) and Haun et al. (1972) to derive the ACs, data from the NTP (1986) study and the Burek et al. (1984) study were used. These studies are more recent and they were peer-reviewed, but the studies of MacEwen et al. and Haun et al. were not. The NTP study showed that repetitive exposures to methylene chloride at 2100 ppm, 6 h/d, 5 d/w, for 13 w caused no histopathology in rats.

TABLE 8-6 Temporal Pattern of Methylene Chloride's Hepatic Effects in Mice

Exposure Time	Triglyceride Level	Histopathology
3 d	No significant change	No changes
1 w	No significant change	Small fat droplets
2 w	240% of control's	Increase in fat-droplet size
3 w	420% of control's	Fatty changes, enlarged nuclei, small autophagic vacuoles
4 w	190% of control's	Fatty changes, enlarged nuclei, small autophagic vacuoles
10 w	140% of control's	Fatty changes, enlarged nuclei, large autophagic vacuoles

7-d AC based on liver toxicity
= 90-d NOAEL × 1/species factor
= 2100 ppm × 1/10
= 210 ppm.

The NTP (1986) study also showed that exposures to rats at 1000, 2000, or 4000 ppm, 6 h/d, 5 d/w, for 2 y could lead to cytoplasmic vacuolization, hemosiderosis, and focal granulomatous inflammation in liver. Burek et al. (1984) found that a similar 2-y exposure of rats to methylene chloride produced cytoplasmic vacuolization, indicative of fatty liver, at as low as 500 ppm, so the LOAEL for non-neoplastic hepatotoxicity is 500 ppm.

30-d AC based on liver toxicity
= 2-y LOAEL × 1/NOAEL factor × 1/species factor
= 500 ppm × 1/10 × 1/10
= 5 ppm.

180-d AC based on liver toxicity
= 2-y LOAEL × 1/NOAEL factor × 1/species factor × time adjustment
= 500 ppm × 1/10 × 1/10 ×
(6 h/d × 5 d/w × 104 w)/(24 h/d × 180 d)
= 3.6 ppm.

No evidence of liver toxicity has been found in the literature for acute methylene chloride exposures; therefore, 1-h and 24-h ACs based on liver toxicity are not derived.

Carcinogenicity

A 2-y exposure to methylene chloride at 0, 2000, and 4000 ppm produced 3 of 50, 30 of 48, and 41 of 48 cases of lung tumors and 3 of 50, 16 of 48, and 40 of 48 cases of liver tumors, respectively, in female $B6C3F_1$ mice in the NTP study (1986). Instead of using the airborne methylene chloride concentrations to calculate the 10^{-4} tumor dose, it is better to use the doses of active metabolite produced by the glutathione transferase pathway in the lung and liver, as estimated by a physiologically based pharmacokinetic model (Andersen et al., 1987). According to this pharmacokinetic model, 2000 and 4000 ppm of airborne methylene chloride are equivalent to 123 and 256 mg of methylene chloride metabolized per liter of lung per exposure day. Similarly, 2000 and 4000 ppm are equivalent to 851 and 1811 mg of methylene chloride metabolized per liter of liver per exposure-day. By substituting these values in the linearized multistage model using GLOBAL86 (Howe and Crump, 1986), 0.011 mg of methylene chloride metabolized per liter of lung per day and 0.24 mg of methylene chloride metabolized per liter of liver per day are the lower 95% confidence limit of the dose that will yield a 10^{-4} lung and liver tumor risk, respectively. Based on the pharmacokinetic model (Andersen et al., 1987), 0.011 mg/L lung and 0.24 mg/L liver are equivalent to about 6 and 12 ppm of methylene chloride for humans, respectively. The lower concentration of 6 ppm is used in the risk assessment.

The continuous exposure concentration to get a lung tumor risk of 10^{-4}
 = 6 ppm × (6 h/d × 5 d/w)/(24 h/d × 7 d/w)
 = 1.1 ppm.

Instead of the physiologically based pharmacokinetics model, EPA (1990) used the body-surface-area ratio to extrapolate the tumor data in mice to humans. With the linearized multistage model, EPA estimated that a continuous lifetime exposure to methylene chloride at 0.02 mg/m^3 or 5.8 × 10^{-3} ppm would produce an excess tumor risk of 1 in 10,000 in humans. In comparison, EPA's estimate is 190 times more conservative than the risk assessment estimate based on the physiologically based pharmacokinetics model.

According to the Committee on Toxicology, setting $k = 3$ (the number of stages in the carcinogenic process affected by methylene chloride), $t = 25,550$ d (lifetime of 70 y), and $t_o = 10,950$ d (an initial exposure age of 30 y), the adjustment factor for a near instantaneous exposure is calculated to be 26,082 (NRC, 1992).

24-h exposure level that would produce a 10^{-4} excess tumor risk
= 1.1 ppm × 26,082
= 29,000 ppm.

Similarly, by setting $k = 3$, $t = 25,550$ d, and $t_o = 10,950$ d, the adjustment factor for estimating the 7-d exposure concentration that would yield the same excess tumor risk as that for a lifetime exposure is 3728 (NRC, 1992).

7-d exposure level that would produce a 10^{-4} excess tumor risk
= 1.1 ppm × 3728
= 4100 ppm.

With a similar approach, 871 and 146.7 are calculated to be the adjustment factors for converting a lifetime exposure concentration to 30-d and 180-d exposure concentrations for the same excess tumor risk (NRC, 1992).

30-d exposure level that would produce a 10^{-4} excess tumor risk
= 1.1 ppm × 871
= 960 ppm.

180-d exposure level that would produce a 10^{-4} excess tumor risk
= 1.1 ppm × 146.7
= 160 ppm.

Establishment of SMAC Values

The ACs for the three toxic end points are listed in Table 8-7. The lowest AC for each exposure duration is selected to be the SMAC. As a result, the 1-h, 24-h, 7-d, 30-d, and 180-d SMACs are set at 100, 35, 15, 5, and 3 ppm, respectively.

No adjustments of the SMACs are needed for any microgravity-induced physiological changes. The reason is that the in-flight hemoglobin concentrations obtained in Skylabs were higher than the preflight values by only 10%, so the carbon monoxide produced from methylene chloride metabolism is not going to be significantly more toxic in-flight than on earth.

TABLE 8-7 End Points and Acceptable Concentrations

End Point	Exposure Data	Species and Reference	Uncertainty Factors			Acceptable Concentrations, ppm				
			NOAEL	Time	Species	1 h	24 h	7 d	30 d	180 d
CO formation	COHb data from several sources	Human (Astrand et al., 1975; DiVincenzo and Kaplan, 1981; Stewart et al., 1972; Peterson, 1978; Ratney et al., 1974; Putz et al., 1979; Andersen et al., 1991)	—	—	—	100	—[a]	—	—	—
	PB-PK model data	Human (Andersen et al., 1991)	—	PB-PK	—	—	35	14	14	14
Liver toxicity	NOAEL at 2100 ppm, 6 h/d, 5 d/w, 13 w	Rat (National Toxicology Program, 1986)	—	—	10	—	—	210	—	—
	LOAEL at 500 ppm, 6 h/d, 5 d/w, 2 y	Rat (Burek et al., 1984)	10	HR[b]	10	—	—	—	5	3.6
Carcinogenesis	2-y study	Mouse (National Toxicology Program, 1986)	—	COT[c]	—	—	29,000	4100	960	160
SMAC						100	35	15	5	3

[a]Extrapolation to these exposure durations produces unacceptable uncertainty in the values.
[b]HR = Haber's rule.
[c]Calculated based on COT's equation (NRC, 1985) derived from Crump and Howe's multistage carcinogenicity model and using a lifetime cancer risk of 10^{-4}. This model was not used to calculate acceptable concentrations for exposures shorter than 24 h.

REFERENCES

ACGIH. 1986. Methylene chloride. In Documentation of the Threshold Limit Values and Biologic Exposure Indices. American Conference of Governmental Industrial Hygienists, Cincinnati, Ohio.

Ahmed, A.E., and M.W. Anders. 1978. Metabolism of dihalomethanes to formaldehyde and inorganic halide. I. In vitro studies. Drug Metab. Dispos. 4:357-361.

Andersen, M.E., M.G. MacNaughton, H.J. Clewell III, and D.J. Paustenbach. 1987. Physiologically based pharmacokinetics and the risk assessment process for methylene chloride. Toxicol. Appl. Pharmacol. 87:185-205.

Andersen, M.E., H.J. Clewell, M.L. Gargas, M.G. MacNaughton, R.H. Reitz, R.J. Nolan, and M.J. McKenna. 1991. Physiologically based pharmcokinetic modeling with dichloromethane, its metabolite, carbon monoxide, and blood carboxyhemoglobin in rats and humans. Toxicol. Appl. Pharmacol. 108:14-27.

Astrand, I., P. Ovrum, and A. Carlsson. 1975. Exposure to methylene chloride. I. Its concentration in alveolar air and blood during rest and exercise and its metabolism. Scand. J. Work Environ. Health 1:78-94.

Balmer, M.F., F.A. Smith, L.J. Leach, and C.L. Yulie. 1976. Effects in the liver of methylene chloride inhaled alone and with ethyl alcohol. Am. Ind. Hyg. Assoc. J. 37:345-352.

Burek, J., K.D. Nitschke, T.J. Bell, D.L. Wackerle, R.C. Childs, J.E. Beyer, D.A. Dittenber, L.W. Rampy, and M.J. McKenna. 1984. Methylene chloride: A two-year inhalation toxicity and oncogenicity study in rats and hamsters. Fundam. Appl. Toxicol. 4:30-47.

Carlsson, A., and M. Hultengren. 1975. Exposure to methylene chloride. III. Metabolism of ^{14}C-labelled methylene chloride in rat. Scand. J. Work Environ. Health 1:104-108.

Casanova, M., D.F. Deyo, and H. d'A. Heck. 1992. Dichloromethane (methylene chloride): Metabolism to formaldehyde and formation of DNA-protein cross-links in B6C3F$_1$ mice and Syrian golden hamsters. Toxicol. Appl. Pharmacol. 114:162-165.

DiVincenzo, G.D., and C.J. Kaplan. 1981. Uptake, metabolism, and elimination of methylene chloride vapor by humans. Toxicol. Appl. Pharmacol. 59:130-140.

DiVincenzo, G.D., F.J. Yanno, and B.D. Astill. 1972. Human and canine exposure to methylene chloride vapor. Am. Ind. Hyg. Assoc. J. 33:125-135.

EPA. 1990. Methylene chloride. In Integrated Risk Information System. Office of Research and Development, Environmental Criteria and Assessment Office, U.S. Environmental Protection Agency, Cincinnati, Ohio.

Foster, J.R., P.M. Hext, and T. Green. 1986. The bronchiolar Clara cell: Selective cytotoxicity with methylene chloride. Pp. 2755-2766 in the Proceedings of the 11th International Congress on Electron Microscopy, Aug. 31-Sept. 7, 1986, Kyoto, Japan. Tokyo: Japanese Society of Electron Microscropy.

Foster, J.R., T. Green, L.L. Smith, R.W. Lewis, P.M. Hext, and I. Wyatt. 1992. Methylene chloride: An inhalation study to investigate pathological and biochemical events occurring in the lungs of mice over an exposure period of 90 days. Fundam. Appl. Toxicol. 18:376-388.

Friedlander, B., T. Hearne, and S. Hall. 1978. Epidemiologic investigation of employees chronically exposed to methylene chloride. Mortality analysis. J. Occup. Med. 20:657-666.

Gamberale, F., G. Annwall, and M. Hultengren. 1975. Exposure to methylene chloride. II. Psychological Functions. Scand. J. Work Environ. Health 1:95-103.

Gocke, E., M.-T. King, K. Eckhardt, and D. Wild. 1981. Mutagenicity of cosmetics ingredients licensed by the European Communities. Mutat. Res. 90:91-109.

Hardin, B.D., and J.M. Manson. 1980. Absence of dichloromethane teratogenicity with inhalation exposure in rats. Toxicol. Appl. Pharmacol. 52:22-28.

Haun, C.C., E.H. Vernot, K.I. Darmer, and S.S. Diamond. 1972. Continuous animal exposure to low levels of dichloromethane. Proceedings of the Third Annual Conference on Environmental Toxicology. AMRL-TR-72-130. Wright-Patterson Air Force Base, Dayton, Ohio.

Heppel, L., P.A. Neal, E.L. Perrin, M.L. Orr, and V.T. Porterfield. 1944. Toxicology of dichloromethane. I. Studies on effects of daily inhalation. J. Ind. Hyg. Toxicol. 26:8-16.

Howe, R.B., and K.S. Crump 1986. GLOBAL86. A Computer Program to Extrapolate Quantal Animal Toxicity Data to Low Doses. Clement Associates, Inc., Ruston, La.

Jongen, W.M.F., G.M. Alink, and J.H. Koeman. 1978. Mutagenic effect of dichloromethane on *Salmonella typhimurium.* Mutat. Res. 56:245-248.

Kim, Y.C., and G.P. Carlson. 1986. The effect of an unusual work shift on chemical toxicity. I. Studies on the exposure of rats and mice to dichloromethane. Fundam. Appl. Toxicol. 6:162-171.

Kubic, V.L., M.W. Anders, R.R. Engel, C.H. Barlow, and N.S. Caughey. 1974. Metabolism of dihalomethanes to carbon monoxide. I. In vivo studies. Drug Metab. Dispos. 2:53-57.

Laham, S., M. Potvin, and K. Schrader. 1978. Toxicological studies on dichloromethane, a solvent simulating carbon monoxide poisoning. Toxicol. Eur. Res. 1:63-73.

MacEwen, J.D., E.H. Vernot, and K.I. Darmer, Jr. 1972a. Continuous animal exposure to dichloromethane [abstract]. Proceedings of the Annual Conference on Environmental Toxicology. AMRL-TR-72-28. Wright-Patterson Air Force Base, Dayton, Ohio.

MacEwen, J.D., E.H. Vernot, and C.C. Haun. 1972b. Continuous animal exposure to dichloromethane. AMRL-TR-72-28. Aerospace Medical Research Laboratory, Wright-Patterson Air Force Base, Dayton, Ohio.

McKenna, M.J., J.A. Zempel, and W.H. Braun. 1982. The pharmacokinetics of inhaled methylene chloride in rats. Toxicol. Appl. Pharmacol. 65:1-10.

Merck Index. 1989. P. 954 in Merck Index, 11th Ed., S. Budavari, M.J. O'Neil, A. Smith, and P.E. Heckelman, eds. Rahway, N.J.: Merck & Co.

Moskowitz, S., and H. Shapiro. 1952. Fatal exposure to methylene chloride vapor. AMA Arch. Ind. Hyg. Occup. Med. 6:116-123.

NASA. 1989. Air Analysis Results Obtained by the JSC Toxicology Laboratory, Toxicology Group, Biomedical Operations and Research Branch, Johnson Space Center, NASA, Houston, Tex.

NRC. 1986. Criteria and Methods for Preparing Emergency Exposure Guidance Level (EEGL), Short-Term Public Emergency Guidance Level (SPEGL), and Continuous Exposure Guidance Level (CEGL) Documents. Washington, D.C.: National Academy Press.

NRC. 1992. Guidelines for Developing Spacecraft Maximum Allowable Concentrations for Space Station Contaminants. Washington, D.C.: National Academy Press.

NTP. 1986. Toxicology and Carcinogenesis Studies of Dichloromethane (Methylene Chloride) in F344/N Rats and B6C3F$_1$ Mice (Inhalation Studies). NTP Tech. Rep. Series No. 306. National Institute of Environmental Health Sciences, Research Triangle Park, N.C.

Peterson, J.E. 1978. Modelling the uptake, metabolism and excretion of dichloromethane by man. Am. Ind. Hyg. Assoc. J. 39:41-47.

Putz, V.R., B.L. Johnson, and J.V. Setzer. 1979. A comparative study of the effects of carbon monoxide and methylene chloride on human performance. J. Environ. Pathol. Toxicol. 2:97-112.

Ratney, R.S., D.H. Wegman, and H.B. Elkins. 1974. In vivo conversion of methylene chloride to carbon monoxide. Arch. Environ. Health 28:223-226.

Sahu, S., D. Lowther, and A. Ulsamer. 1980. Biochemical studies on pulmonary response to inhalation of methylene chloride. Toxicol. Lett. 7:41-45.

Savolainen, H., K. Kruppa, P. Pfaffli, and H. Kivisto. 1981. Dose-related effects of dichloromethane on rat brain in short-term inhalation exposure. Chem.-Biol. Interact. 34:315-322.

Schwetz, B.A., B.K.H. Leong, and P.J. Gehring. 1975. The effect of maternally inhaled trichloroethylene, perchloroethylene, methyl chloroform, and methylene chloride on embryonal and fetal development in mice and rats. Toxicol. Appl. Pharmacol. 32:84-96.

Stewart, R.D., T.N. Fisher, M.J. Hosko, J.E. Peterson, E.D. Baretta, and H.C. Dodd. 1972. Experimental human exposure to methylene chloride. Arch. Environ. Health 25:342-348.

Thilager, A., and V. Kumaroo. 1983. Induction of chromosome damage by methylene chloride in CHO cells. Mutat. Res. 116:361-367.

Thomas, A.A., M.K. Pinkerton, and J.A. Warden. 1972. Effects of low level dichloromethane exposure on the spontaneous activity of mice. Proceedings of the Third Annual Conference on Environmental Toxicology. AMRL-TR-72-130. Wright-Patterson Air Force Base, Dayton, Ohio.

Weinstein, R.S., and S.S. Diamond. 1972. Hepatotoxicity of dichloromethane with continuous inhalation exposure at a low dose level. Proceedings of the Third Annual Conference on Environmental Toxicology. AMRL-TR-72-130. Wright-Patterson Air Force Base, Dayton, Ohio.

Weinstein, R.S., D.D. Boyd, and K.C. Back. 1972. Effects of continuous inhalation of dichloromethane in the mouse: Morphologic and functional observations. Toxicol. Appl. Pharmacol. 23:660-679.

Winneke, G. 1981. The neurotoxicity of dichloromethane. Neurobehav. Toxicol. Teratol. 3:391-395.

Winneke, G., and G.G. Fodor. 1976. Dichloromethane produces narcotic effect. Occup. Health Safety 45(2):34-49.

Wong, K.L. 1990. Carbon monoxide. Pp. 61-90 in Spacecraft Maximum Allowable Concentrations for Selected Airborne Contaminants, Vol. 1. Washington, D.C.: National Academy Press.

B9 Methyl Ethyl Ketone

King Lit Wong, Ph.D.
Johnson Space Center Toxicology Group
Biomedical Operations and Research Branch
Houston, Texas

PHYSICAL AND CHEMICAL PROPERTIES

Methyl ethyl ketone (MEK) is a flammable, colorless liquid with an acetone-like odor (ACGIH, 1991).

Synonyms:	2-butanone, methyl acetone
Formula:	$CH_3COCH_2CH_3$
CAS number:	78-93-3
Molecular weight:	72
Boiling point:	79.6°C
Melting point:	-85.9°C
Vapor pressure:	71.2 mm Hg at 20°C
Conversion factors	1 ppm = 2.94 mg/m^3
at 25°C, 1 atm:	1 mg/m^3 = 0.34 ppm

OCCURRENCE AND USE

MEK is used as a solvent in synthetic resins manufacturing and in the surface-coating industry (ACGIH, 1991). We are not aware of any use of MEK in the spacecraft, but MEK has been found in the cabin atmosphere during several space-shuttle missions, ranging from 0.3 to 69 ppb (Huntoon, 1993). Off-gassing is probably the source of MEK in space shuttles. Based on the off-gassing data in the Spacelab, it was estimated that 3.8 g of MEK will be generated in the space station each day (J. Perry, Marshall Space Flight Center, personal commun., 1989).

PHARMACOKINETICS AND METABOLISM

Absorption

Most disposition studies of MEK were performed on humans, so the following review refers to human results unless specified otherwise. The average pulmonary uptake rate of MEK was 1.05 mg/min in workers exposed to MEK at a concentration of 100 ppm (Perbellini et al., 1984). In resting male volunteers exposed to MEK at 200 ppm for 4 h, the expired MEK concentration was about 53% of the inhaled MEK all through the exposure (Liira et al., 1988a). It remained unchanged at 53% when the volunteers performed three 10-min, 100-watt-workload exercises during the 4-h MEK exposure beginning at 5, 95, and 225 min (Liira et al., 1988a). At the end of the 4-h MEK exposure at 200 ppm, the venous blood concentration reached 80 μmol/L (5.8 μg/mL) in the resting volunteers, and the venous blood concentration was 80% higher in the volunteers who did the exercises (Liira et al., 1988a). In another study, during a 4-h exposure to MEK at 100 or 200 ppm, the venous blood concentrations of MEK reached about 1.8 or 3.5 μg/mL, respectively, at 4 h (Liira et al., 1988a; Dick et al., 1988). The venous blood concentrations at 4 h were approximately 10-30% higher than those at 2 h (Brown et al., 1987; Liira et al., 1988a; Dick et al., 1988).

Metabolism and Excretion

In guinea pigs, MEK is reduced to 2-butanol or oxidized to 3-hydroxy-2-butanone and 2,3-butanediol (DiVincenzo et al., 1976). In human volunteers exposed to MEK at 200 ppm, Liira et al. (1988a) discovered 2,3-butanediol in the urine. In contrast, Perbellini et al. (1984) did not find 2-butanol or 2,3-butanediol in the urine of workers exposed to MEK at concentrations equal to or lower than 300 mg/m^3 (100 ppm); they found only 3-hydroxy-2-butanone in the urine. Liira et al. (1988a) reported that 3% of the absorbed MEK was exhaled unchanged and 2%

was excreted as 2,3-butanediol in the urine during the 4 h of exposure and 20 h after exposure. Miyasaka et al. (1982) found that only approximately 0.1% of MEK absorbed during an inhalation exposure was excreted unchanged in the urine after the exposure in humans. These data suggest that most of the absorbed MEK apparently entered into intermediary metabolism (Liira et al., 1988a), so that, had these investigators used radioactively labeled MEK, they would have found that most of the MEK absorbed is eliminated from the body as CO_2. Using a physiologically based pharmacokinetic model, Liira et al. (1990a) estimated that MEK metabolism would be saturated at an airborne MEK concentration of 100 ppm at rest or 50 ppm during exercise. The saturable metabolism of MEK explains why the MEK concentrations in blood tend to vary nonlinearly with exposure concentrations above 100 ppm: the peak MEK concentrations in blood during a 4-h exposure of resting human volunteers at 25, 200, or 400 ppm were 0.3 μg/mL, 7.5 μg/mL, or 23.0 μg/mL, respectively (Liira et al., 1990a).

In humans exposed to MEK at a concentration of 200 ppm, MEK was cleared from the blood in two exponential phases (Liira et al., 1988a). The initial phase had a 30-min half-life and the second phase had an 81-min half-life. One and a half hours after a 4-h MEK exposure at 200 ppm, the venous blood concentration of MEK dropped from 3.5 μg/mL to 1.0 μg/mL, and it decreased to below detectable concentrations 20 h after exposure in 22 volunteers (Dick et al., 1988). In guinea pigs given MEK intraperitoneally at 450 mg/kg (a much higher dose than the estimated dose of 28 mg/kg used in the inhalation study by Liira et al. (1988a)), MEK was cleared from the blood monoexponentially with a much longer half-life of 270 min (DiVincenzo et al., 1976).

TOXICITY SUMMARY

Acute and Short-Term Toxicity

Mucosal Irritation

The major acute toxicity of MEK is mucosal irritation. Nelson et al. (1943) exposed 10 human subjects for 3-5 min to MEK or to 1 of 15 other organic solvents and told them to rank the effects on the eye, nose, and throat as either no irritation, slightly irritating, or very irritating. They were not told of the exposure concentrations. Unfortunately, Nelson et al. did not report whether they analytically determined the exposure concentrations, and it appears that the exposures were done using nominal concentrations. In the study, MEK at a concentration of 100 ppm resulted in slight nose and throat irritation. Mild eye irritation was reported by some subjects at 200 ppm. Most, if not all of the subjects, rejected the vapor at 300 ppm. In a mouse model of sensory irritation, a 5-min exposure to MEK at 10,745 ppm reduced the respiratory rate by 50% (De Ceaurriz et al., 1981).

Subchronic and Chronic Toxicity

In subchronic exposures, MEK is known to cause death, hepatic effects, and central-nervous-system (CNS) impairment.

Neural Studies

Although methyl *n*-butyl ketone is neurotoxic, MEK has been found not to cause peripheral neuropathy in rats. A continuous exposure of rats to MEK at 1125 ppm, 24 h/d. for up to 55 d in one study (Saida et al., 1976) or a repetitive exposure to MEK at 700 ppm, 8 h/d, 5 d/w, for 16 w in another (Duckett et al., 1979) failed to cause any peripheral neuropathy. An exposure of five rats to MEK at 10,000 ppm in the first few days and 6000 ppm thereafter for 8 h/d, 7 d/w, for 7 w resulted in no neurological signs (Altenkirch et al., 1978). During the MEK exposure, all the rats were excited in the initial several minutes, but they became somnolent within 5-10 min (Altenkirch et al., 1978). In the seventh week, all the rats died of severe bronchopneumonia as determined by gross pathological and histological examinations (Altenkirch et al., 1978). Electron microscopic examination revealed no alterations of the sciatic, tibial, peroneal, and sural nerves (Altenkirch et al., 1978).

Mortality

As mentioned earlier, all five rats died after an exposure to MEK for 8 h/d, 7 d/w, for 7 w (Altenkirch et al., 1978). The exposure concentrations were 10,000 ppm in the first several days and 6000 ppm for the balance of the 7 w. The 100% mortality was again obtained when the experiment was repeated (Altenkirch et al., 1978). The mortality at 6000-ppm MEK seen in that study was probably of little significance at lower exposure concentrations, because a 90-d study showed that exposures of rats to MEK at 1250, 2500, or 5000 ppm, 6 h/d, 5 d/w, for 13 w failed to cause any increase in mortality compared with the control group (Cavender et al., 1983).

Hepatic Effects

A 90-d exposure to MEK at 5000 ppm, 6 h/d, 5 d/w did not change the morphology of any tissues, but it increased the liver weight and the liver-to-body-weight ratio (Cavender et al., 1983). A similar exposure at 2500 ppm increased the liver weight in the female rats but not in the male rats, and it had no effect on the liver-to-body-weight ratio (Cavender et al., 1983). It can be concluded that a subchronic MEK exposure could have a mild effect on the liver. In the same study, MEK at 5000 ppm decreased the SGPT slightly, and it increased the serum potassium, glucose, and alkaline phosphatase levels in the females. A similar exposure to MEK at 1250 ppm failed to elicit any toxic effects (Cavender et al., 1983).

CNS Effects

Like most organic solvents, MEK might also impair the CNS at relatively high concentrations (Patty et al., 1935; Altenkirch et al., 1978). In an exposure of rats to MEK at 6000 ppm, 8 h/d, 7 d/w for 7 w, it was discovered that the rats developed transient excitation in the first few minutes of exposure followed by somnolence (Altenkirch et al., 1978). After the somnolence set in, the rats were still arousable. An exposure of guinea pigs to MEK at 10,000 ppm for 13.5 h resulted in incoordination in 90 min and narcosis in about 250 min (Patty et al., 1935).

Carcinogenicity

No carcinogenicity bioassay of inhaled MEK was found in the literature (EPA, 1990). A study with only 10 C3H/He mice per group showed that a cutaneous application, twice a week for 1 y, of 50 mL of a solution containing 25% MEK resulted in no skin tumors, whereas a skin tumor was found in 1 of 10 mice after a similar application of 29% MEK (Horton, 1965). The animal data to assess the carcinogenicity of MEK are inadequate..

Likewise, there are insufficient epidemiological data to determine MEK's carcinogenicity, if any, in humans. Generally, uncertainties on the exposure history of the study populations make it difficult to interpret the epidemiological data reported in the literature.

In a retrospective cohort mortality study of 14,067 aircraft manufacturing workers who have spent, on the average, 15.8 y in the facility, there was no significant excess of mortality from cancer of various organs (Garabrandt et al., 1988). These workers were exposed to various substances on their jobs, including aluminum alloy dusts, welding fumes, methylene chloride, trichloroethylene, MEK, lubricating oils and greases, and metal cutting fluids (Garabrandt et al., 1988). Forty-seven percent of their jobs had a potential for MEK exposure (Garabrandt et al., 1988). In another retrospective cohort mortality study, the standardized mortality ratio (SMR) for all cancers was much lower than unity in 1008 male oil-refinery workers, and the SMR for prostate cancer was 1.82, which was not statistically significant (Wen et al., 1985). These workers have been working in a lubricating-dewaxing process with exposures to various solvents, primarily MEK and toluene, at concentrations below OSHA's standards (Wen et al., 1985).

In an historical prospective study of 446 men who had worked in two MEK dewaxing plants, where they had been followed for 13.9 y on the average, there was a statistically significant excess of deaths from buccal cavity and pharynx cancers and fewer deaths from lung cancer

(Alderson and Rattan, 1980). The investigators concluded that "there is no clear evidence of a cancer hazard in these workers." Finally, childhood leukemia might be related to parents' occupational exposure to MEK. A case-control study of children less than 10 y old showed a statistically significant increase of leukemogenic risk in children whose fathers had been occupationally exposed, after the children's birth, to MEK, chlorinated solvents, spray paint, dyes, pigments, and cutting oil (Lowengart et al., 1987). Because of exposure to a wide mixture of chemicals, these epidemiology studies failed to demonstrate carcinogenicity specifically associated with MEK in humans.

Genotoxicity

MEK was not mutagenic for several strains of *Salmonella typhimurium* (Douglas et al., 1980; Florin et al., 1980), but it induced aneuploidy in *Saccharomyces cerevisiae* strain M (Zimmermann et al., 1985). MEK was not found to be genotoxic in the mouse lymphoma assay, unscheduled DNA synthesis assay, and micronucleus assay (O'Donoghue et al., 1988).

Developmental Toxicity

MEK exposure at a concentration of 3000 ppm for 7 h/d on gestation days 6-15 was not teratogenic in rats (Deacon et al., 1981). However, a similar MEK exposure was mildly teratogenic in mice because a concentration-related trend of increases in misaligned sternebrae was found, but no increase in any single malformation was seen at exposure concentrations up to 3000 ppm (Schwetz et al., 1991). MEK exposure at 3000 ppm for 7 h/d on gestation days 6-15 is slightly toxic to the fetus by delaying ossification in the rat and reducing fetal weight in the mouse (Deacon et al., 1981; Schwetz et al., 1991). It also decreased the body-weight gain and increased the water consumption in the pregnant rats (Deacon et al., 1981), and it increased the maternal liver-to-body-weight ratio in the mouse (Schwetz et al., 1991). It can be concluded that MEK has only very slight developmental toxicity.

Interaction with Other Chemicals

Ethanol has been found to affect MEK metabolism. In humans, ingestion of ethanol at 0.8 g/kg just before a 4-h exposure to MEK at 200 ppm resulted in increased blood concentrations of MEK and 2-butanol and decreased blood concentrations of 2,3-butanediol (Liira et al., 1990b).

Although MEK is not neurotoxic, it potentiates the neurotoxicity of *n*-hexane. A weekly exposure of rats to *n*-hexane at 10,000 ppm, 8 h/d, 7 d/w led to slight weakness and severe paresis in some of the rats in the eighth week (Altenkirch et al., 1978). A similar exposure to a mixture of 1000-ppm MEK with 9000-ppm *n*-hexane hastened the neurotoxicity development by the third week and caused a higher percent of the rats to be inflicted with slight weakness and severe paresis (Altenkirch et al., 1978). Exposure to the MEK and *n*-hexane mixture also caused hypersalivation not produced by an exposure to *n*-hexane alone (Altenkirch et al., 1978).

An explanation for the potentiation is that co-exposure to MEK and *n*-hexane reduced the clearance of methyl *n*-butyl ketone, *n*-hexane's neurotoxic metabolite, in rats (Shibata et al., 1990a), although MEK also inhibited the oxidation of *n*-hexane to 2,5-hexanedione, another neurotoxic metabolite (Shibata et al., 1990b). These findings were made by the same investigators by exposing rats for 8 h to *n*-hexane at a concentration of 2000 ppm plus MEK at concentrations of 0, 200, 630, or 2000 ppm (Shibata et al, 1990a,b). Because MEK reduced the clearance of methyl *n*-butyl ketone, it is reasonable that an exposure of rats to 1125-ppm MEK and 225-ppm methyl *n*-butyl ketone, 24 h/d for 35 or 55 d resulted in more severe nerve injuries than an exposure to methyl *n*-butyl ketone alone at 225 ppm (Saida et al., 1976).

In addition to affecting *n*-hexane's metabolism, MEK also has an influence on xylene's metabolism. Exposure of male volunteers to MEK at 200 ppm and *m*-xylene at 100 ppm for 4 h resulted in higher xylene concentrations in the blood than an exposure to *m*-xylene alone at 100 ppm, because MEK inhibited *m*-xylene metabolism (Liira et al., 1988b). However, *m*-xylene did not change the blood concentration of MEK and the urinary excretion of 2,3-butanediol, an MEK metabolite (Liira et al., 1988b).

In conclusion, the interaction findings indicate that NASA needs to be aware of the possibility that ethanol ingestion might potentiate MEK's toxicity and MEK inhalation could potentiate the toxicity of n-hexane, methyl n-butyl ketone, and m-xylene during combined exposure. Due to the large number of possible combinations of these chemicals, MEK's SMACs are not set based on the interaction information.

TABLE 9-1 Toxicity Summary[a]

Concentration, ppm	Exposure Duration	Species	Effects	Reference
100[b]	3-5 min	Human	Slight noise and throat irritation	Nelson et al., 1943
200[b]	3-5 min	Human	Mild eye irritation in some subjects	Nelson et al., 1943
200	4 h	Human	No effects on choice reaction time, visual vigilance, auditory tone tracking and discrimination, memory scanning, and postural sway test	Dick et al., 1988
200	4 h	Human	No effects on pattern recognition, choice reaction time, and visual vigilance	Dick et al., 1984
200	4 h	Human	No effects on reaction time, pattern discrimination, and visual vigilance	Dick et al., 1984
300[b]	3-5 min	Human	Vapor was "conclusively rejected"	Nelson et al., 1943
3300	N.S.[c] (a few breaths)	Human	Moderately irritating to the eyes and nose; moderate-to-strong odor	Patty et al., 1935
10,000	N.S. (a few breaths)	Human	Intolerable from irritation to the eyes and nose	Patty et al., 1935
100	24 h/d, 7 d	Baboon	No effect on the accuracy of pressing the correct level to get a banana pellet in response to stimuli; but the response was slowed on the second and third days of exposure, which returned to normal by the last 2 d of exposure	Geller et al., 1979
300	8 h/d, 7 d	Rat	No effect on alkaline phosphatase activities in serum and leukocyte	Li et al., 1986
700	8 h/d, 5 d/w, 16 w	Rat	No clinical or pathological manifestations of peripheral neuropathy	Duckett et al., 1979
800	7 h/d, 5 d/w, 4 w	Rat	Increased liver weight/body weight; depressed the hepatic metabolism of androstenedione; no effect on cytochrome P-450 concentrations in liver	Toftgard et al., 1981

Dose	Exposure	Species	Effects	Reference
1000	7 h/d, d 6-15 of gestation	Rat	No increase in malformation, resorption, or preimplantation loss; no effect on ossification in the fetus; no effect on maternal weight gain, food and water consumption, appearance, or behavior	Deacon et al., 1981
1000	7 h/d, d 6-15 of gestation	Mouse	No maternal or developmental toxicity	Schwetz et al., 1991
1125	24 h/d; 16, 25, 35, or 55 d	Rat	No increases of nerve inpouchings, denuded fibers, or swollen axons	Saida et al., 1976
1250	6 h/d, 5 d/w, 13 w	Rat	No changes in liver weight, liver-weight/body-weight ratio, serum chemistry, and histology	Cavender et al., 1983
2500	6 h/d, 5 d/w, 13 w	Rat	Liver weight increased by 11% in females (not in males); no changes in liver-weight/body-weight ratio and serum chemistry; no histopathology	Cavender et al., 1983
3000	7 h/d, d 6-15 of gestation	Rat	Delayed ossification in the fetus; no increases in major malformation, resorption, or preimplantation loss; decreased body-weight gain and increased water consumption in the dams; no changes in maternal liver weight or food consumption	Deacon et al., 1981
3000	7 h/d, d 6-15 of gestation	Mouse	No significant increase in any single malformation; a significant concentration-dependent increase of misaligned sternebrae; no increase in resorptions; reduction in fetal body weight; slight increase in maternal liver-weight/body-weight ratio	Schwetz et al., 1991
3300	13.5 h	Guinea pig	No signs of eye and nose irritation; no lacrimation, incoordination, narcosis, gasping, or death	Patty et al., 1935
5000	6 h/d, 5 d/w, 13 w	Rat	Increased liver weight (27% in males and 18% in females) and liver-weight/body-weight ratio; increased mean corpuscular hemoglobin; in females (not in males): small increases in serum K^+, glucose, alkaline phosphatase levels; small decrease in SGPT; no histopathology	Cavender et al., 1983

TABLE 9-1 (Continued)

Concentration, ppm	Exposure Duration	Species	Effects	Reference
8000	8 h	Rat	3/6 rats died	Smyth et al., 1962
10,000 in the first few days, 6,000 thereafter	8 h/d, 7 d/w, 7 w	Rat	Transient excitation in the first few min of exposure, followed by arousable somnolence. 5/5 rats died in w 7 with severe bronchopneumonia; no nerve injury or neurological signs	Altenkirch et al., 1978
10,000	13.5 h	Guinea pig	Rubbing the nose (2 min into exposure); squinting (4 min); lacrimation (40 min); incoordination (90 min); narcosis (240-280 min)	Patty et al., 1935
33,000	13.5 h	Guinea pig	Squinting and nose rubbing (1 min into exposure); lacrimation (4 min); incoordination (20-30 min); narcosis (50-90 min); gasping (180 min); death (200-260 min); no corneal opacity	Patty et al., 1935
100,000	13.5 h	Guinea pig	Squinting and nose rubbing (beginning of exposure); lacrimation (1 min); incoordination (4 min); narcosis (10 min); gasping (20-30 min); corneal opacity (30 min), which disappeared 8 d after exposure; death (45-55 min) with slight brain congestion and marked congestion of other systemic organs	Patty et al., 1935

[a]Only the more important results are included.
[b]It is unknown whether the concentration was nominal or analytical.
[c]N.S. = not specified.

METHYL ETHYL KETONE

TABLE 9-2 Exposure Limits Set by Other Organizations

Organization	Concentration, ppm
ACGIH's TLV	200 (TWA)
ACGIH's STEL	300
OSHA's PEL	200 (TWA)

TLV = threshold limit value. TWA = time-weighted average. STEL = short-term exposure limit. PEL = permissible exposure limit.

TABLE 9-3 Spacecraft Maximum Allowable Concentrations

Duration	ppm	mg/m^3	Target Toxicity
1 h	50	150	Mucosal irritation
24 h	50	150	Mucosal irritation
7 d[a]	10	30	Mucosal irritation
30 d	10	30	Mucosal irritation
180 d	10	30	Mucosal irritation

[a]The current 7-d SMAC = 20 ppm.

RATIONALE FOR ACCEPTABLE CONCENTRATIONS

To set MEK's SMACs, the maximum acceptable concentrations (ACs) for the various exposure durations, according to the different toxic end points, are compared. The noncarcinogenic end points produced by MEK include mucosal irritation (Nelson et al., 1943), hepatomegaly (Cavender et al., 1983), and CNS impairment (Altenkirch et al., 1978). The NRC's Committee on Toxicology advised that MEK's SMACs need not be set using hepatomegaly as an end point because of doubtful clinical significance of a mild increase in liver weight absent any histological hepatic damages. Accordingly, ACs are derived for only mucosal irritation and CNS impairment.

Mucosal Irritation

A 3-5 min exposure at 100-ppm MEK produced slight nose and throat irritation, but not eye irritation, in 10 human subjects (Nelson et al., 1943). There are no data on MEK's irritancy at 1 h. The best time-response data on mucosal irritants are those gathered by Weber-Tschopp et al. (1977) with acrolein. They showed that the nose irritation at 0.3-ppm acrolein increased from no irritation in the first few minutes to mild irritation at 40 min, and the degree of throat irritation went from no irritation in the first few minutes to very mild irritation at 40 min. The severity of nasal and throat irritation did not change from 40 to 60 min of the 1-h exposure (Weber-Tschopp et al., 1977). Assuming that MEK's irritancy develops with time like that of acrolein, the nose and throat irritation produced by 100-ppm MEK at 1 h probably would be only mild to moderate.

There are no time-response data on MEK's mucosal irritation. However, if the exposure to MEK at 100 ppm were prolonged to more than 1 h, the degree of nose and throat irritation would most likely remain mild to moderate because mucosal irritation, being a surface response, is not expected to increase with time after the first hour. This is supported by the time-response data of two other irritants, acrolein and formaldehyde. The mucosal irritancy of acrolein did not increase from 40 to 60 min into a 1-h exposure (Weber-Tschopp et al., 1977), and the mucosal irritancy in volunteers exposed to formaldehyde at 3 ppm was mild to moderate at 1 or 3 h (Sauder et al., 1986; Green et al., 1987).

The 1-h and 24-h ACs based on mucosal irritation should be lower than 100 ppm, because NASA should not subject the crew to more than mild mucosal irritation in a 1-h or 24-h contingency. MEK's irritancy appears to have quite a steep concentration-response curve. Although all the subjects found a 3-5 min exposure to MEK at 300 ppm not tolerable, only slight nose and throat irritation was felt at 100 ppm (Weber-Tschopp et al., 1977). Therefore, a factor of 3 applied to the mildly to moderately irritating concentration of 100 ppm should yield a nonirritating concentration of 33 ppm. Since some degree of nose and throat irritation is acceptable in a 1-h or 24-h contingency, the 1-h and 24-h ACs should be set slightly higher than the nonirritating concentration of 33 ppm, so the 100-ppm concentration is divided by an extrapolation factor of 2 instead of 3.

1-h and 24-h ACs based on mucosal irritation
= mildly to moderately irritating level × 1/extrapolation factor
= 100 ppm × ½
= 50 ppm.

For the 7-d, 30-d, and 180-d ACs, it is important that mucosal irritation be prevented completely, so an extrapolation factor of 3 is used to derive a nonirritating concentration from a mildly to moderately irritating concentration. To compensate for the relatively small number of subjects used in getting the mildly to moderately irritating concentration, a safety factor is also used.

7-d, 30-d, and 180-d ACs based on mucosal irritation
= mildly to moderately irritating level × 1/extrapolation factor × 1/small n factor
= 100 ppm × 1/3 × (square root of n)/10
= 100 ppm × 1/3 × (square root of 10)/10
= 10 ppm.

CNS Impairment

Dick et al. (1984, 1988) showed that a 4-h exposure of 137 human volunteers to MEK at a concentration of 200 ppm had no effects on the choice reaction time, visual vigilance, auditory tone tracking, auditory discrimination, memory scanning, and postural sway test. Thus, the NOAEL for CNS impairment is 200 ppm for a 4-h MEK exposure.

Classical pharmacokinetics analysis is used instead of the conservative Haber's rule to derive ACs for MEK on the basis of CNS impairment. A pharmacokinetic technique is used to predict an MEK concentration in blood during an inhalation exposure of humans to MEK. By assuming that MEK's CNS effect is dependent on its concentration in the blood, the pharmacokinetic technique can predict an acceptable exposure concentration based on the prevention of CNS impairment. The assumption is probably valid for two reasons. First, only 2% of the absorbed MEK is excreted in the urine as 2,3-butanediol in human subjects within a day of an MEK exposure (Liira et al., 1988a). This suggests that only a very small fraction of absorbed MEK is converted into organic metabolites in the body, with the majority going into intermediary metabolism (Liira et al., 1988a). Second, even when MEK is converted into organic metabolites, such as 3-hydroxy-2-butanone and 2,3-butanediol, they are more polar than MEK and should be eliminated much faster than MEK. From the data of Liira et al. (1988a) on the rate of urinary excretion of 2,3-butanediol in human subjects after an

MEK exposure, it can be estimated that the elimination half-life of 2,3-butanediol is about 1 h, which is less than MEK's elimination half-life of 81 min determined in that study. Therefore, the organic metabolites are not expected to contribute much to the CNS effect of MEK.

According to classical pharmacokinetics, during a continuous exposure in which the subject is exposed to the same dose throughout, the blood concentration would reach 87.5% or 99% of the final steady-state value at 3 or 7 elimination half-lives, respectively, past the start of the exposure (Gibaldi and Perrier et al., 1975). It follows that the blood concentration, expressed in percent of the final steady-state value that is reached at a given time, expressed in number of elimination half-lives, after the start of a continuous exposure, can be calculated with the following formula:

Concentration reached at time $t = 100\% - (100\%/2^n)$,

where $t = n \times$ elimination half-life.

Although the absorption rate might not stay constant in an inhalation exposure (in fact as the body gradually absorbs a vapor, the absorption rate should decrease with time of exposure), to simplify the analysis and to err slightly on the conservative side, the absorption rate is assumed to be constant throughout a continuous inhalation exposure. With MEK, for instance, the expired MEK concentration was about 53% of the inhaled MEK concentration all through a 4-h exposure of human subjects at 200 ppm, indicating that the MEK absorption rate remained constant through at least 4 h of an inhalation exposure. It is, therefore, safe to assume that the blood concentration of the vapor will be no higher than the concentration predicted by the formula above.

According to the formula, when an individual inhales a constant concentration of a vapor for time t equaled to 3 times the chemical's elimination half-life, the blood concentration will be 87.5%, or about 90%, of the steady-state concentration. The half-life of the elimination phase of MEK is 81 min in humans (Liira et al., 1988a). In other words, the blood concentration at 3 × 81 min, or 4 h, into an MEK inhalation exposure is about 90% of the steady-state concentration. It can be calculated with the formula that the MEK concentrations in blood achieved in a 4-h inhalation exposure should be 37% higher than those in a 2-h exposure. Because measurements done in 70 male and female human subjects exposed to MEK at 100 or 200 ppm showed that the 4-h blood concentration was, on the average, 35% higher than that at 2 h, the formula predicts the blood concentration quite well (Brown et al., 1987).

As predicted by the formula, the blood concentration will reach 99% of the steady-state concentration if the inhalation exposure is extended to time t equaled to 7 elimination half-lives. That means at 7×81 min, or 9.5 h, into an inhalation exposure to a fixed airborne MEK concentration, the MEK concentration in blood would practically reach steady state. If MEK's CNS effect is proportional to its blood concentration, as assumed above, an airborne MEK concentration acceptable for 9.5 h should also be acceptable for up to 180 d.

Dick et al. (1984, 1988) showed that a 4-h exposure to MEK at 200 ppm does not cause CNS impairment. Since the elimination half-life of MEK is 81 min, 4 h is 3 elimination half-lives. As shown previously, at 3 elimination half-lives into an inhalation exposure, the blood concentration should be about 90% of the final steady-state concentration. So the data of Dick et al. can be interpreted to mean that a blood MEK concentration of 90% the steady-state blood concentration produced by a long-term continuous exposure to MEK at 200 ppm does not cause CNS impairment. Therefore, the steady-state blood concentration of MEK produced by a long-term continuous exposure to MEK at 180 ppm, which is equal to 90% of 200 ppm, should also be devoid of CNS effects. Because it only takes an MEK exposure of 9.5 h for the blood concentration to reach steady state, the ACs based on CNS impairment for 24 h, 7 d, 30 d, and 180 d are all estimated to be 180 ppm.

The 1-h AC can be derived as follows: The pharmacokinetic formula shows that, during an inhalation exposure to MEK, the MEK concentration in blood reached in 1 h should be 40% of that reached in 4 h. So a 1-h exposure to twice the 4-h no-observed-adverse-effect level (NOAEL) of 200 ppm should also yield an MEK blood concentration that would not impair the CNS. Therefore, 400 ppm, which is twice the 4-h NOAEL, is selected as the 1-h AC on the basis of CNS impairment.

Establishment of SMAC Values

The various ACs are given in Table 9-4. Comparison of the ACs for mucosal irritation and CNS impairment shows that the 1-h, 24-h, 7-d, 30-d, and 180-d SMACs are set at 50, 50, 10, 10, and 10 ppm, respectively. Because neither irritation nor CNS impairment is expected to be affected by microgravity-induced physiological changes, no microgravity adjustments are needed for the SMACs.

TABLE 9-4 End Points and Acceptable Concentrations

End Point	Exposure Data	Species and Reference	Uncertainty Factors				Acceptable Concentrations, ppm				
			NOAEL	Time	Small n		1 h	24 h	7 d	30 d	180 d
Mucosal irritation	LOAEL at 100 ppm, 3-5 min	Human (n = 10) (Nelson et al., 1943)	—	2	—		50	50	—	—	—
	LOAEL at 100 ppm, 3-5 min	Human (n = 10) (Nelson et al., 1943)	3	—	10/(sq. rt. 10)		—	—	10	10	10
CNS depression	NOAEL at 200 ppm, 4 h	Human (n = 137) (Dick et al., 1984, 1988)	—	1/2	—		200	—	—	—	—
	NOAEL at 200 ppm, 4 h	Human (n = 137) (Dick et al., 1984, 1988)	—	1/0.9	—		—	180	180	180	180
SMAC							50	50	10	10	10

REFERENCES

ACGIH. 1991. Documentation of Threshold Limit Values and Biologic Monitoring Indices. Threshold Limit Values Committee, American Conference of Governmental Industrial Hygienists, Cincinnati, Ohio.

Alderson, M.R., and N.S. Rattan. 1980. Mortality of workers on an isopropyl alcohol plant and two MEK dewaxing plants. Br. J. Ind. Med. 37:85-89.

Altenkirch, H., G. Stoltenburg, and H.M. Wagner. 1978. Experimental studies on hydrocarbon neuropathies induced by methyl-ethyl-ketone. MEK. J. Neurol. 219:159-170.

Brown, W.D., J.V. Setzer, R.B. Dick, F.C. Phipps, and L.K. Lowry. 1987. Body burden profiles of single and mixed solvent exposures. J. Occup. Med. 29:877-883.

Cavender, F.L., H.W. Casey, H. Salem, J.A. Swenberg, and E.J. Gralla. 1983. A 90-day vapor inhalation toxicity study of methyl ethyl ketone. Fund. Appl. Toxicol. 3:264-270.

De Ceaurriz, J.C., J.C. Micillino, P. Bonnet, and J.P. Guenier. 1981. Sensory irritation caused by various industrial airborne chemicals. Toxicol. Lett. 9:137-143.

Deacon, M.M., M.D. Pilny, J.A. John, B.A. Schwetz, F.J. Murray, H.O. Yakel, and R.A. Kuna. 1981. Embryo- and fetotoxicity of inhaled methyl ethyl ketone in rats. Toxicol. Appl. Pharmacol. 59:620-622.

Dick, R.B., J.V. Setzer, R. Wait, M.B. Hayden, B.J. Taylor, B. Tolos, V. Putz-Anderson. 1984. Effects of acute exposure of toluene and methyl ethyl ketone on psychomotor performance. Int. Arch. Occup. Environ. Health 54:91-109.

Dick, R.B., W.D. Brown, J.V. Setzer, B.J. Taylor, and R. Shukla. 1988. Effects of short duration exposures to acetone and methyl ethyl ketone. Toxicol. Lett. 43:31-49.

DiVincenzo, G.D., C.J. Kaplan, and J. Dedinas. 1976. Characterization of the metabolites of methyl n-butyl ketone, methyl iso-butyl ketone, and methyl ethyl ketone in guinea pig serum and their clearance. Toxicol. Appl. Pharmacol. 36:511-522.

Douglas, G.R., E.R. Nestmann, E.G. Lee, and J.C. Miller. 1980. Mutagenic activity in pulp mill effluents. Environ. Impact Health Effects 3:865-880.

Duckett, S., L.J. Streletz, R.A. Chambers, M. Auroux, and P. Galle.

1979. 50 ppm MnBK subclinical neuropathy in rats. Experientia 35:1365-1367.
EPA. 1990. 2-Butanone. In Integrated Risk Information System, Office of Health and Environmental Assessment, U.S. Environmental Protection Agency, Washington, D.C.
Florin, I., L. Rutberg, M. Curvall, and C.R. Engell. 1980. Screening of tobacco smoke constituents for mutagenicity using the Ames test. Toxicology 18:219-232.
Garabrandt, D.H., J. Held, B. Langholz, and L. Bernstein. 1988. Mortality of aircraft manufacturing workers in Southern California. Am. J. Ind. Med. 13:683-693.
Geller, I., E. Gause, H. Kaplan, and R.J. Hartmann. 1979. Effects of acetone, methyl ethyl ketone and methyl isobutyl ketone on a match-to-sample task in the baboon. Pharmacol. Biochem. Behav. 11401-406.
Gibaldi, M., and D. Perrier. 1975. Two-compartment model. P. 71 in Pharmacokinetics. New York: Marcel Dekker.
Green, D.J., L.R. Sauder, T.J. Kulle, and R. Bascom. 1987. Acute response to 3.0 ppm formaldehyde in exercising healthy nonsmokers and asthmatics. Am. Rev. Resp. Dis. 135:1261-1266.
Horton, A.W. 1965. Carcinogenesis of the skin. III. The contribution of elemental sulfur and of organic sulfur compounds. Cancer Res. 25:1759-1763.
Huntoon, C.L. 1993. Summary Report of Preflight and Postflight Atmospheric Analyses for STS-26 through STS-41. JSC Memo. No. SD4-93-021. Space and Life Sciences Directorate, Johnson Space Center, NASA, Houston, Tex..
Li, G.L., S.N. Yin, T. Watanabe, H. Nakatsuka, M. Kasahara, H. Abe, and M. Ikeda. 1986. Benzene-specific increase in leukocyte alkaline phosphatase activity in rats exposed to vapors of various organic solvents. J. Toxicol. Environ. Health 19:581-589.
Liira, J., V. Riihimaki, and P. Pfaffli. 1988a. Kinetics of methyl ethyl ketone in man: absorption, distribution and elimination in inhalation exposure. Int. Arch. Occup. Environ. Health 60:195-200.
Liira, J., V. Riihimaki, K. Engstrom, and P. Pfaffli. 1988b. Coexposure of man to *m*-xylene and methyl ethyl ketone. Kinetics and metabolism. Scand. J. Work Environ. Health 14:322-327.
Liira, J., G. Johanson, and V. Riihimaki. 1990a. Dose-dependent kinet-

ics of inhaled methylethylketone in man. Toxicol. Lett. 50:195-201.
Liira, J., V. Riihimaki, and K. Engstrom. 1990b. Effects of ethanol on the kinetics of methyl ethyl ketone in man. Br. J. Ind. Med. 47:235-240.
Lowengart, R.A., J.M. Peters, C. Cicioni, J. Buckley, L. Bernstein, S. Preston-Martin, and E. Rappaport. 1987. Childhood leukemia and parents occupational and home exposures. J. Nat. Cancer Inst. 79:39-46.
Miyasaka, M., M. Kumai, A. Koizumi, T. Watanabe, K. Kurasako, K. Sato, and M. Ikeda. 1982. Biological monitoring of occupational exposure to methyl ethyl ketone by means of urinanalysis for methyl ethyl ketone itself. Int. Arch. Occup. Environ. Health 50:131-137.
Nelson, K.W., J.F. Ege, Jr., M. Ross, L.E. Woodman, and L. Silverman. 1943. Sensory response to certain industrial solvent vapors. J. Ind. Hyg. Toxicol. 25:282-285.
O'Donoghue, J.L., S.R. Haworth, R.D. Cullen, P.E. Kirby, T. Lawlor, E.J. Moran, R.D. Phillips, D.L. Putnam, A.M. Roers-Back, and R.S. Slesinski. 1988. Mutagenicity studies on ketone solvents: methyl ethyl ketone, methyl isobutyl ketone, and isophorone. Mutat. Res. 206:149-161.
Patty, F.A., H.H. Schrenk, and W.P. Yant. 1935. Acute response of guinea pigs to vapors of some new commercial organic compounds. VIII. Butanone. Public Health Rep. 50:1217-1228.
Perbellini, L., F. Brugnone, P. Mozzo, V. Cocheo, and D. Caretta. 1984. Methyl ethyl ketone exposure in industrial workers. Uptake and kinetics. Int. Arch. Occup. Environ. Health 54:73-81.
Sauder, L.R., M.D. Chatham, D.J. Green, and T.J. Kulle. 1986. Acute pulmonary response to formaldehyde exposure in healthy nonsmokers. J. Occup. Med. 28:420-424.
Saida, K, J.R. Mendell, and H.S. Weiss. 1976. Peripheral nerve changes induced by methyl n-butyl ketone and potentiation by methyl ethyl ketone. J. Neuropathol. Exp. Neurol. 35:207-225.
Schwetz, B.A., T.J. Mast, R.J. Weigel, J.A. Dill, and R.E. Morrissey. 1991. Developmental toxicity of inhaled methyl ethyl ketone in Swiss mice. Fundam. Appl. Toxicol. 16:742-748.
Shibata, E., J. Huang, N. Hisanaga, Y. Ono, I. Saito, and Y. Takeuchi. 1990a. Effects of MEK on kinetics of n-hexane metabolites in serum. Arch. Toxicol. 64:247-250.

Shibata, E., J. Huang, Y. Ono, N. Hisanaga, M. Iwata, I. Saito, and Y. Takeuchi. 1990b. Changes in urinary n-hexane metabolites by co-exposure to various concentrations of methyl ethyl ketone and fixed n-hexane levels. Arch. Toxicol. 64:165-168.

Smyth, H.F., C.P. Carpenter, C.S. Weil, U.C. Pozzani, and J.A. Striegel. 1962. Range-finding toxicity data: List VI. Amer. Ind. Hyg. Assoc. J. 23:95-107.

Toftgard, R., O.G. Nilsen, and J.-A. Gustafsson. 1981. Changes in rat liver microsomal cytochrome P-450 and enzymatic activities after inhalation of n-hexane, xylene, methyl ethyl ketone and methylchloroform for four weeks. Scand. J. Work Environ. Health 7:31-37.

Weber-Tschopp, A., T. Fischer, R. Gierer, and E. Grandjean. 1977. [Experimentally induced irritating effects of acrolein on men.] Int. Arch. Occup. Environ. Health 40:117-130.

Wen, C.P., S.P. Tsai, N.S. Weiss, R.L. Gibson, O. Wong, and W.A. McClellan. 1985. Long-term mortality study of oil refinery workers. IV. Exposure to the lubricating-dewaxing process. J. Natl. Cancer Inst. 74:11-18.

Zimmermann, F.K., V.W. Mayer, I Schael, and M.A. Resnick. 1985. Acetone, methyl ethyl ketone, ethyl acetate, acetonitrile and other polar aprotic solvents are strong inducers of aneuploidy in *Saccharomyces cerevisiae*. Mutat. Res. 149:339-351.

B10 Nitromethane

King Lit Wong, Ph.D.
Johnson Space Center Toxicology Group
Biomedical Operations and Research Branch
Houston, Texas

PHYSICAL AND CHEMICAL PROPERTIES

Nitromethane is a colorless liquid with a fruity odor (Sax, 1984).

Synonyms:	Nitrocarbol
Formula:	CH_3NO_2
CAS number:	75-52-5
Molecular weight:	61
Boiling point:	101.2°C
Melting point:	-29°C
Vapor pressure:	27.8 mm Hg at 20°C
Conversion factors at 25°C, 1 atm:	1 ppm = 2.62 mg/m^3
	1 mg/m^3 = 0.38 ppm

OCCURRENCE AND USE

Nitromethane is used as a solvent, an intermediate in the chemical synthesis of pharmaceuticals and pesticides, as a rocket fuel, and a stabilizer for halogenated alkanes (EPA, 1985; ACGIH, 1986). Nitromethane is not known to be used in spacecraft, but it has been predicted to be off-gassed in the Space Station Freedom (Leban and Wagner, 1989).

PHARMACOKINETICS AND METABOLISM

No pharmacokinetic data have been found in the literature search. Because nitromethane has been found to be toxic in rabbits after oral or inhalation exposures, but not after cutaneous applications on closely clipped abdominal skin (the liquid nitromethane applied was allowed to remain on the skin until totally evaporated) (Machle et al., 1940), it could be inferred that nitromethane is absorbed via the intestinal and respiratory tracts and nitromethane might not be very well absorbed through the skin.

There are no in vivo data on nitromethane metabolism. However, a few studies have been performed with liver microsomal preparations. Incubation of nitromethane with rat liver microsomes under oxidative conditions has been shown to produce nitrite in one study (Marletta and Stuehr, 1983) and to form nitrite and formaldehyde in a stoichiometric fashion in another study (Sakurai et al., 1980). Anaerobic incubation of rat liver microsomes with nitromethane produced only formaldehyde, but not nitrite (Sakurai et al., 1980).

There is no study on the excretory pathways of nitromethane. However, nitromethane has been detected in the urine of mice not exposed to nitromethane (Miyashata and Robinson, 1980).

TOXICITY SUMMARY

In in vitro systems, nitromethane is known to interact with certain heme proteins. Incubation of cytochrome P-450, myoglobin, or hemoglobin with nitromethane has been shown to form nitro complexes of these heme proteins (Mansuy, 1977a,b). Nevertheless, it is unknown whether these interactions contribute to any toxicity in the body during nitromethane exposures.

Acute and Short-Term Toxicity

Nitromethane is known to produce certain symptoms and death in laboratory animals. However, whether nitromethane is toxic to the liver is a controversy.

Lethality

The oral LD_{50} of nitromethane in mice is 1.44 g/kg (Weatherby, 1955). There are very few inhalation studies with nitromethane. Machle et al. exposed two rabbits and two guinea pigs per group to nitromethane vapor at concentrations ranging from 0.1% to 5% in cumulative exposure durations of 0.25-140 h (the cumulative exposure durations above 6 h consisted of daily 6-h exposures) (Machle et al., 1940). The lethal responses of the two rabbits and two guinea pigs combined varied with concentration × time (C × T) (Machle et al., 1940). Closer scrutiny of the data shows that the lethal response followed Haber's rule only for exposures at concentrations of 0.10% or higher. No mortality was recorded in the group exposed cumulatively to 0.05% nitromethane for 140 h, which is equivalent to a C×T of 7 %-h. The fact that the group exposed at 0.05% for 140 h was an outlier was apparent after the mortality data were tabulated (see Tables 10-1 and 10-2).

Miscellaneous Symptoms

Without reference to specific exposure concentrations, Machle et al. (1940) reported that restlessness, slight mucosal irritation (no occular

TABLE 10-1 Mortality in Single Exposures (Machle et al., 1940)

Exposure Concentration, %	Exposure Time, h	C × T	Mortality
3	0.25	0.75	0/4
1	1	1	0/4
3	0.5	1.5	1/4
0.50	3	1.5	1/4
2.25	1	2.25	1/4
3	1	3	2/4
1	3	3	2/4
0.50	6	3	2/4
5	1	5	4/4
1	6	6	4/4
3	2	6	4/4

TABLE 10-2 Mortality in Repetitive Exposures (Machle et al., 1940)

Exposure Concentration, %	Exposure Time, h	C × T	Mortality
0.25	12	3	3/4
0.10	30	3	2/4
0.05	140	7	0/4

discharge, coughing, or sneezing), slight narcosis, salivation, weakness, ataxia, incoordination, circus movements, convulsions, and twitching were produced by the exposures. In addition, the animals appeared uncomfortable or ill. They reported that it took exposures at 3% or 5% nitromethane for more than 1 h or 1% nitromethane for 5 h or more to cause CNS symptoms in rabbits and guinea pigs.

Hepatic Toxicity

Weatherby (1955) showed that an oral administration of nitromethane at 500 mg/kg in dogs produced marked fatty changes in the periportal and midzonal regions of the liver 32 h after the administration. However, Dayal et al. (1989) reported that nitromethane injected intraperitoneally at 550 mg/kg failed to cause any liver histopathology or any changes in the plasma levels of alanine transaminase, aspartate transaminase, and sorbitol dehydrogenase in mice. Similarly, no evidence of hepatic toxicity has been found with inhalation exposure of nitromethane. Lewis et al. (1979) exposed 10 rats to nitromethane at 98 or 745 ppm, 7 h/d, for 2 d. The level of serum glutamic-pyruvic transaminase (alanine transaminase) did not increase and no histopathology was detected in various tissues, including the liver, in these animals. Therefore, there is no evidence that nitromethane, at least via inhalation exposures, is hepatotoxic.

Subchronic and Chronic Toxicity

Hepatic Toxicity

Subchronic nitromethane exposures might result in liver changes, but there is no evidence of any liver toxicity when animals inhaled nitromethane repetitively. Subcutaneous injections of nitromethane at 290 mg/kg every other day for 18 d in rats led to an 80% reduction of hepatic histidase activity and two- to three-fold increases in the histidine levels in the plasma and liver (Wang and Lee, 1973). There were no changes in the levels of other amino acids in the plasma and liver, the liver protein content, and the body-weight gain in these rats. Because it is doubtful whether increases in the histidine levels in the plasma and liver have any adverse health impacts, the SMACs are not set according to these biochemical changes.

Weatherby (1955) exposed male rats at 0.1% or 0.25% nitromethane in drinking water for 15 w. Cells with enlarged, prominent nuclei and cytoplasm more granular than in the controls in both the 0.1% and 0.25% groups were observed in the liver. In the rats exposed at 0.25% nitromethane, numerous lymphocytes were observed in the periportal regions of the liver. In both nitromethane groups, the light microscopic morphology of other tissues appeared normal. The National Research Council Subcommittee on SMACs doubted the clinical significance of these minor light microscopic changes in the liver. The subcommittee advised that NASA disregard the hepatic end point in setting nitromethane's SMACs, especially because Lewis et al. (1979) failed to find any increase in alanine transaminase or histopathology in the livers of rats exposed to nitromethane at a concentration of 745 ppm, 7 h/d, 5 d/w, for 10 d or 1, 3, or 6 mo. The hepatic findings of Weatherby (1955) also had doubtful meaning in inhalation exposures of nitromethane because Griffin (1990) also did not detect any increases in the level of alanine transaminase, aspartate transaminase, bilirubin or protein, or any hepatic histopathology in rats exposed to nitromethane at 100 or 200 ppm, 7 h/d, 5 d/w, for 2 y.

Thyroid Toxicity

Exposures of rats to nitromethane have been shown to result in thyroid toxicity. An exposure of 10 rats at 98 or 745 ppm, 7 h/d, 5 d/w, for 6 mo resulted in an increase in thyroid weight by 19% or 26%, respectively, but no change in the thyroxine level in plasma (Lewis et al., 1979). In contrast, an exposure of five rabbits to nitromethane at 98 ppm, 7 h/d, 5 d/w, for 6 mo caused a 48% reduction in the thyroxine level in plasma, but no change in the thyroid weight. However, a similar exposure of five rabbits at 745 ppm increased the thyroid weight by 38% and reduced the plasma thyroxine level by 52%. No thyroid weight or thyroxine level changes were detected in 10 rats exposed at 98 or 745 ppm, 7 h/d, 5 d/2, for 2 d, 10 d, 1 mo, or 3 mo. Similarly, no thyroid weight or thyroxine level changes were found in five rabbits exposed to nitromethane at 98 ppm, 7 h/d, 5 d/w, for 1 or 3 mo (Lewis et al., 1979). The thyroid data of Lewis et al. are summarized in Table 10-3.

Unfortunately, in the 2-y study conducted by Griffin (1990), he did not weigh the thyroid or measure the serum levels of thyroid hormones in rats exposed to nitromethane at up to 200 ppm. Despite a close to 50% reduction in thyroxine level and a 20-40% increase in thyroid weight caused by a 6-mo exposure to nitromethane at 98 or 745 ppm in rabbits, the subcommittee on SMACs advised that nitromethane's SMACs need not be set to prevent the thyroid toxicity. The subcommittee's reasoning is that the thyroid changes are not clinically significant in humans and that quite a number of chemicals known to produce thyroid changes of similar magnitude in rodents do not produce any thyroid disease in humans. It should be noted that nitromethane's effect on thyroxine is opposite to that of microgravity. Studies done by NASA showed that the blood concentrations of thyroxine and thyroid-stimulating hormone were increased on the day astronauts returned to earth after several Skylab and space-shuttle missions (Huntoon et al., 1989).

TABLE 10-3 Thyroid Data from Rats and Rabbits

Animal and Parameter	Nitromethane Concentration			
	98 ppm		745 ppm	
	3 mo	6 mo	3 mo	6 mo
Rats				
Thyroxine level	±	±	±	±
Thyroid weight	±	+19%	±	+26%
Rabbits				
Thyroxine level	±	−48%	±	−52%
Thyroid weight	±	+19%	±	+38%

Hematological Changes

As mentioned earlier, nitromethane is metabolized in vitro to nitrite, so there is a possibility that nitromethane produces methemoglobin. Despite the fact that nitrite was found in the lung, heart, kidney, and spleen of rats exposed to nitromethane at 13,000 ppm for 6 h, no methemoglobin was detected (Dequidt et al., 1973). Similarly, Lewis et al. (1979) did not find methemoglobinemia in rats and rabbits exposed to nitromethane at 745 ppm, 7 h/d, 5 d/w, for 6 mo.

In the study of rats and rabbits exposed to nitromethane for 7 h/d, 5 d/w, the hematocrit and hemoglobin levels were reduced 5-12% in rats exposed at 745 ppm for 10 d or 1, 3, or 6 mo, but not in the rats exposed for 2 d (Lewis et al., 1979). The rabbits appeared to be less sensitive than rats to nitromethane's effects in causing anemia because a 13% reduction in hemoglobin levels was found in the rabbits only at 1 mo and not after 3 or 6 mo of repetitive exposures to nitromethane at 745 ppm. In addition, repetitive exposures of rabbits to nitromethane at 745 ppm for 1, 3, or 6 mo did not decrease the hematocrit.

There is evidence that chronic nitromethane inhalation exposures are devoid of any hematological effects at up to 200 ppm. Griffin (1990) exposed rats to nitromethane at 100 or 200 ppm, 7 h/d, 5 d/w, for 2 y and found no changes in the red- or white-blood-cell counts, hemoglobin concentration, hematocrit, and platelet count.

Pulmonary Toxicity

Lewis et al. (1979) showed that nitromethane caused pulmonary damage in one group of rabbits at one time point. They observed moderate to moderately severe focal lung hemorrhages and congestion, associated with interstitial lung edema, in rabbits exposed to nitromethane at 745 ppm, 7 h/d, 5 d/w, for 1 mo. The interstitial pulmonary edema evidently must not be severe because no increase in the wet-weight-to-dry-weight ratio was found in the lung of these rabbits. These changes in the pulmonary morphology were absent in the rabbits exposed to nitromethane at 98 ppm. No lung histopathology was detected in groups of 10 rats each exposed at 745 ppm, 7 h/d, 5 d/w, for 2 or 10 d or 1, 3, or 6 mo. The apparent pulmonary toxicity discovered in these rabbits was not relied on in setting the SMACs for three reasons. First, only five rabbits were exposed at 745 ppm for 1 mo. For lung morphological evaluation, something highly susceptible to artifacts, more animals are required to make the result more reliable. Second, the lung changes were seen in the rabbit only at 1 mo but not at 3 or 6 mo. It is highly suspicious that the morphological changes resolved after further repetitive nitromethane exposures. Third, the lung changes have never been reported by other investigators, including Weatherby (1955), who exposed rabbits, rats, and dogs to nitromethane, and Griffin (1990), who exposed rats to nitromethane at up to 200 ppm for 2 y (Machle et al., 1940).

Effects on Reproductive Function

Nitromethane did not produce reproductive toxicity in female rats. Intraperitoneal injections of nitromethane with 0.5 mL (equivalent to 46 mg) every third day, begun 1 w before the rats were bred and continued throughout gestation, failed to produce any changes in the percentage of successful matings, litter size, pup death rate, birth weights, or maternal behavior (Whitman et al., 1977). The negative finding of reproductive toxicity in rats exposed to nitromethane does not prove that nitromethane is devoid of adverse effects on reproduction. According to the subcommittee on SMACs, the rat is not an ideal model to test for some classes of reproductive toxicity, and both male and female effects can be difficult to detect in the absence of histopathological examination. This is particularly germane in an experiment such as this where treatment began only 1 w before mating.

Carcinogenesis

Although an exposure to 2-nitropropane at 207 ppm, 7 h/d, 5 d/w, for only 6 mo produced liver tumors in 10 out of 10 rats, a similar exposure to nitromethane at 745 ppm failed to show any carcinogenicity (Lewis et al., 1979). The lack of carcinogenic response to nitromethane has been proven by Griffin (1990) in rats exposed to nitromethane at 200 or 100 ppm, 7 h/d, 5 d/w, for 2 y.

Genotoxicity

Nitromethane does not appear to be genotoxic. An exposure of *Salmonella typhimurium* (strains TA98, TA100, and TA102) to nitromethane at up to 200 μmol per plate did not increase the mutation frequency (Dayal et al., 1989).

Developmental Toxicity

No data on nitromethane's developmental toxicity were found in the Toxline and Toxlit databases of the National Library of Medicine.

Interaction with Other Chemicals

No data on nitromethane's interaction with other chemicals were found.

TABLE 10-4 Toxicity Summary

Concentration, ppm	Exposure Duration	Species	Effects	Reference
98	7 h/d, 5 d/w; 1, 3, or 6 mo	Rabbit	No histopathology, anemia, or methemoglobinemia; no changes in prothrombin time, SGPT, body-weight gain, or thyroid weight; 48% reduction in thyroxine level at 6 mo (not at other time points)	Lewis et al., 1979
98	7 h/d, 5 d/w; 2 or 10 d or 1, 3, or 6 mo	Rat	No histopathology, anemia, or methemoglobinemia; no changes in prothrombin time, SGPT, body-weight gain, or thyroxine level; 19% increase in thyroid weight at 6 mo (not at other time points)	Lewis et al., 1979
100 or 200	7 h/d, 5 d/w, 2 y	Rat	No changes in body weight in males, but dropped slightly in females starting at 9 mo; no visible signs of toxicity; no significant effects on mortality, organ weights of selective organs (liver, kidney, brain, heart, and lung), serum chemistry (Na, K, aspartate transaminase, alanine transaminase, bilirubin, protein, blood urea nitrogen, and creatinine), or histopathology	Griffin, 1990
500	6 h/d, 23 d	Rabbit, guinea pig (n = 2)	No deaths	Machle et al., 1940
745	7 h/d, 5 d/w, 2 d	Rat	Red-blood-cell (RBC) count increased by 7%; no changes in hematocrit, hemoglobin level, prothrombin time; no methemoglobinemia and histopathology; no changes in SGPT, thyroxine level, thyroid weight, and body-weight gain	Lewis et al., 1979
745	7 h/d, 5 d/w, 10 d	Rat	RBC count and hemoglobin level decreased by 7%; hematocrit decreased by 5%; no changes in prothrombin time, SGPT, thyroxine level, thyroid weight, and body-weight gain; no methemoglobinemia and histopathology	Lewis et al., 1979

745	7 h/d, 5 d/w, 1 mo	Rat	Hemoglobin level and hematocrit decreased by 5-6%; no changes in RBC count, prothrombin time, SGPT, thyroxine level, thyroid weight, and body-weight gain; no methemoglobinemia or histopathology	Lewis et al., 1979
745	7 h/d, 5 d/w, 1 mo	Rabbit	Hemoglobin level reduced 13%; no changes in hematocrit, RBC count, or prothrombin time; thyroxine reduced 44%; no changes in thyroid weight, body-weight gain, or SGPT; no methemoglobinemia; in the lung, focal areas of moderate hemorrhages, congestion, and interstitial edema; no histopathology in other tissues	Lewis et al., 1979
745	7 h/d, 5 d/w, 3 mo	Rabbit	No changes in hemoglobin level, hematocrit, RBC count, prothrombin time, thyroxine level, thyroid weight, body weight gain, or SGPT; no methemoglobinemia or histopathology	Lewis et al., 1979
745	7 h/d, 5 d/w, 3 or 6 mo	Rat	Retarded body-weight gain; hemoglobin level reduced 12%, hematocrit reduced 7%; no changes in RBC count, prothrombin time, SGPT, and thyroxine level; thyroid weight increased 26% at 6 mo, but not 3 mo; no methemoglobinemia or histopathology	Lewis et al., 1979
745	7 h/d, 5 d/w, 6 mo	Rabbit	Thyroxine level reduced 52% and thyroid weight increased 38%; no changes in body weight gain, hemoglobin level, hematocrit, RBC count, prothrombin time, and SGPT; no methemoglobinemia or histopathology	Lewis et al., 1979
1000	6 h/d, 5 d	Rabbit, guinea pig (n = 2)	Deaths: 0/2 rabbits and 2/2 guinea pigs	Machle et al., 1940
2500	6 h/d, 2 d	Rabbit, guinea pig (n = 2)	Deaths: 2/2 rabbits and 1/2 guinea pigs	Machle et al., 1940
5000	3 h	Rabbit, guinea pig	Deaths: 0/2 rabbits and 1/2 guinea pigs	Machle et al., 1940

TABLE 10-1 (Continued)

Concentration, ppm	Exposure Duration	Species	Effects	Reference
5000	6 h	Rabbit, guinea pig (n = 2)	Deaths: 1/2 rabbits and 1/2 guinea pigs	Machle et al., 1940
10,000	1 h	Rabbit, guinea pig (n = 2)	No deaths	Machle et al., 1940
10,000	3 h	Rabbit, guinea pig (n = 2)	Deaths: 0/2 rabbits and 2/2 guinea pigs	Machle et al., 1940
10,000	6 h	Rabbit, guinea pig (n = 2)	Deaths: 2/2 rabbits and 2/2 guinea pigs	Machle et al., 1940
22,500	1 h	Rabbit, guinea pig (n = 2)	Deaths: 0/2 rabbits and 1/2 guinea pigs	Machle et al., 1940
30,000	0.25 h	Rabbit, guinea pig (n = 2)	No deaths	Machle et al., 1940
30,000	0.5 h	Rabbit, guinea pig (n = 2)	Deaths: 0/2 rabbits and 1/2 guinea pigs	Machle et al., 1940
30,000	1 h	Rabbit, guinea pig (n = 2)	Deaths: 0/2 rabbits and 2/2 guinea pigs	Machle et al., 1940
30,000	2 h	Rabbit, guinea pig (n = 2)	Deaths: 2/2 rabbits and 2/2 guinea pigs	Machle et al., 1940
50,000	1 h	Rabbit, guinea pig (n = 2)	Deaths: 2/2 rabbits and 2/2 guinea pigs	Machle et al., 1940

[a]Only inhalation data are included.
[b]N.S. = not specified.

NITROMETHANE 343

TABLE 10-5 Exposure Limits Set by Other Organizations

Organization	Concentration, ppm
ACGIH's TLV	100 (TWA)
NIOSH's IDLH	1000
OSHA's PEL	100 (TWA)

TLV = threshold limit value. TWA = time-weighted average. IDLH = immediately dangerous to life and health. PEL = permissible exposure limit. TWA = time-weighted average.

TABLE 10-6 Spacecraft Maximum Allowable Concentrations

Duration	ppm	mg/m^3	Target Toxicity
1 h	25	65	Anemia
24 h	15	40	Anemia
7 d[a]	7	18	Anemia
30 d	7	18	Anemia
180 d	5	13	Anemia

[a]There was no 7-d SMAC.

RATIONALE FOR ACCEPTABLE CONCENTRATIONS

The SMACs are set to protect the astronauts against the following toxic end points: CNS symptoms and anemia. For a given exposure duration, the SMACs are set by selecting the lowest acceptable concentration (AC) among the two toxic end points. An acceptable concentration for a toxic end point for a given exposure duration is derived from the no-observed-adverse-effect level (NOAEL) for that end point and that duration. Because all the NOAELs are estimated from animal data, an interspecies factor of 10 is applied on all of them. Other than the AC for anemia, no microgravity safety factor is needed for other toxic end points because they are not known to be affected by microgravity.

Miscellaneous Symptoms

Machle et al. (1940) reported that inhalation exposures of rabbits or guinea pigs led to slight mucosal irritation, but they did not report the exposure concentration at which they detected mucosal irritation in the animals. Similarly, both the ACGIH (1986) and Lewis et al. (1979) commented that nitromethane is a mucosal irritant not supported by any quantitative data. The SMACs are not set based on mucosal irritation because there are no data on the irritating concentrations of nitromethane. Lewis et al., however, did state that the current permissible exposure limit (PEL) of 100 ppm is sufficiently low to prevent respiratory irritation based on animal data and industrial health experience. Because the SMACs are set below 100 ppm, it appears that mucosal irritation will not be a problem if Lewis et al. are correct.

Machle et al. (1940) also stated, without reference to any specific exposure concentration, that nitromethane could produce restlessness, narcosis, ataxia, incoordination, and convulsions in the rabbits and guinea pigs they exposed. Since Lewis et al. (1979) did not mention these symptoms in rats exposed to nitromethane at 98 or 745 ppm in a 90-d study, the NOAEL for CNS symptoms is estimated to be 745 ppm.

1-h, 24-h, 7-d, 30-d, and 180-d ACs based on CNS symptoms
= 90-d NOAEL × 1/species factor
= 745 ppm × 1/10
= 75 ppm.

The 180-d AC based on CNS symptoms is set equal to the 30-d AC because if CNS symptoms fail to develop in an exposure at 75 ppm for 30 d, it is highly unlikely that the symptoms will occur when the exposure is extended to 180 d. The reason is that the CNS should equilibrate with nitromethane in blood in 30 d.

Anemia

Exposures of rats to nitromethane at 98 ppm, 7 h/d, 5 d/w, caused no anemia in 2 d to 6 mo, but a similar exposure at 745 ppm produced anemia, as shown in Table 10-7, in 10 d to 6 mo (Lewis et al., 1979).

The 7% increase in the RBC count seen in the rat after 2 d of exposure to nitromethane at 745 ppm (Lewis et al., 1979) is not considered important because it did not repeat itself in all the later time points. Be-

TABLE 10-7 Changes in Red Blood Cells in Rats Exposed at 745 pp (Lewis et al., 1979)

Parameter	2 d	10 d	1 mo	3 mo	6 mo
Hematocrit	±	−5%	−5%	−7%	−7%
Hemoglobin concentration	±	−7%	−6%	−12%	−12%
RBC count	+7%	−7%	±	±	±

cause no anemia was found in the rat after a 2-d exposure at 745 ppm, 745 ppm is chosen to be the 1-h NOAEL for anemic effects.

To derive the acceptable concentrations based on nitromethane's induction of anemia, a microgravity safety factor of 3 is applied because microgravity is known to reduce the RBC mass by 10-20% (Huntoon et al., 1989). Three is chosen because 5 has been used for microgravity-induced arrhythmia in the derivation of the ACs of other compounds. The anemic effect is not as serious as the arrhythmic effect for the reason that anemia is usually not life threatening, and some arrhythmia could be. So a smaller safety factor is justified for the anemic effect than that for the arrhythmic effect.

With the microgravity safety factor and the traditional interspecies extrapolation factor, the 1-h and 24-h ACs are derived from the fact that a 2-d exposure of rats to nitromethane at 745 ppm at 7 h/d did not produce anemia in the study by Lewis et al. (1979).

1-h AC based on anemic effects
= 2-d NOAEL × 1/species factor × 1/microgravity factor
= 745 ppm × 1/10 × 1/3
= 25 ppm.

Haber's rule is used to derive the 24-h AC.

24-h AC based on anemic effects
= 2-d NOAEL × time adjustment × 1/species factor
× 1/microgravity factor
= 745 ppm × (7 h/d × 2 d)/24 h × 1/10 × 1/3
= 745 ppm × 0.58 × 1/10 × 1/3
= 15 ppm.

Lewis et al. (1979) showed that an exposure to nitromethane at 98 ppm, 7 h/d, 5 d/w, for 10 d, 1 mo, or 6 mo failed to produce any changes in the hematocrit, hemoglobin concentration, and RBC count in rats. However, based on Griffin's (1990) finding that an exposure of rats to nitromethane at 200 ppm, 7 h/d, 5 d/w for 2 y, did not produce any changes in the hematocrit, hemoglobin concentration, and RBC count, 200 ppm is selected to be the NOAEL for anemia in a continuous exposure lasting 7, 30, or 180 d.

7-d and 30-d ACs based on anemic effects
= 2-y NOAEL × 1/species factor × 1/microgravity factor
= 200 ppm × 1/10 × 1/3
= 7 ppm.

180-d AC based on anemic effects
= 2-y NOAEL × time adjustment × 1/species factor
  × 1/microgravity factor
= 200 ppm × (7 h/d × 5 d/w × 104 w)/(24 h/d × 180 d)
  × 1/10 × 1/3
= 200 ppm × 0.84 × 1/10 × 1/3
= 5.6 ppm.

Establishment of SMAC Values

From the comparison of the various ACs at each time point, the 1-h, 24-h, 7-d, 30-d, and 180-d SMACs are set at 25, 15, 7, 7, and 5 ppm, respectively. As reported in the Toxicity Summary, acute exposures of rabbits and guinea pigs to nitromethane by inhalation have been known to cause death, according to Machle et al (1940). However, those mortality data are not relied on in setting ACs because Machle et al. used only two rabbits and two guinea pigs per combination of exposure concentration and duration. It is of interest, nevertheless, to assess whether exposures at the short-term SMACs are likely to be lethal using the benchmark dose approach as follows (Beck et al., 1993).

The mortality data of rabbits and guinea pigs are combined and grouped together in terms of C × T, which is the product of exposure concentration in percent and exposure duration in hours. But the data of the outlier group (animals exposed at 0.05% nitromethane, 6 h/d for more than 23 d) are excluded. The mortality rates are plotted against the logarithms of C × T of the exposure groups. Via probit analysis according to Finney (1971), the lower 95% confidence limit of the LD_{50} is estimated to be 0.49 %-h.

Acceptable C × T based on lethality for acute exposure
 = Benchmark dose × 1/species factor
 = 0.49 %-h × 1/10
 = 490 ppm-h.

Acceptable concentration for a 24-h exposure
 = 490 ppm-h × 1/24 h
 = 20 ppm.

Acceptable concentration for a 1-h exposure
 = 490 ppm.

Because the 24-h and 1-h SMACs are lower than 20 and 490 ppm, respectively, it is highly unlikely that any deaths will result from exposures at these SMACs.

TABLE 10-8 End Points and Acceptable Concentrations

End Point	Exposure Data	Species and Reference	Uncertainty Factors			Acceptable Concentrations, ppm				
			Time	Species	Micro-gravity	1 h	24 h	7 d	30 d	180 d
CNS symptoms	NOAEL at 745 ppm, 7 h/d, 5 d/w, 90 d	Rat (Lewis et al., 1979)	—	10	—	75	75	75	75	75
Anemia	NOAEL at 745 ppm, 7 h/d, 2 d	Rat (Lewis et al., 1979)	—	10	3	25	—	—	—	—
	NOAEL at 745 ppm, 7 h/d, 2 d	Rat (Lewis et al., 1979)	HR[a]	10	3	—	15	—	—	—
	NOAEL at 200 ppm, 7 h/d, 5 d/w, 2 y	Rat (Griffin, 1990)	—	10	3	—	—	7	7	—
	NOAEL at 200 ppm, 7 h/d, 5 d/w, 2 y	Rat (Griffin, 1990)	HR	10	3	—	—	—	—	5
SMAC						25	15	7	7	5

[a]HR = Haber's rule.

REFERENCES

ACGIH. 1986. Nitromethane. P. 439 in Documentation of the Threshold Limit Values and Biological Exposure Indexes. American Conference of Governmental Industrial Hygienists, Akron, Ohio.
Beck, B.D., R.B. Conolly, M.L. Dourson, D. Guth, D. Hattis, C. Kimmel, and S.C. Lewis. 1993. Symposium overview. Improvements in quantitative noncancer risk assessment. Fundam. Appl. Toxicol. 20:1-14.
Dayal, R., A. Gescher, E.S. Harpur, I. Pratt, and J.K. Chipman. 1989. Comparison of the hepatotoxicity in mice and the mutagenicity of three nitroalkanes. Fundam. Appl. Toxicol. 13:341-348.
Dequidt, J., P. Vasseur, and J. Potencier. 1973. [Experimental toxicological study of some nitroparaffins. 4. Nitromethane.] Bull. Soc. Pharm. Lillie 1973(1):29-35.
EPA. 1985. Health and Environmental Effects Profile for Nitromethane. EPA/600/X-85/116. U.S. Environmental Protection Agency. Cincinnati, Ohio.
Finney, D.J. 1971. Probit Analysis. Cambridge, U.K.: Cambridge University Press.
Griffin, T.B. 1990. Chronic Inhalation of Nitromethane. Lab. Rep. 850902. Coulston International, Inc., White Sands Research Center, Alamogordo, N.M.
Huntoon, C.L., P.C. Johnson, and N.M. Cintron. 1989. Hematology, immunology, endocrinology, and biochemistry. Pp. 222-239 in Space Physiology and Medicine, A.E. Nicogossian, ed. Philadelphia: Lea & Febiger.
Leban, M.I., and P.A. Wagner. 1989. Space Station Freedom Gaseous Trace Contaminant Load Model Development. SAE Technical Paper Series 891513. Warrendale, Pa.: Society of Automotive Engineers.
Lewis, T.R., C.E. Ulrich, and W.M. Busey. 1979. Subchronic inhalation toxicity of nitromethane and 2-nitropropane. J. Environ. Pathol. Toxicol. 2:233-249.
Machle, W., E.W. Scott, and J. Treon. 1940. The physiological response of animals to some simple mononitroparaffins and to certain derivatives of these compounds. J. Ind. Hyg. Toxicol. 22:315-332.
Mansuy, D., P. Gans, J.C. Chottard, and J.F. Bartoli. 1977a. Nitrosoalkanes as iron(II) ligands in the 455-nm-absorbing cyto-

chrome P-450 complexes formed from nitroalkanes in reducing conditions. Eur. J. Biochem. 76:607-615.

Mansuy, D., J.C. Chottard, and G. Chottard. 1977b. Nitrosoalkanes as iron(II) ligands in the hemoglobin and myoglobin complexes formed from nitroalkanes in reducing conditions. Eur. J. Biochem. 76:617-623.

Marletta, M.A., and D.J. Stuehr. 1983. Biologically-derived reactive oxygen and the oxidation of simple nitrogenous compounds. Pp. 56-61 in Oxy Radicals. Their Scavenger System. Vol. 1. Proceedings of the International Conference on Superoxide and Superoxide Dismutase, G. Cohen and R.A. Greenwald, eds. New York: Elsevier Biomedical.

Miyashata, K., and A.B. Robinson. 1980. Identification of compounds in mouse urine vapor by gas chromatography and mass spectrometry. Mech. Age. Dev. 13:177-184.

Sakurai, H., G. Hermann, H.H. Ruf, and V. Ullrich. 1980. The interaction of aliphatic nitro compounds with the liver microsomal monooxygenase system. Biochem. Pharmacol. 29:341-345.

Sax, I. 1984. P. 74 in Dangerous Properties of Industrial Materials. New York: Van Nostrand Reinhold.

Wang, M.-L., and S.-C. Lee. 1973. Alteration of plasma and liver amino acid concentrations in rats treated with nitromethane. J. Chin. Biochem. Soc. 2:25-30.

Weatherby, J.H. 1955. Observations on the toxicity of nitromethane. Arch. Ind. Health 11:102-106.

Whitman, R., B.A. Maher, and R. Abeles. 1977. Deficits in discrimination and maze learning resulting from maternal histidinemia in rats. J. Abnorm. Physchol. 86:659-661.

B11 2-Propanol

John T. James, Ph.D., and Harold L. Kaplan, Ph.D.
Johnson Space Center Toxicology Group
Biomedical Operations and Research Branch
Houston, Texas

PHYSICAL AND CHEMICAL PROPERTIES

Isopropyl alcohol is a colorless, volatile liquid at room temperature (Rowe and McCollister, 1982).

Synonyms: Isopropanol, 2-propanol
Formula: $CH_3CHOHCH_3$
CAS number: 67-63-0
Molecular weight: 60.09
Boiling point: 82.5°C
Melting point: -89.5°C
Specific gravity: 0.79
Vapor pressure: 44 mm Hg at 25°C
Solubility: Miscible with water and most organic solvents
Conversion factors at 25°C, 1 atm: 1 ppm = 2.45 mg/m^3
1 mg/m^3 = 0.41 ppm

OCCURRENCE AND USE

Isopropyl alcohol (IPA) is used commercially in the manufacture of acetone, as a solvent, and in skin lotions, cosmetics, and pharmaceuticals (Rowe and McCollister, 1982). It is widely used by the public in a 70% solution in water as rubbing alcohol to reduce fever and as a disinfectant.

The odor threshold in air is 22 ppm (Amoore and Hautala, 1983). Isopropanol is consistently found in air samples taken during shuttle flights at concentrations in the range of 0.1 to 10 mg/m^3 (James et al., 1994). It originates from flight-hardware off-gassing and the use of IPA as a disinfectant and cleaner.

PHARMACOKINETICS AND METABOLISM

Absorption

The pharmacokinetics and metabolism of IPA have been studied in animals using oral and intravenous (i.v.) administration and, to a limited extent, using inhalation. Acute toxicity inhalation studies show that IPA at high concentrations (near 20,000 ppm) is absorbed by the lungs rapidly and in sufficient quantities to cause central-nervous-system (CNS) depression and lethality within a few hours (Starrek, 1938 (cited by Lehman and Flurry, 1943; Rowe and McCollister, 1982); Carpenter et al., 1949). Apparently, sufficient IPA can be absorbed by rats in 4 h at 2000 ppm to induce what the authors report as slight anesthesia effects (Nakaseko et al., 1991a). Studies of anesthetized dogs given IPA injections into isolated segments of their alimentary tract showed that absorption is more rapid from the intestine than from the stomach and occurs rapidly (82% complete in 30 min) (Wax et al., 1949). In four workers exposed at a mean concentration of 410 ppm, the alveolar air contained an average of only 100 ppm, whereas nine workers exposed at an average of 140 ppm had an alveolar air concentration of only 56 ppm (Folland et al., 1976). This suggests a respiratory absorption of 60-75% in these concentration ranges. In a study of printing-plant workers exposed to IPA in a range of concentrations up to 260 ppm, the ratio of the alveolar concentration to the ambient concentration averaged 0.4, suggesting an uptake of about 60% (Brugnone et al., 1983).

Distribution

Once IPA reaches the blood, it is distributed to the spinal fluid and brain, liver, kidney, and skeletal muscle in dogs (Wax et al., 1949).

2-PROPANOL 353

The results did not suggest any pattern of preferential uptake by any of these organs, nor did the distribution depend on concentration of IPA.

Excretion

The exact metabolic fate of IPA has not been established. Perfusion experiments with isolated rabbit liver demonstrated that the liver converts 30-50% of IPA to acetone (Ellis, 1952). Also in rabbits, IPA is conjugated with glucuronic acid, but only 10% of the intragastrically (i.g.) administered dose was accounted for in the urine as the glucuronide (Kamil et al., 1953). Acetone (19 ppm) was found in the expired air of four workers exposed to IPA at an average of 410 ppm, and an acetone concentration of 7.5 ppm was found in nine workers exposed at an average of 140 ppm (Folland et al., 1976). In another study involving 12 printing-plant workers exposed to IPA at 3-260 ppm for 7 h the elimination of acetone from the lungs reached a steady state in 6-7 h, suggesting that blood concentrations of IPA and acetone had also reached steady-state concentrations (Brugnone et al., 1983).

Metabolism

Limited data are available on the metabolism of IPA in animals and in humans. At the end of a 4-h exposure of rats to IPA at 8000 ppm, the blood concentrations of IPA and acetone were nearly equal; however, at exposures at 500 to 1000 ppm, the acetone-to-IPA ratio was about 3 (Laham et al., 1980). Twenty-four hours after ending the 8-h exposure to 8000 ppm, the blood concentrations of IPA and acetone were 0.008 and 0.110 mg/mL, respectively. In another study, after i.v. administration of 1 or 2 g/kg to pigeons, rats, rabbits, cats, and dogs, the rate of disappearance of alcohol from the blood during the 6 h after equilibration depended on the dose as well as the species (Lehman et al., 1945). In contrast to the linear disappearance of ethanol, the disappearance of IPA from the blood of dogs given large fractions of the fatal dose was more rapid a few hours after administration than at times approaching 24 h (Lehman et al., 1944). In two humans, who had ingested large doses of rubbing alcohol, and reportedly were heavy users of alcoholic beverages, the concentrations of blood IPA were consistent with an exponential model with half-lives of 155 and 187 min (Daniel et al., 1981).

Studies with enzyme inhibitors indicate that oxidation of IPA to acetone is catalyzed by alcohol dehydrogenase (ADH). Pyrazole, an inhibitor of ADH and catalase, reduced the clearance of IPA from the blood of rats and slowed the rate of acetone production (Lester and Benson, 1970; Nordmann et al., 1973). In contrast, pre-exposure of rats to 3-amino-1,2,4-triazole, an inhibitor of catalase, did not result in blood alcohol or acetone concentrations that were different from concentrations in control animals (Nordmann et al., 1973).

TOXICITY SUMMARY

Acute Toxicity

Isopropyl alcohol vapor is irritating to the eyes and upper respiratory tract, and is a CNS depressant at higher concentrations. At 400 ppm, human volunteers exposed for 3-5 min experienced mild irritation of the eyes, nose, and throat (Nelson et al., 1943). At 800 ppm, the effects were not severe but were considered unsuitable for an 8-h workday. Most subjects estimated 200 ppm as the highest concentration satisfactory for 8 h.

No reports of CNS depressant effects in humans solely from inhalation of IPA vapor were found in the scientific literature; however, many cases of acute poisoning in humans from ingestion of IPA have been reported (Adelson, 1962; King et al., 1970). Common manifestations are nausea, vomiting, headache, and varying degrees of CNS depression, soon followed by coma with or without shock (Nelson et al., 1943). When shock is present, death might occur within the first 24 h. In a study of seven subjects given 10-15 cc IPA orally, Fuller and Hunter (1927) found increased blood pressure, sensations of warmth, and various degrees of dizziness and numbness. Tolerance was evident by reduced responses on successive days of administration.

Animals exposed to vapors of IPA exhibit signs of CNS depression, with the severity dependent on the concentration and duration of exposure. In mice, ataxia was produced in 12-26 min at a concentra-

2-PROPANOL 355

tion of 24,000 ppm and was increasingly delayed with decreasing concentrations until, at 3250 ppm, 180-195 min of exposure were required (Starrek, 1938). Prostration of mice occurred in 37-46 min at a concentration of 24,000 ppm but required almost 6 h at 3250 ppm. The onset of narcosis ranged from 100 min at a concentration of 24,000 ppm to almost 8 h at 3250 ppm. Exposure at 12,800 ppm for 200 min or at 19,200 ppm for 160 min caused death in mice (Weese, 1928 (cited by Rowe and McCollister, 1982)). Mice exposed for 8 h at 2050 ppm did not show adverse effects (Starrek, 1938). The ability of IPA to induce CNS depression in rats appears to be comparable to its ability in mice. In rats, an 8-h exposure to IPA vapor at 16,000 ppm resulted in the deaths of four of six animals (Smyth and Carpenter, 1948). In rats exposed to IPA vapor for 4 h at 400, 2000, 4000, or 12,000 ppm, the CNS effects (reduced reaction to sound and dragging hind legs) were obvious only at the highest concentration. In the 2000- and 4000-ppm groups, the authors reported reduced activity and the possibility that anesthesia-type effects might have appeared in some rats during exposures as low as 2000 ppm (Nakaseko et al., 1991a).

The anesthetic dose (elimination of corneal reflect) and lethal dose of IPA by slow i.v. infusion were 2.5 and 6.5 g/kg, respectively, in rabbits and 2.7 and 4.1 g/kg, respectively, in dogs (Lehman and Chase, 1944). Although the anesthetic dose for the two species is almost identical, the anesthetic dose is about 40% of the fatal dose for the rabbit and 65% for the dog.

The principal effect of IPA administered intragastrically to rats, rabbits, and dogs was depression of the CNS, with salivation, retching, and vomiting (Lehman and Chase, 1944). The LD_{50} values for the rat, rabbit, and dog were 5.3, 5.1, and 4.9 g/kg, respectively. Surviving animals recovered rapidly from the depressant effects completely returned to normal behavior. In another study, the oral LD_{50} in young adult rats was 4.7 g/kg, and the first observable toxic signs were at a dose of 2.4 g/kg (Kimura et al., 1971).

Subchronic and Short-Term Toxicity

Rats were unaffected except for slight intoxication when exposed intermittently over a week (total number of exposure hours not given) to air supposedly saturated with IPA vapor (Macht, 1922). Mice exposed to IPA at 10,900 ppm, 4 h/d, for a total of 123 h of exposure were narcotized but survived (Weese, 1928). Slight, reversible fatty changes were observed in the liver. Pregnant rats were exposed to IPA at 10,000, 7000, and 3500 ppm, 7 h/d, for 19 d. No effects were observed clinically at 3500 ppm; however, unsteady gait was observed at the end of early 7000-ppm exposures but was not noticeable after the later exposures (Nelson et al., 1988). The rats were narcotized at the end of early 10,000-ppm exposures, but the effect diminished in later exposures. Since the goal of the study was teratogenic effects, the dams were not subjected to pathology evaluation.

Daily ingestion by human volunteers of IPA at 2.6 or 6.4 mg/kg in a flavored syrup diluted with water for 6 w did not result in adverse signs or symptoms (Wills et al., 1969). There were also no significant changes in clinical chemistry measurements of blood or urine, BSP excretion, optical properties of the eyes, or the general well-being of the subjects.

Chronic Toxicity

Rats exposed to IPA vapor at 8.4 ppm, 24 h/d, for 3 mo showed alterations in reflexes, enzyme activities, BSP retention, leukocyte count, total nucleic acids, urine coproporphyrin, and morphology of the lung, liver, spleen, and CNS (Baykov et al., 1974). At 1.0 ppm, there were lesser changes in some of these end points, and at 0.27 ppm, there were no alterations. These findings are difficult to evaluate in view of their variance with other data and inadequate details on experimental design and statistical analysis.

Ingestion by rats of IPA in drinking water at 0.5-5% for 27 w did not result in any definite toxic signs other than a slight retardation of growth and the probable accidental death of some animals (Lehman and Chase, 1944). There were no gross or microscopic abnormalities in the brain, pituitary, lungs, heart, liver, spleen, kidneys, or adrenals.

Ingestion by three dogs of IPA in drinking water at 4% for 1 h daily (average of 1.3 g/kg/d) over 6 mo produced inebriation for 3-5 h daily but otherwise normal, healthy behavior (Lehman et al., 1945). At the end of 6 mo, an i.v. test dose indicated the development of tolerance manifested by a decreased response to IPA and an increased rate of removal of alcohol from the blood. Histopathological changes were limited to the kidneys of one and the brains of two of the three dogs.

In contrast to the report by Baykov et al. (1974), 3-mo intermittent exposures of rats to IPA were without adverse effect up to 1000 ppm (Nakaseko et al., 1991b). Rats were exposed 4 h/d, 5 d/w, and were evaluated for hematological, clinical chemistry, and organ-weight changes. At concentrations of 4000 and 8000 ppm, mucous membrane irritation was evident, and a decrease in red-blood-cell (RBC) count occurred. At 8000 ppm, there was an increase in liver enzymes, and liver and spleen weights were reduced. Using 8-h daily exposures, 5 d/w for 20 w, the authors found delayed transmission velocity in the tail peripheral nerve in rats exposed at 8000 ppm but not in rats exposed at 1000 ppm (Nakaseko et al., 1991b).

Reproductive and Developmental Toxicity

The reproductive and developmental effects were studied in rats given 2.5% IPA in their drinking water over two generations (Lehman et al., 1945). First-generation males and females had retarded growth early in life, but nearly caught up with the control group by the 13th week. The second-generation rats showed no retardation of growth in either sex. The study is certainly incomplete by modern standards; however, the authors conclude that IPA does not produce deleterious effects on reproductive function or embryonic development.

Developmental effects have been reported in rats, but only at IPA concentrations that are toxic to the dams (Nelson et al., 1988). There were increased resorptions and decreased fetal weights from dams exposed at 10,000 ppm, 7 h/d, for 19 d. These results might have been due to IPA's being administered excessively early in gestation. Only decreased fetal weights were seen after exposures to IPA at 7000 ppm, and no effects were seen after exposures at 3500 ppm. The dams were narcotized at 10,000 ppm and showed mild CNS effects at 7000 ppm.

Genotoxicity

Recently, a Tier 1 (Toxic Substances Control Act) mutagenicity evaluation has been completed in response to recognition that regulatory risk assessment on IPA was hampered by insufficient data. In eight strains of *Salmonella* tested with and without S-9 activation, IPA was not found to be mutagenic (Zeigler et al., 1992). In vitro sister chromatid exchange assays using V79 cells, with and without S-9 activation, were also negative for mutagenic activity (Vonder Hude et al., 1987). Further demonstration of the lack of mutagenic activity of IPA was reported in a Chinese hamster ovary gene mutation assay and in a bone-marrow micronucleus test (Kapp et al., 1993).

Carcinogenicity

In the early 1940s, suspicion was raised that a carcinogen was involved in the industrial process used to produce isopropanol. Eventually, studies in mice showed that isopropyl oil, a waste product of the production, was most likely the cause of the observed human cancers (Weil et al., 1952). No tumorigenic activity was observed in mice exposed to IPA vapor at 4000 ppm, 3-7 h/d, 5 d/w, for 5-8 mo or when IPA was administered by skin painting or subcutaneous injection (Weil et al., 1952). Epidemiological studies of workers in isopropyl alcohol plants in the United States and United Kingdom have not shown a significant excess of mortalities or malignant diseases (Alderson and Rattan, 1986).

Interaction with Other Chemicals

Isopropyl alcohol has been shown to potentiate the toxicity of several halogenated hydrocarbons. In rats, intubation of IPA at 2.3 g/kg for 16-18 h before exposure to carbon tetrachloride vapor at 1000 ppm caused significantly greater serum SGOT activity than that produced by carbon tetrachloride alone (Cornish and Adefuin, 1967). Pre-exposure of rats with IPA before intraperitoneally injected carbon tetrachloride resulted in increased serum SGPT activity, hepatic triglyceride content, and total serum bilirubin and in decreased hepatic glucose-6-phosphatase activity (Traiger and Plaa, 1971). Suggested mechanisms for the potentiation include lysosomal alterations, changes in the endoplasmic reticulum, stimulation of drug-metabolizing enzymes, and increased sensitivity of hepatocytes to carbon tetrachloride (Coté et al., 1974).

In mice, pre-exposure with IPA or acetone by gavage potentiated the hepatotoxic response to chloroform, 1,1,2-trichloroethane, and trichloroethylene, as measured by serum SGPT activity, but not to 1,1,1-trichloroethane (Traiger and Plaa, 1974).

In an industrial exposure of workers, toxic effects, including renal failure and hepatitis, were attributed to the potentiation of carbon tetrachloride toxicity by IPA, following the inhalation of vapors of the two chemicals (Folland et al., 1976).

TABLE 11-1 Toxicity Summary

Concentration	Exposure Duration	Species	Effects	Reference
400 or 800 ppm	3-5 min	Human (n = 10)	At 400 ppm, mild irritation of eyes, nose, throat; at 800 ppm, irritation not severe but unsuitable for 8 h	Nelson et al., 1943
2.6 or 6.4 mg/kg/d	7 d/w, 6 w	Human (n = 8)	No toxic signs or symptoms, changes in blood or urine chemistries, BSP excretion, or general well-being	Wills et al., 1969
8.4 ppm	24 h/d, 3 mo	Rat	Changes in reflexes, enzyme activities, BSP retention, leukocyte count, organ morphology (results of questionable significance)	Baykov et al., 1974
400 ppm	4 h	Rat	Redness of nose, auricle, eyelid	Nakaseko et al., 1991a
1000 ppm	8 h/d, 5 d/w, 20 w	Rat	Body-weight reduction first week only	Nakaseko et al., 1991b
2000 ppm	4 h	Rat	Redness of nose, auricle, eyelid; loss of active motion	Nakaseko et al., 1991a
2050 ppm	8 h	Mouse	No toxic effects	Starrek, 1943
3000 ppm	3-7 h/d, 5 d/w, 5-8 mo	Mouse	No tumorigenic activity	Weil et al., 1952
3250 ppm	8 h	Mouse	Ataxia after 180-195 min, prostration after 6 h, narcosis after 8 h	Starrek, 1943
3500 ppm	7 h/d, 19 d	Rat (pregnant)	No clinical effects	Nelson et al., 1988
4000 ppm	4 h/d, 5 d/w, 13 w	Rat	Decreased RBC count after 12 w	Nakaseko et al., 1991b
7000 ppm	7 h/d, 19 d	Rat (pregnant)	Gait disturbance, reduced weight gain	Nelson et al., 1988
8000 ppm	4 h/d, 5 d/w, 13 w	Rat	Decreased RBC count, increased liver enzymes, reduced liver weight	Nakaseko et al., 1991b
8000 ppm	8 h/d, 5 d/w, 20 w		Peripheral nerve conduction velocity decreases	Nakaseko et al., 1991b
			Narcosis	Nelson et al., 1988

Dose	Species	Exposure	Effect	Reference
10,900 ppm	Mouse	4 h/d, 123 h total exposure	Narcosis without lethality; slight, reversible fatty changes in liver	Weese, 1928
12,000 ppm	Rat	4 h	Bleeding of nose, gait disturbance, dyspnea, coma	Nakaseko et al., 1991a
12,800 or 19,200 ppm	Mouse	160 or 200 min	Death after 200 min at 12,800 and 160 min at 19,200	Weese, 1928
16,000 ppm	Rat	8 h	Death of 4/6	Smyth et al., 1948
24,000 ppm	Mouse	2 h	Ataxia after 12-26 min, prostration after 37-46 min, narcosis after 100 min	Starrek, 1938
0.5-5% oral	Rat	24 h/d in drinking water	Slight retardation of growth, no histopathological changes	Lehman et al., 1944
4% oral	Dog	1 h/d, 6 mo in drinking water	Average intake 1.3 g/kg/d, daily inebriation for 3-5 h, otherwise normal behavior, histopathology in kidneys (1/3) and brain (2/3); tolerance to IPA test dose at 6 mo	Lehman et al., 1945
2 g/kg oral	Mouse	1 exposure	Increased elevated SGPT response to chloroform, trichloroethylene, and 1,1,2-trichloroethane	Traiger et al., 1974
2 g/kg oral	Rat	1 exposure	Ataxia and potentiation of response to carbon tetrachloride (increased SGPT, hepatic triglyceride, serum bilirubin and decreased hepatic glucose-6-phosphatase activity)	Traiger et al., 1971
2.3 g/kg	Rat	1 exposure	Increased elevated SGOT response to carbon tetrachloride	Cornish et al., 1967
2.4, 4.7 g/kg oral	Rat (young adult)	1 exposure	Lowest toxic dose 2.4 g/kg, LD_{50} 4.7 g/kg	Kimura et al., 1971
2.5, 6.5 g/kg i.v.	Rabbit	1 exposure	Anesthetic dose (elimination of corneal reflex) 2.5 g/kg, lethal dose 6.5 g/kg	Lehman et al., 1944
2.7, 4.1 g/kg i.v.	Dog	1 exposure	Anesthetic dose (elimination of corneal reflex) 2.7 g/kg, lethal dose 4.1 g/kg	Lehman et al., 1944
5.3, 5.1, 4.9 g/kg i.g.	Rat, rabbit, dog	1 exposure	LD_{50} values in each species, respectively; CNS depression in all species and retching and vomiting in dogs	Lehman et al., 1944

TABLE 11-2 Exposure Limits Set by Other Organizations

Organization	Concentration, ppm
ACGIH's TLV	400 (TWA)
ACGIH's STEL	500
OSHA's PEL	400 (TWA)
OSHA's STEL	500
NIOSH's REL	400 (TWA)
NIOSH's STEL	500
NIOSH's IDLH	12,000
NRCs 1-h EEGL	400
NRC's 24-h EEGL	200
NRC's 90-d CEGL	1

TLV = threshold limit value. TWA = time-weighted average. STEL = short-term exposure limit. PEL = permissible exposure limit. REL = recommended exposure limit. IDLH = immediately dangerous to life and death. EEGL = emergency exposure guidance level. CEGL = continuous exposure guidance level.

TABLE 11-3 Spacecraft Maximum Allowable Concentrations

Duration	ppm	mg/m^3	Target Toxicity
1 h	400	1000	CNS depression, irritation
24 h	100	240	CNS depression, irritation, hepatoxicity
7 d[a]	60	150	CNS depression, irritation, hepatoxicity
30 d	60	150	CNS depression, irritation, peripheral nerve damage
180 d	60	150	CNS depression, irritation

[a]Previous 7-d SMAC was 50 ppm.

RATIONALE FOR ACCEPTABLE STANDARDS

CNS depression should be of primary concern in setting SMAC values for IPA vapor; however, hepatoxicity must also be considered because of early reports of fatty liver in mice (Weese, 1928) and liver enzyme elevation after prolonged exposures at high concentrations (Nakaseko et al., 1991b). One investigator has also reported conduction decreases in peripheral nerves after prolonged high exposure

(Nakaseko et al., 1991b). Additionally, the vapor has the potential to cause irritation of the eyes and upper respiratory passages. Although mild irritation might be acceptable under emergency short-term conditions, long-term SMACs should protect against all adverse effects during prolonged exposure. Finally, in setting acceptable concentrations (ACs), the Baykov et al. (1974) report will be disregarded because it is not consistent with the weight of evidence from other studies. Guidelines from the National Research Council's Committee on Toxicology have been used to structure the rationale for astronaut exposure limits (NRC, 1984).

CNS Depression

A combination of animal inhalation studies, structure-activity arguments, and human-worker studies were used to derive ACs to protect against CNS effects. Based on graphical data derived from mice exposed to various IPA concentrations up to 8 h, a no-observed-adverse-effect level (NOAEL) for ataxia (first instability) could be estimated (Lehman and Flury, 1943). For a 1-h exposure, that value was 12,000 ppm (30 mg/L); however, clinically evident ataxia is not a sensitive end point for CNS effects. The question is what factor would be suitable to correct the ataxia NOAEL to a true NOAEL? A factor of 2 appears to be suitable based on two arguments. First, it was noted that increasing the 1-h IPA concentration to 22,000 ppm (approximately doubling the 12,000 ppm value) resulted in more serious CNS effects; specifically, the mice became prone. That result suggests that halving the 12,000-ppm value would greatly reduce the magnitude of any CNS effects. Second, ethanol appears to have approximately the same capacity as IPA to induce CNS effects in mice by inhalation (ethanol at 13,300 ppm in 100 min induces ataxia) (Rowe and McCollister, 1982). The CNS effects of ethanol are such that doubling the concentration from 0.06-0.08% to 0.12-0.15% causes the CNS effect to increase from "beginning of uncertainty" to "stupor" (Rowe and McCollister, 1982). Using this NOAEL correction factor of 2, the 1-h AC based on CNS effects was calculated as follows:

1-h AC based on CNS effects
= 12,000 ppm × 1/10 (species factor) × ½ (NOAEL)
= 600 ppm.

This value is consistent with rat data from Nakeseko et al. (1991b) in which they report that animals exposed to IPA at 8000 ppm did not show effects of anesthesia resulting from the first half of 4-h (or 8-h) exposures. In 4-h exposures, Nakeseko et al. (1991a) clearly observed CNS effects in less than 1 h at a concentration of 12,000 ppm, but no definite CNS effects were reported after a 4-h exposure at 2000 ppm.

To protect against CNS depression for exposure periods longer than 1 h, an exposure concentration must be found that avoids accumulation of blood concentrations of IPA or acetone that is associated with CNS depression. In nine workers exposed during a 7-h work shift to average concentrations between 40 and 200 ppm (average 110 ppm), Brugnone et al. (1983) made the following observations:

- The uptake of IPA from the air was about 60%.
- IPA was not detectable in blood or urine (limit of detection was 1 mg/L).
- In three workers exposed to IPA at 193-202 ppm, the blood acetone average was 7.2-8.2 mg/L.
- Blood acetone concentrations with IPA exposures were highly correlated.
- The slope of the regression lines (IPA exposure vs. blood acetone concentration) suggested a steady state after 3 h of exposure.

These data indicate that any CNS effects would correlate with blood acetone concentrations rather than IPA concentrations. In 22 subjects exposed to acetone at 250 ppm, the blood concentrations after 4 h averaged 15 mg/L, and there were slight psychomotor performance decrements; however, at exposures to acetone at 125 ppm (plus methyl ethyl ketone at 100 ppm) the blood acetaldehyde concentration averaged 10 mg/L, and no performance decrements could be deleted (Dick et al., 1988). This latter blood acetone concentration (10 mg/L) is slightly above that seen in the three most heavily exposed workers (7-8 mg/L) from the Brugnone IPA study. Since a steady state is reached between IPA exposure and blood acetone after 3 h of exposure, it may be concluded that 100-ppm IPA is a safe concentration to avoid CNS effects even during prolonged exposure.

This conclusion is consistent with mouse data showing no ataxia at 2050 ppm after an 8-h exposure (Starrek, 1938). Using the factor of 2 to correct the ataxia NOAEL to a true NOAEL (see discussion above) and noting that accumulation after 8 h would be negligible, the AC to avoid CNS effects for long exposures was calculated as follows:

AC based on CNS effects
= 2050 ppm × 1/10 (species factor) × ½ (NOAEL)
= 100 ppm.

Peripheral Nerve Damage

Prolonged exposure of a total of 800 h induced decreases in peripheral nerve conduction velocity in rats exposed to IPA at 8000 ppm but not in those exposed at 1000 ppm (Nakaseko et al., 1991b). This finding was used to set a 30-d (720 h) AC as follows:

30-d AC based on peripheral nerve damage
= 1000 ppm × 1/10 (species factor)
= 100 ppm.

Irritation

Exposures of approximately 10 human subjects to IPA at concentrations of 200, 400, and 800 ppm for 3-5 min resulted in mild irritation at 400 ppm and the conclusion, by the test subjects, that 200 ppm would be satisfactory for 8 h (Nelson et al., 1943). Even though the study leaves some doubt about how such conclusions were reached, it appears that 400 ppm would be acceptable for 1-h exposures because tolerance to the irritating sensation of alcohols develops quickly. The 200-ppm value would be more suitable for a 24-h exposure where the risk of irritation would have to be minimal. Irritation would not be acceptable for long-term exposure; hence, a long-term NOAEL was calculated using the recommended approach for studies involving a small number (n = 10) of test subjects. Specifically,

AC based on irritation
= 200 ppm × (square root of 10)/10 (small n factor)
= 60 ppm.

Hence, the 7-, 30-, and 180-d ACs to prevent irritation from IPA were all set at 60 ppm.

Hepatotoxicity

Liver injury is not usually thought of as an important effect of IPA exposure; however, an early study involving repeated inhalation exposure of mice reported reversible fatty changes in the liver (Weese, 1928). A lowest-observed-adverse-effect level (LOAEL) for liver effects was a cumulative exposure of 123 h (in 4-h segments) at 11,000 ppm. Based on this observation, the 24-h AC was 110 ppm with uncertainty factors of 10 for LOAEL to NOAEL and 10 for species extrapolation. This may be extended to 7 d by reducing the value by 168 ÷ 123 to give 80 ppm. Extending estimates to longer times was based on the 4000-ppm NOAEL reported in rats exposed 20 h/w for 13 w (Nakeseko et al., 1991b).

The 7-d AC based on liver injury calculated that way gave an AC of 400 ppm. The 30-d AC was calculated by applying a time factor based on Haber's rule (720 ÷ 260) to derive an AC of 140 ppm.

RECOMMENDATIONS FOR ADDITIONAL RESEARCH

Many of the IPA toxicity studies, particularly the inhalation studies, were completed 40 or more years ago. Over that time, many new regulatory guidelines have been developed to provide data that are more useful for setting human-exposure standards. In addition, methods of measuring CNS effects in animals, and performance decrements in humans have improved considerably over the past 40 years. Hence, standardized inhalation studies in rodents, including a long-term, continuous-exposure study with multiple end-point assessments (to determine the accuracy of Baykov et al., 1974), are needed to resolve questions of dose-response and target organs. Short-term human inhalation studies are needed to better understand IPA irritancy and tolerance phenomena. Finally, metabolic studies are needed to better define the adsorption, metabolism, and fate of inhaled IPA.

End Point	Exposure Data	Species and Reference	Uncertainty Factors				Acceptable Concentrations, ppm				
			NOAEL	Time	Species	Space-flight	1 h	24 h	7 d	30 d	180 d
CNS depression	NOAEL at 12,000 ppm, 1 h	Mouse (Starrek, 1938)	2[a]	1	10	1	600	—	—	—	—
	NOAEL at 2050 ppm, 8 h	Mouse (Starrek, 1938)	2[a]	1	10	1	—	100	100	100	100
	NOAEL at 100 ppm, no performance decrement based on blood acetone	Human (Brugnone et al., 1983; Dick et al., 1988)	—	1	1	1	—	100	100	100	100
Peripheral nerve damage	NOAEL at 1000 ppm, 40 h/w, 20 w	Rat (Nakeseko et al., 1991b)	—	1	10	1	—	—	—	100	—
Irritation	LOAEL (mild) at 400 ppm, 3-5 min	Human (Nelson et al., 1943)	—	1	1	1	400	—	—	—	—
	NOAEL[b] at 200 ppm, 8 h	Human (Nelson et al., 1943)	—	1	1	1	—	200	—	—	—
	NOAEL[b] (n = 10) at 200 ppm, 8 h	Human (Nelson et al., 1943)	3	1	1	1	—	—	60	60	60
Hepato-toxicity	LOAEL at 11,000 ppm, 123 h in 4-h intervals	Mouse (Weese, 1928)	10	1 or HR[c]	10	1	—	110	80	—	—
	NOAEL at 4000 ppm, 20 h/w, 13 w	Rat (Nakeseko et al., 1991b)	1	HR	10	1	—	—	400	140	—
SMAC							400	100	60	60	60

[a] A factor of 2 used because the measured NOAEL was for ataxia, which is not a sensitive indicator of CNS effects.
[b] This NOAEL was from the test subjects impression that an exposure of 200 ppm could be tolerated for 8 h.
[c] HR = Haber's rule.

> **NOTE ADDED IN PROOF**
>
> As this document was going to press, two studies were published that address many of the recommendations above. Slauter et al. (Slauter, R.W., D.P. Coleman, N.F. Gaudette, R.H. McKee, L.W. Masten, T.H. Gardiner, D.E. Strother, T.R. Tyler, and A.R. Jeffcoat. 1994. Fundam. Appl. Toxicol 23: 407-420) reported valuable data on the distribution, metabolism, and excretion of IPA in rats and mice exposed for 6 h at a concentration of 500 or 5000 ppm. Burleigh-Flayer et al. (Burleigh-Flayer, H.D., M.W. Gill, D.E. Strother, L.W. Masten, R.H. McKee, T.R. Tyler, T. Gardiner. 1994. Fundam. Appl. Toxicol 23:421-428) showed that repeated exposures to IPA for 13 w produced clearly toxic effects in rats and mice only at a concentration of 5000 ppm. Nephropathy was found in male rats at lower exposures, but this type of lesion has questionable relevance to human risk assessment. Clinical signs of CNS effects (narcosis, ataxia, and hypoactivity) were reported in some mice during, but not immediately after, exposures at 1500 ppm; however, later exposures of mice at 2500 ppm did not fully confirm the presence of CNS effects such as ataxia at 1500 ppm (H.D. Burleigh-Flayer, Busby Run Research Center, Union Carbide Chemicals and Plastics Co., Export, Pa., personal commun., 1995). Additonal new studies on IPA will be published soon and a re-evaluation of the SMACs in light of the new data should be completed at that time.

REFERENCES

Adelson, L. 1962. Fatal intoxication with isopropyl alcohol (rubbing alcohol). Am. J. Clin. Pathol. 38:144-151.

Alderson, M.R., and N.S. Rattan. 1986. Mortality of workers on an isopropyl alcohol plant and two MEK dewaxing plants. Br. J. Ind. Med. 37:85-89.

Amoore, J.E., and E. Hautala. 1983. Odor as an aid to chemical safety: Odor thresholds compared with threshold limit values and volatiles for 214 industrial chemicals in air and water dilution. J. Appl. Toxicol. 3:272-290.

Baykov, B.K., O.Y. Gorlova, Y.V. Novikov, T.V. Yudina, and A.N. Sergeyev. 1974. [Hygienic standardization of the daily average maximum admissible concentrations of propyl and isopropyl alcohols in the atmosphere.] Gig. Sanit. 4:6-13.

Brugnone, F., L. Perbellini, P. Apostoli, M. Bellorri, and D. Caretta. 1983. Isopropanol exposure: Environmental and biological monitoring in a printing works. Br. J. Ind. Med. 40:160-168.

Carpenter, C.P., H.F. Smyth, Jr., and U.C. Pozzani. 1949. The assay of acute vapor toxicity, and the grading and interpretation of results on 96 chemical compounds. J. Ind. Hyg. Toxicol. 31:343-346.

Cornish, H.H., and J. Adefuin. 1967. Potentiation of carbon tetrachloride toxicity by aliphatic alcohols. Arch. Environ. Health 14: 447-449.

Coté, M.G., G.J. Traiger, and G.L. Plaa. 1974. Effect of isopropanol-induced potentiation of carbon tetrachloride on rat hepatic ultrastructure. Toxicol. Appl. Pharmacol. 30:14-25.

Daniel, D.R., B.H. McAnalley, and J.C. Garriott. 1981. Isopropyl alcohol metabolism after acute intoxication in humans. J. Anal. Toxicol. 5:110-112.

Dick, R.B., W.D. Brown, J.V. Selzer, B.J. Taylor, and R. Shukla. 1988. Effects of short-duration exposures to acetone and methyl ethyl ketone. Toxicol. Lett. 43:31-49.

Ellis, F.W. 1952. The role of the liver in the metabolic disposition of isopropyl alcohol. J. Pharmacol. Exp. Ther. 105:427-436.

Folland, D.S., W. Schaffner, H.E. Ginn, O.B. Crofford, and D.R. McMurray. 1976. Carbon tetrachloride toxicity potentiated by isopropyl alcohol. JAMA 236:1853-1856.

Fuller, H.C., and O.B. Hunter. 1927. Isopropyl alcohol-an investigation of its physiologic properties. J. Lab. Clin. Med. 12:326-349.

James, J.T., T.F. Limero, H.J. Leano, J.F. Boyd, and P.A. Covington. 1994. Volatile organic contaminants found in the habitable environment of the Space Shuttle: STS-26 to STS-55. Aviat. Space Environ. Med. 65:851-857.

Kamil, I.A., J.N. Smith, and R.T. Williams. 1953. Studies in detoxication, 46. The metabolism of aliphatic alcohols, the glucuronic acid conjugation of acyclic aliphatic alcohols. Biochem. J. 53:129-136.

Kapp, R.W., D.J. Marino, T.H. Gardiner, L.W. Masten, R.H. McKee, T.R. Taylor, J.L. Ivell, and R.R. Young. 1993. In vitro and in vivo assays of isopropanol for mutagenicity. Environ. Mol. Mutagen 22:93-100.

Kimura, E.T., D.M. Ebert, and P.W. Dodge. 1971. Acute toxicity and limits of solvent residue or sixteen organic solvents. Toxicol. Appl. Pharmacol. 19:699-704.
King, L.H., K.P. Breadley, and D.L. Shires. 1970. Hemodialysis for isopropanol alcohol poisoning. JAMA 211:1855.
Laham, S., M. Potvin, K. Schrader, and I. Marino. 1980. Studies on inhalation toxicity of 2-propanol. Drug Chem. Toxicol. 3:343-360.
Lehman, A.J., and H.F. Chase. 1944. The acute and chronic toxicity of isopropyl alcohol. J. Lab. Clin. Med. 29:561-567.
Lehman, A.J., H. Schwerma, and E. Richards. 1944. Isopropyl alcohol: rate of disappearance from the bloodstream of dogs after intravenous and oral administration. J. Pharmacol. Exp. Ther. 82:196-201.
Lehman, A.J., H. Schwerma, and E. Richards. 1945. Isopropyl alcohol: Acquired tolerance in dogs, rate of disappearance from the bloodstream in various species, and effects on successive generation of rats. J. Pharmacol. Exp. Ther. 85:61-69.
Lehman, K.B., and F. Flury. 1943. Pp. 207-208 in Toxicology and Hygiene of Industrial Solvents. Baltimore, Md.: Williams & Wilkins.
Lester, D., and G.D. Benson. 1970. Alcohol oxidation in rats inhibited by pyrazole, oximes, and amides. Science 169:282-284.
Macht, D.I. 1922. Pharmacological examination of isopropyl alcohol. Arch. Int. Pharmacodyn. Ther. 26:285-289.
Nakaseko, H., K. Teramoto, and S. Horiguchi. 1991a. Toxicity of isopropyl alcohol. Part I: Single inhalation exposure in rats. Jpn. J. Ind. Health 33:198-199.
Nakaseko, H., K. Teramoto, and S. Horiguchi. 1991b. Toxicity of isopropyl alcohol. Part 2: Repeated inhalation exposure in rats. Jpn. J. Ind. Health 33:200-201.
Nelson, K.W., J.F. Ege, Jr., M. Ross, L.E. Woodman, and L. Silverman. 1943. Sensory response to certain industrial solvent vapors. J. Ind. Hyg. Toxicol. 25:282-285.
Nelson, B.K., W.S. Brightwell, D.R. MacKenzie, A. Khan, J.R. Burg, and W.W. Weigel. 1988. Teratogenicity of n-propanol and isopropanol administered at high inhalation concentrations to rats. Food Chem. Toxicol. 26:247-254.

Nordmann, R., Y. Giudicelli, F. Beauge, M. Clement, C. Ribiere, H. Roach, and J. Nordmann. 1973. Studies on the mechanism involved in the isopropanol-induced fatty liver. Biochim. Biophys. Acta 326:1-11.

NRC. 1984. Emergency and Continuous Exposure Limits for Selected Airborne Contaminants, Vol. 2. Washington, D.C.: National Academy Press.

Rowe, V.K., and S.B. McCollister. 1982. Alcohols. Pp. 4527-4708 in Patty's Industrial Hygiene and Toxicology, Vol. 2C, 3rd Rev. Ed. New York: John Wiley & Sons.

Smyth, H.F., Jr., and C.P. Carpenter. 1948. Further experience with the range finding test in the industrial toxicology laboratory. J. Ind. Hyg. Toxicol. 30:63-68.

Starrek, E. 1938. Dissertation. Würzburg, Germany.

Traiger, G.J., and G.L. Plaa. 1971. Differences in the potentiation of carbon tetrachloride in rats by ethanol and isopropanol pretreatment. Toxicol. Appl. Pharmacol. 20:105-112.

Traiger, G.J., and G.L. Plaa. 1974. Chlorinated bycarbon toxicity-potentiation by isopropyl alcohol and acetone. Arch. Environ. Health 28:276-278.

Von der Hude, W., M. Scheutwinkel, U. Gramlich, B. Fibler, and A. Basler. 1987. Genotoxicity of three-carbon compounds evaluated in the SEC test in vitro. Environ. Mutagen 9:401-410.

Wax, J., F.W. Ellis, and A.J. Lehman. 1949. Absorption and distribution of isopropyl alcohol. J. Pharmacol. Exp. Ther. 97:229-237.

Weese, H. 1928. [Comparative studies of the efficacy and toxicity of the vapors of lower aliphatic alcohols.] Arch. Exp. Pathol. Pharmakol. 135:118-130.

Weil, C.S., H.F. Smyth, Jr., and T.W. Nale. 1952. Quest for a suspected industrial carcinogen. Arch. Ind. Hyg. Occup. Med. 5:535-547.

Wills, J.H., E.M. Jameson, and F. Coulston. 1969. Effects on man of daily ingestion of small doses of isopropyl alcohol. Toxicol. Appl. Pharmacol. 15:560-565.

Zeigler, E., B. Anderson, S. Haworth, T. Lawlor, and K. Mortelmans. 1992. Salmonella mutagenicity tests. V. Results from the testing of 311 chemicals. Environ. Mol. Mutagen 19(Suppl. 21):2-141.

B12 Toluene

Hector D. Garcia, Ph.D.
Johnson Space Center Toxicology Group
Biomedical Operations and Research Branch
Houston, Texas

PHYSICAL AND CHEMICAL PROPERTIES

Toluene is a clear, colorless, non-corrosive, flammable liquid with a sweet, pungent, "aromatic" odor. Values reported for the odor threshold range from 0.2 to 16 ppm (Sandmeyer, 1981).

Synonyms: Antisal, phenyl methane, methacide, methyl benzene, methylbenzol, NCI-C07272, tolueen, toluen, toluol, toluolo, tolu-sol, UN 1294
Formula: C_7H_8
Structure:

$\langle$benzene ring$\rangle$—CH_3

CAS number: 108-88-3
Molecular weight: 92.14
Boiling point: 110.62 °C
Melting point: -95 °C
Liquid density: 0.8869
Vapor pressure: 36.7 mm Hg at 30 °C
Solubility: Insoluble in water
Very soluble in alcohol and ether
Conversion factors at 25 °C, 1 atm: 1 ppm = 3.77 mg/m^3
1 mg/m^3 = 0.265 ppm

OCCURRENCE AND USE

Toluene has been measured in urban air at 0.01-0.05 ppm, probably stemming from production facilities, automobile and coke-oven emissions, gasoline evaporation, and cigarette smoke, and can occur in human respiratory air in smokers and nonsmokers (Sandmeyer, 1981). It is used extensively as a component of gasoline, as a solvent in the chemical, rubber, paint, and drug industries, as a thinner for inks, perfumes and dyes, and as a nonclinical thermometer liquid and suspension solution for navigational instruments (Sandmeyer, 1981). Intentional inhalation of toluene vapors from glue was popular among some youth during the last few decades because of toluene's effects on the central nervous system (CNS). Toluene has been detected in spacecraft air in numerous missions at levels of up to 64 ppm.

PHARMACOKINETICS AND METABOLISM

Absorption

The major route of absorption of toluene is by inhalation. During inhalation, arterial blood concentrations of toluene in humans reach 60% of maximum in 10-15 min (Benignus, 1981a). A linear relationship was found between toluene concentrations in alveolar air and arterial blood in human subjects exposed to toluene at 100, 300, 500, and 714 ppm for 20 min per concentration (Gamberale and Hultengren, 1972). Reports of uptake rates (to 95% of asymptote) in humans exposed at up to 500 ppm vary from 10 to 80 min (Gamberale and Hultengren, 1972; Veulemans and Masschelein, 1978; Benignus, 1981a). Exercise increases the rate of uptake (Benignus, 1981a). Retention, including from cigarette smoke, was 86-96% (Sandmeyer, 1981). In mice, however, Peterson and Bruckner (1978) reported that the arterial blood concentration did not approach maximum values until about 2 h after the onset of exposure to toluene at 4000 ppm and was still rising slightly after 3 h.

Distribution

Inhaled toluene is distributed widely throughout the body, most rapidly into highly vascularized tissues, with fatty tissues acting as reservoirs. During exposure, the concentrations of toluene are highest in the liver, then the brain and blood (Bruckner and Peterson, 1981a).

Excretion

The majority of inhaled toluene is exhaled unchanged. Toluene concentrations fall off rapidly after cessation of exposure, dropping to 30% of maximum in humans in 40 min and near zero levels in 4 h when subjects are at rest (Benignus et al., 1981). In rats, toluene blood concentrations fall to 50% of maximum about 60 min after termination of exposure (Benignus, 1981b). In humans, about 80-85% of total toluene is excreted in the urine as conjugates of benzoic acid. Of this, about 80% is hippuric acid (the glycine conjugate of benzoic acid) and 20% is benzoylglucuronide (the glucuronic acid conjugate of benzoic acid). In 23 male volunteers inhaling toluene vapor for 3 h or 7 h (with a break of 1 h), the excretion of urinary hippuric acid over an 18-h period was equivalent to 68% of the toluene absorbed (Ogata et al., 1971). In humans exposed to toluene at up to 200 ppm, the total amount of hippuric acid excreted was proportional to the total exposure (ppm × h) (Ogata et al., 1971).

Metabolism

Toluene is metabolized in several ways, but mainly it is converted by oxidation of the methyl group to benzoic acid via benzyl alcohol and benzaldehyde in the microsomes of the liver parenchymal cells (Benignus et al., 1981). Benzoic acid is conjugated with glycine to form hippuric acid (80%) and with glucuronic acid to form benzoylglucuronide (20%), which are excreted in the urine (Benignus et al., 1981). Possible ethnic differences in toluene metabolism were reported for Chinese, Turkish, and Japanese solvent workers; the male Japanese excreted almost twice as much hippuric acid as the male Chinese under similar exposure conditions, although the difference was less marked between female Chinese and Japanese workers, and there were no differences in the excretion of *o*-cresol (Inoue et al., 1986).

TOXICITY SUMMARY

Acute and Short-Term Toxicity

Lethality

The LC_{50} in Fischer rats for a 60-min inhalation exposure to toluene vapors is 26,700 ppm, and the LC_{100} for a 60-min inhalation exposure is about 40,000 ppm (Pryor et al., 1978). During inhalation studies in which continuous exposure and fixed concentrations of toluene vapor were used, the LC_{50} was a function of both the concentration of toluene and the duration of exposure. A group of pregnant rats (n = 9) exposed to toluene at 400 ppm continuously for 8 d had a 28% mortality, but no mortality was seen at 266 ppm (n = 10) (Hudák and Ungváry, 1978). A later publication from the same laboratory indicated that rats could be exposed to toluene at 950 ppm for 48 h, starting on d 10 of pregnancy without significant maternal mortality (Ungváry et al., 1982).

Irritation and CNS Effects

In 16 volunteers, 21-32 y old, exposure to toluene at 40 ppm for 6 h has been shown to be a no-observed-adverse-effect level (NOAEL) for irritation and CNS effects, including vigilance, visual perception, psychomotor functions, and higher cortical functions (Andersen et al., 1983). In the same study, 100 ppm was found to cause slight irritation of eyes and nose and increased headaches, dizziness, and a feeling of intoxication; in another study (Gamberale and Hultengren, 1972), 100 ppm was found to be a NOAEL for impairment of reaction time.

In a Danish study, 20 printers previously occupationally exposed for 9-25 y to a mixture of solvents containing toluene at 0-20% and 22 naive control subjects were exposed to toluene at 100 ppm for 6 h. All 42 exposed subjects complained of low air quality, strong odor, fatigue, sleepiness, a feeling of intoxication, and irritation of the eyes, nose, and throat. The exposed naive control subjects showed a statistically significant 1% decrease in manual dexterity, decreased color discrimination, and slightly decreased accuracy in visual perception compared with unexposed naive controls (Bælum et al., 1985).

Male rats and mice exposed to toluene at 12,000 ppm for 5 min exhibited marked depression, but recovered fully after breathing fresh air for 10 min (Bruckner and Peterson, 1981a). A study by Guillot et al. (1982) reported that a 1-h exposure to liquid toluene using several protocols irritated rabbits' eyes, but the irritation was reduced to slight irritation if the eyes were rinsed 30 s after instillation.

Subchronic and Chronic Toxicity

Histopathology, Clinical Chemistry, and Hematology

A well-executed chronic inhalation study in F344 rats performed by the Chemical Industry Institute of Toxicology in 1980 failed to show any adverse effects at the doses tested (Gibson et al., 1983). Groups of 120 male and 120 female F344 rats were exposed to toluene at 30, 100, or 300 ppm (>99.98% pure), 6 h/d, 5 d/w, for 24 mo. An unexposed group of 120 male and 120 female rats served as a control. Clinical chemistry, hematology, and urinalysis tests were conducted at 18 and 24 mo. All parameters measured at the termination of the study were normal except for a dose-related reduction in hematocrit values in females exposed to toluene at 100 and 300 ppm. Forty-one tissues from each animal (5-76 animals per group) sacrificed at 6, 12, 18, and 24 mo were examined grossly and histopathologically. The authors considered the highest dose of 300 ppm to be a NOAEL.

Hepatotoxicity and Nephrotoxicity

Evidence for liver and kidney toxicity by toluene is weak and inconsistent. Rats exposed to toluene at 5000 ppm for 7 h/d for 25 exposures showed reversible increases in kidney and liver weights (Von Oettingen, 1942). Mice exposed at 40,000 ppm for 3 h/d for 40 exposures and rats exposed at 1000 ppm for 6 h/d for 65 exposures showed no evidence of kidney or liver damage. Clinical case reports on humans who have either accidentally or intentionally been exposed to long durations or high levels or both of toluene also show reversible kidney and liver pathology symptoms, but the reports are not consistent in their findings of pathology (Benignus, 1981a).

CNS Effects

Chronic inhalation abuse of pure toluene produces irreversible cerebellar, brain-stem, and pyramidal-tract dysfunction (Spencer and Schaumberg, 1985), but comparable changes have not been found in solvent workers (n = 43) occupationally exposed for 22 y (SD = 7.4) to toluene at a mean concentration of 117 ppm (Bælum et al., 1985; Juntunen et al., 1985).

Rats exposed to toluene at 1000 ppm, 6 h/d, 5 d/w, for 30 d exhibited a small but significant alteration in brain function (flash-evoked potentials) 18-26 h after the last exposure (Dyer et al., 1984).

Lethality

In an extensive study conducted by the National Toxicology Program (NTP) on nonpregnant animals, mice (n = 120) and rats (n = 100) exposed to toluene at 1200 ppm for 2 y had no significant differences in survival compared with unexposed controls (Huff, 1990).

Carcinogenicity

In a 1989 NTP study (Huff, 1990), groups of 60 male and 60 female F344 rats were exposed to toluene at 0, 600, or 1200 ppm and 60 male and 60 female B6C3F$_1$ mice were exposed at 0, 120, 600, or 1200 ppm for 6.5 h/d, 5 d/w, for 2 y. At 15 mo, an interim sacrifice revealed an increased incidence and severity of nonneoplastic lesions of the nasal cavity of exposed rats. Minimal hyperplasia of the bronchial epithelium was seen in 4 of 10 female mice at 1200 ppm. There were also effects on the olfactory and respiratory epithelia of exposed rats. There was no evidence of carcinogenicity for any group of animals in this study.

Genotoxicity

Chromosome studies on peripheral blood lymphocytes of 34 workers of a rotogravure plant found no significant differences between toluene-exposed workers and a group of 34 matched controls (Forni et al., 1971). A similar study found no increases in the frequencies of sister chromatid exchanges (SCEs) or chromosomal aberrations (CAs) (Maki-Paakkanen et al., 1980). No increases in SCEs or CAs were found in vitro in toluene-treated human lymphocytes (Gerner-Smidt and Friedrich, 1978). Toluene was found to be nonmutagenic in the Ames assay (Bos et al., 1981). Toluene was found to be a potent mitotic arrestant in the grasshopper embryo system, but it did not induce an accumulation of colchicine-like mitoses (Liang et al., 1983).

Reproductive Toxicity

Wives (n = 28) of men with high or frequent occupational exposure to toluene had an increased odds ratio of spontaneous abortion compared with 29 referents (controls), but there was no association with congenital malformations (Taskinen et al., 1989). No increases were seen for intermediate or low or rare exposures.

A study examining occupational exposure to chemicals in the workplace compared 301 working women who had recently given birth to a child with congenital defects with 301 matched working women whose most recent child was born normal (McDonald et al., 1987). The authors' analysis indicated that, of nine categories of chemical exposures examined, only in those women exposed to aromatic solvents, primarily toluene, were there a suspicious excess number of cases of defects. Six mothers of children with congenital defects were exposed to toluene and two of these were also exposed to other solvents. The authors state that the statistical significance of this result ($Z = 1.77$; one sided $p \approx 0.04$) might be an overestimate.

Developmental Toxicity

There are conflicting reports of the teratogenicity of toluene. Exposure to toluene at 399 ppm for 24 h/d during various portions of pregnancy was not teratogenic to rats, but there was considerable retardation of fetal development (Hudák and Ungváry, 1978). In mice, however, exposure at 400 ppm from gestation days 6-15 indicated teratogenicity (a significant shift in the rib profile) but no fetotoxicity (Courtney et al., 1986). Continuous exposure at 266 ppm from gestation days 6-15 caused spontaneous abortion in rabbits, while maternal-weight gain decreased (Ungváry and Tátrai, 1985). Toluene was found to be negative in the Chernoff/Kavlock developmental toxicity screen in mice (Taskinen et al., 1989).

The offspring of pregnant mice given toluene at 400 ppm in drinking water and maintained at that concentration after weaning showed a decreased habituation or open-field activity at 35 d of age and depressed rotorod performance at 45-55 d of age, but no change in surface-righting response (Kostas and Hotchin, 1981).

Interaction with Other Chemicals

Toluene and trichloroethylene have been shown to noncompetitively inhibit each other's metabolism in rats. Trichloroethylene suppresses the urinary excretion of hippuric acid, a main metabolite of toluene, and toluene reduces the amount of urinary total trichloro-compounds, the metabolites of trichloroethylene (Ikeda, 1974). Ikeda also reported suppression of benzene and styrene oxidation in vivo by co-administration of toluene in rats (Ikeda et al., 1972). Benzene and toluene mutually inhibit each other's metabolism, with toluene more effectively inhibiting benzene metabolism than the reverse (Purcell et al., 1990). A drastic increase in toluene concentration in the blood of rats was reported after combined inhalation with acetone (Freundt and Schneider 1986). Inhalation of toluene at 954 ppm by pregnant rats combined with acetylsalicylic acid at 500 mg/kg resulted in increased maternal and fetal toxicity and teratogenicity (Tátrai et al., 1979).

Simultaneous exposure of human volunteers to toluene and ethanol has been shown to reduce the urinary excretion of hippuric acid and *o*-cresol, metabolites of toluene, to less than half of the value for exposure to toluene alone (Døssing et al., 1984).

TABLE 12-1 Toxicity Summary

Concentration, ppm	Exposure Duration	Species	Effects	Reference
7-112	Occupational (rotogravure)	Human (n = 32)	NOAEL for SCE and chomosome aberrations	Maki-Paakkanen et al. 1980
40	6 h/d, 4 d	Human (n = 16), ages 21-32 y	NOAEL for irritation of eyes and nose, lung function, nasal mucus flow, visual perception, vigilance, psychomotor functions, higher cortical functions	Andersen et al., 1983
100	6.5 h	Human (n = 41), ages 29-50 y	Strong odor, fatigue, sleepiness, inebriation; irritation of eyes, nose, and throat; decreased manual dexterity, decreased color discrimination; decreased accuracy in visual perception	Bælum et al., 1985
100	6 h/d, 4 d	Human (n = 16), ages 21-32	Odor; irritation of eyes and nose; NOAEL for lung function, nasal mucus flow, visual perception, vigilance, psychomotor functions, higher cortical functions	Andersen et al. 1983
117	Occupational, 22 y	Human (n = 43)	NOAEL for neurological functions, neuroanatomical changes, and neurophysiology	Juntunen et al., 1985
"High"	Occupational, frequent	Human (n = 301)	Congenital defects in children	McDonald et al., 1987
70	18 h	Monkey (n = 1)	Reduced shock avoidance	Weiss et al., 1979
100	6 h/d, 4 d	Human (n = 12), ages 20-35	NOAEL for impairment of reaction time	Gamberale et al., 1972
300	6 h/d, 4 d	Human (n = 12), ages 20-35	Impairment of reaction time	Gamberale et al., 1972
300	6 h/d, 5 d/w, 24 mo	Rat	NOAEL for histopathology	Gibson et al., 1983

TABLE 12-4 (Continued)

Concentration, ppm	Exposure Duration	Species	Effects	Reference
400	4 h	Pigeon	Increased rate of key pecking	Weiss et al., 1979
500	6 h/d, 5 d/w, 6 mo	Rat	Increased brain-cell volume; altered neurotransmitter levels	Ladefoged et al., 1991
1000	1 h	Monkey	Self-administration of toluene (toluene "sniffing")	Weiss et al., 1979
1000	8 h/d, 3 mo	Rat (n = 8)	No change in catecholamine content of sympathetic neurons and adrenal medulla	Alho et al., 1986
1000	6 h/d, 5 d/w, 30 d	Rat	Altered latencies for flash-evoked potentials (CNS)	Dyer et al., 1984
1200	6.5 h/d, 5 d/w, 2 y	Rat	NOAEL for lethality	Huff, 1990
2000	48 h, continuous	Rat (n = 8)	Reduction in catecholamine content of sympathetic neurons	Alho et al., 1986

TABLE 12-2 Exposure Limits Set by Other Organizations

Organization	Concentration, ppm
ACGIH's TLV	100 (TWA) (proposed TLV = 50 ppm)
ACGIH's STEL	150
OSHA's PEL	200 (TWA) (8 h/d, 40 h/2, lifetime)
OSHA's PEL	300 (ceiling)
OSHA's 10-min PEL	500 (ceiling)
NIOSH's 1-h REL	100 (TWA)
NIOSH's 10- min REL	200 (ceiling)
NRCs 1-h EEGL	200
NRC's 24-h EEGL	100
NRC's 90-d CEGL	20 (continuous)

TLV = threshold limit value. TWA = time-weighted average. STEL = short-term exposure limit. PEL = permissible exposure limit. REL = recommended exposure limit. EEGL = emergency exposure guidance level. CEGL = continuous exposure guidance level.

TABLE 12-3 Spacecraft Maximum Allowable Concentrations

Duration	ppm	mg/m^3	Target Toxicity
1 h	16	60	Neurotoxicity
24 h	16	60	Neurotoxicity
7 d[a]	16	60	Neurotoxicity, irritation
30 d	16	60	Neurotoxicity, irritation
180 d	16	60	Neurotoxicity, irritation

[a]Temporary 7-d SMAC was set at 20 ppm.

RATIONALE FOR ACCEPTABLE CONCENTRATIONS

The SMAC values listed above were set based on the lowest acceptable concentration (AC) for any adverse effect at each exposure duration using guidelines established jointly by the National Research Council and NASA (NRC, 1992). The evidence and logic used to determine the ACs for each adverse effect and exposure duration are documented below.

Irritation

A NOAEL of 40-ppm toluene vapor for irritation of the eyes and nose during a 6-h exposure was established in 16 young male volunteers. Because irritation is dependent on concentration but not on exposure duration, the ACs for all exposure durations from 7 d to 180 d were based on the 40-ppm NOAEL, adjusting for the low number of subjects by a factor equal to one-tenth the square root of the number of subjects tested:

7-d, 30-d, and 180-d ACs based on irritation
= NOAEL × 1/small n factor
= 40 ppm × (square root of 16)/10
= 16 ppm.

Some irritation is acceptable for short-term SMACs; therefore, the irritancy ACs for 1 h and 24 h were set equal to the LOAEL:

1-h and 24-h ACs based on irritation
= LOAEL
= 100 ppm.

Neurotoxicity

A variety of CNS effects in humans have been reported in the literature, and ACs were calculated separately for each set of end points or experiment. These ACs were not adjusted for duration of exposure because the CNS depressant effects of toluene have been shown to be reversible within a few minutes of cessation of exposure (Bruckner and Peterson, 1981b), except possibly after chronic exposure to high concentrations, indicating that the effects are dependent on blood concentrations but not exposure duration. Pharmacokinetic experiments have shown that arterial blood concentrations reach 60% of maximum in 10-

15 min and decrease rapidly after cessation of exposure (Benignus et al., 1981).

Although there are a number of reports in the literature that have examined the effects on humans of exposure to toluene vapors at concentrations in the range of the threshold limit value (TLV) of 100 ppm, the ACs have been set using the study that yields the lowest values of the ACs: In 16 young subjects, exposure at 10 or 40 ppm for 6 h did not result in any adverse effects for any of the end points examined (nasal mucus flow, lung function, subjective response, and psychometric performance), but exposure at 100 ppm produced irritation of eyes and nose and borderline significance in reduced performance in a battery of eight psychometric tests for visual perception, vigilance, psychomotor functions and higher cortical functions (Andersen et al., 1983). Acceptable concentrations for vigilance, visual perception, psychomotor functions, higher cortical functions, headache, dizziness, and a feeling of inebriation were calculated. The ACs for 7 d, 30 d, and 180 d were based on the 40-ppm NOAEL, adjusting for the number of subjects.

7-d, 30-d, and 180-d ACs based on CNS effects
= NOAEL × 1/small n factor
= 40 ppm × (square root of 16)/10
= 16 ppm.

Although headaches, irritation, and insignificant decrements in psychometric tests would be acceptable for short-term contingency exposures, dizziness would not be acceptable, even for brief exposures during contingency operations. Thus, the ACs for 1 h and 24 h were also based on the 40-ppm NOAEL, adjusting for the number of subjects.

1-h and 24-h ACs based on CNS effects
= NOAEL × 1/small n factor
= 40 ppm × (square root of 16)/10
= 16 ppm.

ACs for clinical, neuropsychological, or autonomic nervous system effects that persist after long-term occupational exposure were calculated on the basis of the 117-ppm NOAEL, correcting it for the number of subjects.

180-d AC based on neurotoxicity
= NOAEL × 1/small n factor
= 117 ppm × (square root of 43)/10
= 77 ppm.

ACs for exposure periods shorter than 180 d were not based on these persistent effects because the effects involve anatomical lesions produced after long-term exposures rather than a blood-concentration-dependent functional impairment.

High-frequency hearing loss and appropriate cochlear changes were found in weanling rats exposed to toluene at 1200-1400 ppm, 14 h/d, 7 d/w, for 5 w (Rebert et al., 1982; Pryor and Rebert, 1984). In adult animals, however, no solvent-induced alterations in the structure of brain or peripheral nerves were seen after exposure at 1500 ppm, 6 h/d, 5 d/w, for 6 mo (Spencer et al., 1985). Thus, this finding was not used to set an AC.

One study (Dyer et al., 1984) reported that a 30-d intermittent exposure of rats to toluene at 1000 ppm led to subtle changes in brain-wave activity, which was measured a day after exposure. This report was not used in setting an AC for neurotoxicity for the following reasons: First, the time course for this effect was not studied, and, second, the brain-wave changes were not correlated with any functional deficit. Similarly, the study of mice (Kostas and Hotchin, 1981) exposed to toluene in the womb through 35 d of age at a concentration of 400 ppm in drinking water indicated a CNS effect that could not easily be correlated with effects in humans (that is, decreased habituation or open-field activity at 35 d of age and depressed rotorod performance at 45-55 d of age).

Reproductive Toxicity

There are inconclusive reports concerning toluene's reproductive toxicity. In rabbits, continuous exposure at 266 ppm for gestation days 6-15 caused a decrease in maternal-weight gain and a slight, but not statistically significant, increase in spontaneous abortion (Ungváry and Tátrai, 1985). A study of spontaneous abortion rates in the wives (n = 28) of men who were occupationally exposed to toluene suggested an

TOLUENE 387

increased risk for spontaneous abortion, but the exposure concentrations could not be quantitated (Taskinen et al., 1989). Thus, no AC could be set for reproductive toxicity.

Hepatotoxicity and Nephrotoxicity

Even at high but unmeasured concentrations and long exposure durations in humans (such as chronic glue sniffers), liver and kidney toxicity is variable and reversible. Subchronic intermittent exposure of rats to toluene at 5000 ppm produced reversible weight gains in the liver and kidneys, which is not considered an adverse effect. Thus, an AC was not set for hepatotoxicity and nephrotoxicity.

Lethality

A NOAEL of 1200-ppm toluene was seen in a 2-y exposure of mice (n = 120) and rats (n = 100), 6.5 h/d, 5 d/w (equivalent to 140 d continuous exposure). Thus, applying Haber's rule for 180-d exposures (but not increasing the AC for exposures shorter than the experimental exposure) and using a species extrapolation factor of 10, the 30-d and 180-d ACs are the following:

30-d AC based on lethality
= NOAEL × 1/species factor
= 1200 ppm × 1/10
= 120 ppm.

180-d AC based on lethality
= NOAEL × time adjustment × 1/species factor
= 1200 ppm × (140 d/180 d) × 1/10
= 93 ppm (rounded to 90 ppm).

ACs for shorter exposures are not justified from this data set.
For 1-h exposures, the 1-h rat LC_{50} of 27,000 ppm was reduced by factors of 10 for species extrapolation and 10 for estimating an LC_0 from an LC_{50}.

1-h AC based on lethality
= LC_{50} × 1/species factor × $1/LC_{50}$ to LC_0 factor
= 27,000 ppm × 1/10 × 1/10
= 270 ppm.

Spaceflight Effects

The toxicity of toluene is not expected to be altered by the conditions of spaceflight.

RECOMMENDATIONS

The molecular mechanisms of action of toluene in producing its CNS effects and other toxic effects are poorly understood. Multi-species studies (including humans) are needed to elucidate the role of metabolites in the pharmacokinetics of toluene toxicity and offer results for comparison between species.

The development and validation of neurobehavioral tests, which measure effects relevant to the ability of humans to function competently in space or on earth, are needed to quantitate the effects of toluene exposure. Studies using such validated tests should include measurements at several vapor concentrations and with and without exercise at light and heavy workloads.

Although the data would not be applicable to spaceflight under current NASA guidelines, which prohibit spaceflight by pregnant astronauts, more data are needed on the developmental and reproductive toxicity of inhaled toluene, particularly with regard to teratogenicity and spontaneous abortion. Although some data are available on tests conducted in rats and mice that support suggestive results from epidemiological studies in female workers, these data are not compelling and require replication and extension.

TABLE 12-4 End Points and Acceptable Concentrations

End Point	Exposure Data	Species and Reference	Uncertainty Factors				Acceptable Concentrations, ppm				
			NOAEL	Time	Species	Space-flight	1 h	24 h	7 d	30 d	180 d
Irritation Eye and nose	NOAEL at 40 ppm, 6 h	Human (n = 16) (Andersen et al., 1983)	1	[a]	1	1	100	100	16	16	16
Neurotoxicity Headache, dizziness, inebriation, visual perception, vigilance, psychomotor functions, higher cortical functions	NOAEL at 40 ppm, 6 h	Human (n = 16) (Andersen et al., 1983)	1	[a]	1	1	16	16	16	16	16
Clinical, neurophysiological, neurological, and autonomic function effects	NOAEL at 117 ppm, 8 h/d, 5 d/w, 22 y	Human (n = 43) (Juntunen et al., 1985)	1	—	1	1	—	—	—	—	77
Lethality	NOAEL at 1200 ppm, 6.5 h/d, 5d/w, 2 y	Rat (Huff, 1990)	10	HR[b]	10	1	—	—	—	120	90
	LC_{50} at 27,000 ppm, 1 h	Rat (Pryor, 1978)	10	—	10	1	270	—	—	—	—
SMAC							16	16	16	16	16

[a]Haber's rule was not used to extrapolate from the experimental exposure duration because both irritancy and neurotoxicity are dependent on concentration but not on exposure duration.
[b]HR = Haber's rule.

REFERENCES

Alho, H., H. Tahti, J. Koistinaho, and A. Hervonen. 1986. The effect of toluene inhalation exposure on catecholamine content of rat sympathetic neurons. Med. Biol. 64:285-288.
Andersen, I., G.R. Lundqvist, L. Mølhave, O.F. Pedersen, D.F. Proctor, M. Væth, and D.P. Wyon. 1983. Human response to controlled levels of toluene in six-hour exposures. Scand. J. Work Environ. Health 9:405-418.
Bælum, J., I. Anderson, G.R. Lundqvist, L. Mølhave, O.F. Pedersen, M. Væth, and D.P. Wyon. 1985. Response of solvent-exposed printers and unexposed controls to six-hour toluene exposure. Scand. J. Work Environ. Health 11:271-280.
Benignus, V.A. 1981a. Health effects of toluene: A review. Neurotoxicology 2:526-588.
Benignus, V.A. 1981b. Neurobehavioral effects of toluene: A review. Neurobehav. Toxicol. Teratol. 3:407-415.
Benignus, V.A., K.E. Muller, C.N. Barton, and J.A. Bittikofer. 1981. Toluene levels in blood and brain of rats during and after respiratory exposure. Toxicol. Appl. Pharmacol. 61:326-334.
Bos, R.P., R.M. Brouns, R. Van Doorn, J.L. Theuws, and P.T. Henderson. 1981. Non-mutagenicity of toluene, o-xylene, m-xylene, p-xylene, o-methylbenzylalcohol and o-methylbenzylsulfate in the Ames assay. Mutat. Res. 88:273-279.
Bruckner, J.V., and R.G. Peterson. 1981a. Evaluation of toluene and acetone inhalant abuse. I. Pharmacology and pharmacodynamics. Toxicol. Appl. Pharmacol. 61:27-38.
Bruckner, J.V., and R.G. Peterson. 1981b. Evaluation of toluene and acetone inhalant abuse. II. Model development and toxicology. Toxicol. Appl. Pharmacol. 61:302-312.
Courtney, K.D., J.E. Andrews, J. Springer, M. Menache, T. Williams, L. Dalley, and J.A. Graham. 1986. A perinatal study of toluene in CD-1 mice. Fundam. Appl. Toxicol. 6:145-154.
Døssing, M., J. Bælum, S. Hansen, and G. Lundqvist. 1984. Effects of ethanol, cimetidine, and propanolol on toluene metabolism in man. Int. Arch. Occup. Environ. Health 54:309-315.
Dyer, R.S., K.E. Muller, R. Janssen, C.N. Barton, W.K. Boyes, and V.A. Benignus. 1984. Neurophysiological effects of 30 day chronic

exposure to toluene in rats. Neurobehav. Toxicol. Teratol. 6:363-368.
Forni, A., E. Pacifico, and A. Limonta. 1971. Chromosome studies in workers exposed to benzene or toluene or both. Arch. Environ. Health 22:373-378.
Freundt, K.J., and J.-C. Schneider. 1986. Drastic increase of the m-xylene or toluene concentrations in blood of rats after combined inhalation with acetone. In 27th Spring Meeting of the Deutsche Pharmakologische Gesellschaft [German Pharmacological Society], Mainz, West Germany, March 11-14, 1986.
Gamberale, F., and M. Hultengren. 1972. Toluene exposure. II. Psychophysiological functions. Work Environ. Health 9:131-139.
Gerner-Smidt, P., and U. Friedrich. 1978. The mutagenic effect of benzene, toluene, and xylene studied by the SCE technique. Mutat. Res. 58:313-316.
Gibson, J.E., and J.F. Hardisty. 1983. Chronic toxicity and oncogenicity bioassay of inhaled toluene in Fischer-344 rats. Fundam. Appl. Toxicol. 3:315-319.
Guillot, J.P., J.F. Gonnet, C. Clement, L. Caillard, and R. Truhaut. 1982. Evaluation of the ocular-irritation potential of 56 compounds. Food Chem. Toxicol. 20:573-582.
Hudák, A., and G. Ungváry. 1978. Embryotoxic effects of benzene and its methyl derivatives: toluene, xylene. Toxicology 11:55-63.
Huff, J. 1990. NTP Technical Report on the Toxicology and Carcinogenesis Studies of Toluene (CAS No. 108-88-3) in F344/N Rats and B6C3F1 Mice (Inhalation Studies). NTP TR 371. National Institutes of Health, National Toxicology Program, Research Triangle Park, N.C.
Ikeda, M. 1974. Reciprocal metabolic inhibition of toluene and trichloroethylene in vivo and in vitro. Int. Arch. Arbeitsmed. 33:125-130.
Ikeda, M., H. Otsuji, and T. Imamura. 1972. In vivo suppression of benzene and styrene oxidation by co-administered toluene in rats and effects of phenobarbital. Xenobiotica 2:101-106.
Inoue, O., K. Seiji, T. Watanabe, M. Kasahara, J. Nakatsuka, S. Yin, G. Li, S. Cai, C. Jin, and M. Ikeda. 1986. Possible ethnic difference in toluene metabolism: A comparative study among Chinese, Turkish, and Japanese solvent workers. Toxicol. Lett. 34:167-174.

Juntunen, J., E. Matikainen, M. Antti-Poika, H. Souranta, and M. Valle. 1985. Nervous system effects of long-term occupational exposure to toluene. Acta Neurol. Scand. 75:512-517.

Kostas, J., and J. Hotchin. 1981. Behavioral effects of low-level perinatal exposure to toluene in mice. Neurobehav. Toxicol. Teratol. 3:467-469.

Ladefoged, O., P. Strange, A. Moeller, H.R. Lam, G. Oestergaard, J. Larsen, and P. Arlien-Soeborg. 1991. Irreversible effects in rats of toluene (inhalation) exposure for six months. Pharmacol. Toxicol. (Copenhagen) 68:384-390.

Liang, J.C., T.C. Hsu, and J.E. Henry. 1983. Cytogenetic assays for mitotic poisons. The grasshopper embryo system for volatile liquids. Mutat. Res. 113:467-479.

Maki-Paakkanen, J., K. Husgafvel-Pursiainen, P.L. Kalliomäki, J. Tuominen, and M. Sorsa. 1980. Toluene-exposed workers and chromosome aberrations. J. Toxicol. Environ. Health 6:775-781.

McDonald, J.C., J. Lavoie, R. Cote, and A.D. McDonald. 1987. Chemical exposures at work in early pregnancy and congenital defect: a case-referent study. Br. J. Ind. Med. 44:527-533

NRC. 1992. Guidelines for Developing Spacecraft Maximum Allowable Concentrations for Space Station Contaminants. Washington, D.C.: National Academy Press.

Ogata, M., Y. Takatsuka, K. Tomokuni, and K. Muroi. 1971. Excretion of hippuric acid and *m*- or *p*-methylhippuric acid in the urine of persons exposed to vapours of toluene and *m*- or *p*-xylene in a exposure chamber and in workshops, with specific reference to repeated exposures. Br. J. Ind. Med. 28:382-385.

Peterson, R., and J. Bruckner. 1978. Animal model for studying the toxicological and pharmacological effects resulting from exposure to volatile substances. Final Report. Contract No. 271-75-3067. National Institute on Drug Abuse, Rockville, Md.

Pryor, G.T., and C.S. Rebert. 1984. Hearing loss in rats caused by inhalation of toluene, xylenes, and styrene. Abstract No. 315 in Arbetarskyddsverket. International Conference on Organic Solvent Toxicity, Stockholm, Sweden.

Pryor, G., H.R. Howd, R. Malik, R. Jensen, C. Rebert. 1978. Biomedical studies on the effects of abused inhalant mixtures. Annual Progress Report 2, Contract No. 271-77-3402. National Institute on Drug Abuse, Rockville, Md.

Purcell, K.J., G.H. Cason, M.L. Gargas, M.E. Anderson, and C. C. Travis. 1990. In vivo metabolic interaction of benzene and toluene. Toxicol. Lett. 52:141-152.
Rebert, C.S., S.S. Sorensen, R.A. Howd, and G.T. Pryor. 1982. Toluene-induced hearing loss in rats evidenced by the brain stem auditory-evoked response. Pp. 314-319 in The Toxicology of Petroleum Hydrocarbons. H.N. MacFarland, C.E. Holdsworth, J.A. Macgregor, R.W. Call, and M.L. Kane, eds. American Petroleum Institute, Washington, D.C.
Sandmeyer, E.E. 1981. Aromatic hydrocarbons. Pp. 3283-3291 in Patty's Industrial Hygiene and Toxicology, Vol. 2B, G.D. Clayton and F.E. Clayton, eds. New York: John Wiley & Sons.
Spencer, P.S., and H.H. Schaumberg. 1985. Organic solvent neurotoxicity. Facts and research needs. Scand. J. Work Environ. Health 11:53-60.
Taskinen, H., A. Anttila, M.L. Lindbohm, M. Sallman, and K. Hemminki. 1989. Spontaneous abortions and congenital malformations among the wives of men occupationally exposed to organic solvents. Scand. J. Work Environ. Health 15:345-352.
Tátrai, E., A. Hudák, and G. Ungváry. 1979. [Experimental model for the study of the teratogenic interaction of chemical agents and drugs (toluene and acetylsalicylic acid)]. Morphol. Igazsagugyi Orv. Sz. 19:299-304.
Ungváry, G., and E. Tátrai. 1985. On the embryotoxic effects of benzene and its alkyl derivatives in mice, rats, and rabbits. Arch. Toxicol. 8:425-430.
Ungváry, G., E. Tátrai, M. Lorincz, and G. Barcza. 1982. Joint embryotoxic effect of toluene, an extensively used industrial chemical substance, and acetylsalicylic acid (aspirin). Munkavedelem 28:25-31.
Veulemans, J., and R. Masschelein. 1978. Experimental human exposure to toluene: Factors influencing the individual respiratory uptake and elimination. Int. Arch. Occup. Environ. Health 42:91-103.
Von Oettingen, W., P. Neal, D. Donahue, J. Svirbely, H. Baernstein, A. Monaco, P. Valaer, and J. Mitchell. 1942. The toxicity and potential dangers of toluene, with special reference to its maximum permissible concentration. U.S. Public Health Service Bull. 279:1-50.
Weiss, B., R.W. Wood, and D.A. Macys. 1979. Behavioral toxicology of carbon disulfide and toluene. Environ. Health Perspect. 30: 39-45.